MW00444801

COMPLEX VARIABLES

Second Edition

Robert B. Ash

Department of Mathematics
University of Illinois
Urbana, Illinois

W. Phil Novinger

Mathematics Department
Florida State University, Retired
Tallahassee, Florida

DOVER PUBLICATIONS, INC.
Mineola, New York

Copyright

Bibliographical Note

This Dover edition by R. B. Ash and W. P. Novinger, first published in 2007, is the first publication in book form of the revised edition of *Complex Variables* by Robert B. Ash, originally published by Academic Press, New York, in 1971.

This revised edition is also available from the following web site: http://www.math.uiuc.edu/~r-ash/.

International Standard Book Number
ISBN 13: 978-486-46250-9
ISBN 10: 0-486-46250-1

Manufactured in the United States of America
Dover Publications, Inc., 31 East 2nd Street, Mineola, N.Y. 11501

Preface

This book represents a substantial revision of the first edition which was published in 1971. Most of the topics of the original edition have been retained, but in a number of instances the material has been reworked so as to incorporate alternative approaches to these topics that have appeared in the mathematical literature in recent years.

The book is intended as a text, appropriate for use by advanced undergraduates or graduate students who have taken a course in introductory real analysis, or as it is often called, advanced calculus. No background in complex variables is assumed, thus making the text suitable for those encountering the subject for the first time. It should be possible to cover the entire book in two semesters.

The list below enumerates many of the major changes and/or additions to the first edition.

1. The relationship between real-differentiability and the Cauchy-Riemann equations.

2. J.D. Dixon's proof of the homology version of Cauchy's theorem.

3. The use of hexagons in tiling the plane, instead of squares, to characterize simple connectedness in terms of winding numbers of cycles. This avoids troublesome details that appear in the proofs where the tiling is done with squares.

4. Sandy Grabiner's simplified proof of Runge's theorem.

5. A self-contained approach to the problem of extending Riemann maps of the unit disk to the boundary. In particular, no use is made of the Jordan curve theorem, a difficult theorem which we believe to be peripheral to a course in complex analysis. Several applications of the result on extending maps are given.

6. D.J. Newman's proof of the prime number theorem, as modified by J. Korevaar, is presented in the last chapter as a means of collecting and applying many of the ideas and results appearing in earlier chapters, while at the same time providing an introduction to several topics from analytic number theory.

For the most part, each section is dependent on the previous ones, and we recommend that the material be covered in the order in which it appears. Problem sets follow most sections, with solutions provided (in a separate section).

We have attempted to provide careful and complete explanations of the material, while at the same time maintaining a writing style which is succinct and to the point.

Robert B Ash
W. Phil Novinger

Table Of Contents

Chapter 1: Introduction

Chapter 2: The Elementary Theory

Chapter 3: The General Cauchy Theorem

Chapter 4: Applications of the Cauchy Theory

Chapter 5: Families of Analytic Functions

Chapter 6: Factorization of Analytic Functions

Chapter 7: The Prime Number Theorem

Chapter 1

Introduction

The reader is assumed to be familiar with the complex plane $\mathbb{C}$ to the extent found in most college algebra texts, and to have had the equivalent of a standard introductory course in real analysis (advanced calculus). Such a course normally includes a discussion of continuity, differentiation, and Riemann-Stieltjes integration of functions from the real line to itself. In addition, there is usually an introductory study of metric spaces and the associated ideas of open and closed sets, connectedness, convergence, compactness, and continuity of functions from one metric space to another. For the purpose of review and to establish notation, some of these concepts are discussed in the following sections.

1.1 Basic Definitions

The complex plane $\mathbb{C}$ is the set of all ordered pairs (a, b) of real numbers, with addition and multiplication defined by

$$(a,b)+(c,d)=(a+c,b+d) \qquad \text{and} \qquad (a,b)(c,d)=(ac-bd,ad+bc).$$

If $i = (0,1)$ and the real number a is identified with $(a,0)$, then $(a,b) = a+bi$. The expression $a+bi$ can be manipulated as if it were an ordinary binomial expression of real numbers, subject to the relation $i^2 = -1$. With the above definitions of addition and multiplication, $\mathbb{C}$ is a field.

If $z = a+bi$, then a is called the *real part* of z, written $a = \operatorname{Re} z$, and b is called the *imaginary part* of z, written $b = \operatorname{Im} z$. The *absolute value* or *magnitude* or *modulus* of z is defined as $(a^2+b^2)^{1/2}$. A complex number with magnitude 1 is said to be *unimodular*. An *argument* of z (written $\arg z$) is defined as the angle which the line segment from $(0,0)$ to (a,b) makes with the positive real axis. The argument is not unique, but is determined up to a multiple of 2π.

If r is the magnitude of z and θ is an argument of z, we may write

$$z = r(\cos\theta + i\sin\theta)$$

and it follows from trigonometric identities that

$$|z_1z_2| = |z_1||z_2| \qquad \text{and} \qquad \arg(z_1z_2) = \arg z_1 + \arg z_2$$

(that is, if θ_k is an argument of $z_k, k = 1, 2$, then $\theta_1 + \theta_2$ is an argument of $z_1 z_2$). If $z_2 \neq 0$, then $\arg(z_1/z_2) = \arg(z_1) - \arg(z_2)$. If $z = a + bi$, the *conjugate* of z is defined as $\overline{z} = a - bi$, and we have the following properties:

$$|\overline{z}| = |z|, \qquad \arg \overline{z} = -\arg z, \qquad \overline{z_1 + z_2} = \overline{z}_1 + \overline{z}_2, \qquad \overline{z_1 - z_2} = \overline{z}_1 - \overline{z}_2,$$

$$\overline{z_1 z_2} = \overline{z}_1 \overline{z}_2, \qquad \operatorname{Re} z = (z + \overline{z})/2, \qquad \operatorname{Im} z = (z - \overline{z})/2i, \qquad z\overline{z} = |z|^2.$$

The *distance* between two complex numbers z_1 and z_2 is defined as $d(z_1, z_2) = |z_1 - z_2|$. So $d(z_1, z_2)$ is simply the Euclidean distance between z_1 and z_2 regarded as points in the plane. Thus d defines a metric on $\mathbb{C}$, and furthermore, d is complete, that is, every Cauchy sequence converges. If $z_1, z_2, \ldots$ is sequence of complex numbers, then $z_n \to z$ if and only if $\operatorname{Re} z_n \to \operatorname{Re} z$ and $\operatorname{Im} z_n \to \operatorname{Im} z$. We say that $z_n \to \infty$ if the sequence of real numbers $|z_n|$ approaches $+\infty$.

Many of the above results are illustrated in the following analytical proof of the triangle inequality:

$$|z_1 + z_2| \leq |z_1| + |z_2| \quad \text{for all } z_1, z_2 \in \mathbb{C}.$$

The geometric interpretation is that the length of a side of a triangle cannot exceed the sum of the lengths of the other two sides. See Figure 1.1.1, which illustrates the familiar representation of complex numbers as vectors in the plane.

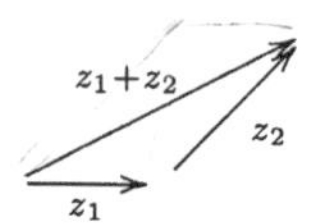

Figure 1.1.1

The proof is as follows:

$$\begin{aligned} |z_1 + z_2|^2 &= (z_1 + z_2)(\overline{z}_1 + \overline{z}_2) = |z_1|^2 + |z_2|^2 + z_1\overline{z}_2 + \overline{z}_1 z_2 \\ &= |z_1|^2 + |z_2|^2 + z_1\overline{z}_2 + \overline{z_1\overline{z}_2} = |z_1|^2 + |z_2|^2 + 2\operatorname{Re}(z_1\overline{z}_2) \\ &\leq |z_1|^2 + |z_2|^2 + 2|z_1\overline{z}_2| = (|z_1| + |z_2|)^2. \end{aligned}$$

The proof is completed by taking the square root of both sides.

If a and b are complex numbers, $[a, b]$ denotes the closed line segment with endpoints a and b. If t_1 and t_2 are arbitrary real numbers with $t_1 < t_2$, then we may write

$$[a, b] = \{a + \frac{t - t_1}{t_2 - t_1}(b - a) : t_1 \leq t \leq t_2\}.$$

The notation is extended as follows. If $a_1, a_2, \ldots, a_{n+1}$ are points in $\mathbb{C}$, a *polygon* from a_1 to a_{n+1} (or a polygon joining a_1 to a_{n+1}) is defined as

$$\bigcup_{j=1}^{n} [a_j, a_{j+1}],$$

often abbreviated as $[a_1, \ldots, a_{n+1}]$.

1.2 Further Topology of the Plane

Recall that two subsets S_1 and S_2 of a metric space are *separated* if there are open sets $G_1 \supseteq S_1$ and $G_2 \supseteq S_2$ such that $G_1 \cap S_2 = G_2 \cap S_1 = \emptyset$, the empty set. A set is *connected* iff it cannot be written as the union of two nonempty separated sets. An open (respectively closed) set is connected iff it is not the union of two nonempty disjoint open (respectively closed) sets.

1.2.1 Definition

A set $S \subseteq \mathbb{C}$ is said to be *polygonally connected* if each pair of points in S can be joined by a polygon that lies in S.

Polygonal connectedness is a special case of path (or arcwise) connectedness, and it follows that a polygonally connected set, in particular a polygon itself, is connected. We will prove in Theorem 1.2.3 that any *open* connected set is polygonally connected.

1.2.2 Definitions

If $a \in \mathbb{C}$ and $r > 0$, then $D(a, r)$ is the open disk with center a and radius r; thus $D(a, r) = \{z : |z - a| < r\}$. The closed disk $\{z : |z - a| \le r\}$ is denoted by $\overline{D}(a, r)$, and $C(a, r)$ is the circle with center a and radius r.

1.2.3 Theorem

If Ω is an open subset of $\mathbb{C}$, then Ω is connected iff Ω is polygonally connected.

Proof. If Ω is connected and $a \in \Omega$, let Ω_1 be the set of all z in Ω such that there is a polygon in Ω from a to z, and let $\Omega_2 = \Omega \setminus \Omega_1$. If $z \in \Omega_1$, there is an open disk $D(z, r) \subseteq \Omega$ (because Ω is open). If $w \in D(z, r)$, a polygon from a to z can be extended to w, and it follows that $D(z, r) \subseteq \Omega_1$, proving that Ω_1 is open. Similarly, Ω_2 is open. (Suppose $z \in \Omega_2$, and choose $D(z, r) \subseteq \Omega$. Then $D(z, r) \subseteq \Omega_2$ as before.)

Thus Ω_1 and Ω_2 are disjoint open sets, and $\Omega_1 \ne \emptyset$ because $a \in \Omega_1$. Since Ω is connected we must have $\Omega_2 = \emptyset$, so that $\Omega_1 = \Omega$. Therefore Ω is polygonally connected. The converse assertion follows because *any* polygonally connected set is connected. ♣

1.2.4 Definitions

A *region* in $\mathbb{C}$ is an open connected subset of $\mathbb{C}$. A set $E \subseteq \mathbb{C}$ is *convex* if for each pair of points $a, b \in E$, we have $[a, b] \subseteq E$; E is *starlike* if there is a point $a \in E$ (called a *star center*) such that $[a, z] \subseteq E$ for each $z \in E$. Note that any nonempty convex set is starlike and that starlike sets are polygonally connected.

1.3 Analytic Functions

1.3.1 Definition

Let $f : \Omega \to \mathbb{C}$, where Ω is a subset of $\mathbb{C}$. We say that f is *complex-differentiable* at the point $z_0 \in \Omega$ if for some $\lambda \in \mathbb{C}$ we have

$$\lim_{h \to 0} \frac{f(z_0 + h) - f(z_0)}{h} = \lambda \tag{1}$$

or equivalently,

$$\lim_{z \to z_0} \frac{f(z) - f(z_0)}{z - z_0} = \lambda. \tag{2}$$

Conditions (3), (4) and (5) below are also equivalent to (1), and are sometimes easier to apply.

$$\lim_{n \to \infty} \frac{f(z_0 + h_n) - f(z_0)}{h_n} = \lambda \tag{3}$$

for each sequence $\{h_n\}$ such that $z_0 + h_n \in \Omega \setminus \{z_0\}$ and $h_n \to 0$ as $n \to \infty$.

$$\lim_{n \to \infty} \frac{f(z_n) - f(z_0)}{z_n - z_0} = \lambda \tag{4}$$

for each sequence $\{z_n\}$ such that $z_n \in \Omega \setminus \{z_0\}$ and $z_n \to z_0$ as $n \to \infty$.

$$f(z) = f(z_0) + (z - z_0)(\lambda + \epsilon(z)) \tag{5}$$

for all $z \in \Omega$, where $\epsilon : \Omega \to \mathbb{C}$ is continuous at z_0 and $\epsilon(z_0) = 0$.

To show that (1) and (5) are equivalent, just note that ϵ may be written in terms of f as follows:

$$\epsilon(z) = \begin{cases} \frac{f(z)-f(z_0)}{z-z_0} - \lambda & \text{if } z \neq z_0 \\ 0 & \text{if } z = z_0. \end{cases}$$

The number λ is unique. It is usually written as $f'(z_0)$, and is called the *derivative* of f at z_0.

If f is complex-differentiable at every point of Ω, f is said to be *analytic* or *holomorphic* on Ω. Analytic functions are the basic objects of study in complex variables.

Analyticity on a nonopen set $S \subseteq \mathbb{C}$ means analyticity on an open set $\Omega \supseteq S$. In particular, f is analytic at a point z_0 iff f is analytic on an open set Ω with $z_0 \in \Omega$. If f_1 and f_2 are analytic on Ω, so are $f_1 + f_2, f_1 - f_2, kf_1$ for $k \in \mathbb{C}$, $f_1 f_2$, and f_1/f_2 (provided that f_2 is never 0 on Ω). Furthermore,

$$(f_1 + f_2)' = f_1' + f_2', \quad (f_1 - f_2)' = f_1' - f_2', \quad (kf_1)' = kf_1'$$

$$(f_1 f_2)' = f_1 f_2' + f_1' f_2, \quad \left(\frac{f_1}{f_2}\right)' = \frac{f_2 f_1' - f_1 f_2'}{f_2^2}.$$

The proofs are identical to the corresponding proofs for functions from $\mathbb{R}$ to $\mathbb{R}$.

Since $\frac{d}{dz}(z) = 1$ by direct computation, we may use the rule for differentiating a product (just as in the real case) to obtain

$$\frac{d}{dz}(z^n) = nz^{n-1}, \ n = 0, 1, \ldots$$

This extends to $n = -1, -2, \ldots$ using the quotient rule.

If f is analytic on Ω and g is analytic on $f(\Omega) = \{f(z) : z \in \Omega\}$, then the composition $g \circ f$ is analytic on Ω and

$$\frac{d}{dz}g(f(z)) = g'(f(z)f'(z)$$

just as in the real variable case.

As an example of the use of Condition (4) of (1.3.1), we now prove a result that will be useful later in studying certain inverse functions.

1.3.2 Theorem

Let g be analytic on the open set Ω_1, and let f be a continuous complex-valued function on the open set Ω. Assume
(i) $f(\Omega) \subseteq \Omega_1$,
(ii) g' is never 0,
(iii) $g(f(z)) = z$ for all $z \in \Omega$ (thus f is 1-1).
Then f is analytic on Ω and $f' = 1/(g' \circ f)$.

Proof. Let $z_0 \in \Omega$, and let $\{z_n\}$ be a sequence in $\Omega \setminus \{z_0\}$ with $z_n \to z_0$. Then

$$\frac{f(z_n) - f(z_0)}{z_n - z_0} = \frac{f(z_n) - f(z_0)}{g(f(z_n)) - g(f(z_0))} = \left[\frac{g(f(z_n)) - g(f(z_0))}{f(z_n) - f(z_0)}\right]^{-1}.$$

(Note that $f(z_n) \neq f(z_0)$ since f is 1-1 and $z_n \neq z_0$.) By continuity of f at z_0, the expression in brackets approaches $g'(f(z_0))$ as $n \to \infty$. Since $g'(f(z_0)) \neq 0$, the result follows. ♣

1.4 Real-Differentiability and the Cauchy-Riemann Equations

Let $f : \Omega \to \mathbb{C}$, and set $u = \operatorname{Re} f, v = \operatorname{Im} f$. Then u and v are real-valued functions on Ω and $f = u + iv$. In this section we are interested in the relation between f and its real and imaginary parts u and v. For example, f is continuous at a point z_0 iff both u and v are continuous at z_0. Relations involving derivatives will be more significant for us, and for this it is convenient to be able to express the idea of differentiability of real-valued function of two real variables by means of a single formula, without having to consider partial derivatives separately. We do this by means of a condition analogous to (5) of (1.3.1).

Convention

From now on, Ω will denote an open subset of $\mathbb{C}$, unless otherwise specified.

1.4.1 Definition

Let $g : \Omega \to \mathbb{R}$. We say that g is *real-differentiable* at $z_0 = x_0 + iy_0 \in \Omega$ if there exist real numbers A and B, and real functions ϵ_1 and ϵ_2 defined on a neighborhood of (x_0, y_0), such that ϵ_1 and ϵ_2 are continuous at (x_0, y_0), $\epsilon_1(x_0, y_0) = \epsilon_2(x_0, y_0) = 0$, and

$$g(x, y) = g(x_0, y_0) + (x - x_0)[A + \epsilon_1(x, y)] + (y - y_0)[B + \epsilon_2(x, y)]$$

for all (x, y) in the above neighborhood of (x_0, y_0).

It follows from the definition that if g is real-differentiable at (x_0, y_0), then the partial derivatives of g exist at (x_0, y_0) and

$$\frac{\partial g}{\partial x}(x_0, y_0) = A, \quad \frac{\partial g}{\partial y}(x_0, y_0) = B.$$

If, on the other hand, $\frac{\partial g}{\partial x}$ and $\frac{\partial g}{\partial y}$ exist at (x_0, y_0) and one of these exists in a neighborhood of (x_0, y_0) and is continuous at (x_0, y_0), then g is real-differentiable at (x_0, y_0). To verify this, assume that $\frac{\partial g}{\partial x}$ is continuous at (x_0, y_0), and write

$$g(x, y) - g(x_0, y_0) = g(x, y) - g(x_0, y) + g(x_0, y) - g(x_0, y_0).$$

Now apply the mean value theorem and the definition of partial derivative respectively (Problem 4).

1.4.2 Theorem

Let $f : \Omega \to \mathbb{C}, u = \operatorname{Re} f, v = \operatorname{Im} f$. Then f is complex-differentiable at (x_0, y_0) iff u and v are real-differentiable at (x_0, y_0) and the *Cauchy-Riemann equations*

$$\frac{\partial u}{\partial x} = \frac{\partial v}{\partial y}, \quad \frac{\partial v}{\partial x} = -\frac{\partial u}{\partial y}$$

are satisfied at (x_0, y_0). Furthermore, if $z_0 = x_0 + iy_0$, we have

$$f'(z_0) = \frac{\partial u}{\partial x}(x_0, y_0) + i\frac{\partial v}{\partial x}(x_0, y_0) = \frac{\partial v}{\partial y}(x_0, y_0) - i\frac{\partial u}{\partial y}(x_0, y_0).$$

Proof. Assume f complex-differentiable at z_0, and let ϵ be the function supplied by (5) of (1.3.1). Define $\epsilon_1(x, , y) = \operatorname{Re} \epsilon(x, y), \epsilon_2(x, y) = \operatorname{Im} \epsilon(x, y)$. If we take real parts of both sides of the equation

$$f(x) = f(z_0) + (z - z_0)(f'(z_0) + \epsilon(z)) \tag{1}$$

we obtain

$$\begin{aligned} u(x, y) = u(x_0, y_0) &+ (x - x_0)[\operatorname{Re} f'(z_0) + \epsilon_1(x, y)] \\ &+ (y - y_0)[-\operatorname{Im} f'(z_0) - \epsilon_2(x, y)]. \end{aligned}$$

It follows that u is real-differentiable at (x_0, y_0) with

$$\frac{\partial u}{\partial x}(x_0, y_0) = \operatorname{Re} f'(z_0), \quad \frac{\partial u}{\partial y}(x_0, y_0) = -\operatorname{Im} f'(z_0). \tag{2}$$

Similarly, take imaginary parts of both sides of (1) to obtain

$$\begin{aligned} v(x, y) = v(x_0, y_0) &+ (x - x_0)[\operatorname{Im} f'(z_0) + \epsilon_2(x, y)] \\ &+ (y - y_0)[\operatorname{Re} f'(z_0) + \epsilon_1(x, y)] \end{aligned}$$

and conclude that

$$\frac{\partial v}{\partial x}(x_0, y_0) = \operatorname{Im} f'(z_0), \quad \frac{\partial v}{\partial y}(x_0, y_0) = \operatorname{Re} f'(z_0). \tag{3}$$

The Cauchy-Riemann equations and the desired formulas for $f'(z_0)$ follow from (2) and (3).

Conversely, suppose that u and v are real-differentiable at (x_0, y_0) and satisfy the Cauchy-Riemann equations there. Then we may write equations of the form

$$\begin{aligned} u(x, y) = u(x_0, y_0) &+ (x - x_0)[\frac{\partial u}{\partial x}(x_0, y_0) + \epsilon_1(x, y)] \\ &+ (y - y_0)[\frac{\partial u}{\partial y}(x_0, y_0) + \epsilon_2(x, y)], \end{aligned} \tag{4}$$

$$\begin{aligned} v(x, y) = v(x_0, y_0) &+ (x - x_0)[\frac{\partial v}{\partial x}(x_0, y_0) + \epsilon_3(x, y)] \\ &+ (y - y_0)[\frac{\partial v}{\partial y}(x_0, y_0) + \epsilon_4(x, y)]. \end{aligned} \tag{5}$$

Since $f = u + iv$, (4) and (5) along with the Cauchy-Riemann equations yield

$$f(z) = f(z_0) + (z - z_0)[\frac{\partial u}{\partial x}(x_0, y_0) + i\frac{\partial v}{\partial x}(x_0, y_0) + \epsilon(z)]$$

where, at least in a neighborhood of z_0,

$$\epsilon(z) = \left[\frac{x - x_0}{z - z_0}\right][\epsilon_1(x, y) + i\epsilon_3(x, y)] + \left[\frac{y - y_0}{z - z_0}\right][\epsilon_2(x, y) + i\epsilon_4(x, y)] \text{ if } z \neq z_0; \ \epsilon(z_0) = 0.$$

It follows that f is complex-differentiable at z_0. ♣

1.5 The Exponential Function

In this section we extend the domain of definition of the exponential function (as normally encountered in calculus) from the real line to the entire complex plane. If we require that the basic rules for manipulating exponentials carry over to the extended function, there is only one possible way to define $\exp(z)$ for $z = x+iy \in \mathbb{C}$. Consider the following sequence of "equations" that exp should satisfy:

$$\begin{aligned}
\exp(z) &= \exp(x+iy) \\
&\text{“}=\text{”}\ \exp(x)\exp(iy) \\
&\text{“}=\text{”}\ e^x\left(1+iy+\frac{(iy)^2}{2!}+\cdots\right) \\
&\text{“}=\text{”}\ e^x\left[\left(1-\frac{y^2}{2!}+\frac{y^4}{4!}-\cdots\right)+i\left(y-\frac{y^3}{3!}+\frac{y^5}{5!}-\cdots\right)\right] \\
&\text{“}=\text{”}\ e^x(\cos y + i\sin y).
\end{aligned}$$

Thus we have only one candidate for the role of exp on $\mathbb{C}$.

1.5.1 Definition

If $z = x+iy \in \mathbb{C}$, let $\exp(z) = e^x(\cos y + i \sin y)$. Note that if $z = x \in \mathbb{R}$, then $\exp(z) = e^x$ so exp is indeed a extension of the real exponential function.

1.5.2 Theorem

The exponential function is analytic on $\mathbb{C}$ and $\frac{d}{dz}\exp(z) = \exp(z)$ for all z.

Proof. The real and imaginary parts of $\exp(x+iy)$ are, respectively, $u(x,y) = e^x \cos y$ and $v(x,y) = e^x \sin y$. At any point (x_0, y_0), u and v are real-differentiable (see Problem 4) and satisfy the Cauchy-Riemann equations there. The result follows from (1.4.2). ♣

Functions such as exp and the polynomials that are analytic on $\mathbb{C}$ are called *entire functions.*

The exponential function is of fundamental importance in mathematics, and the investigation of its properties will be continued in Section 2.3.

1.6 Harmonic Functions

1.6.1 Definition

A function $g : \Omega \to \mathbb{R}$ is said to be *harmonic* on Ω if g has continuous first and second order partial derivatives on Ω and satisfies *Laplace's equation*

$$\frac{\partial^2 g}{\partial x^2} + \frac{\partial^2 g}{\partial y^2} = 0$$

on all of Ω.

After some additional properties of analytic functions have been developed, we will be able to prove that the real and imaginary parts of an analytic function on Ω are harmonic on Ω. The following theorem is a partial converse to that result, namely that a harmonic on Ω is locally the real part of an analytic function.

1.6.2 Theorem

Suppose $u : \Omega \to \mathbb{R}$ is harmonic on Ω, and D is any open disk contained in Ω. Then there exists a function $v : D \to \mathbb{R}$ such that $u + iv$ is analytic on D.

The function v is called a *harmonic conjugate* of u.

Proof. Consider the differential $Pdx + Qdy$, where $P = -\frac{\partial u}{\partial y}$, $Q = \frac{\partial u}{\partial x}$. Since u is harmonic, P and Q have continuous partial derivatives on Ω and $\frac{\partial P}{\partial y} = \frac{\partial Q}{\partial x}$. It follows (from calculus) that $Pdx + Qdy$ is a locally exact differential. In other words, there is a function $v : D \to \mathbb{R}$ such that $dv = Pdx + Qdy$. But this just means that on D we have

$$\frac{\partial v}{\partial x} = P = -\frac{\partial u}{\partial y} \quad \text{and} \quad \frac{\partial v}{\partial y} = Q = \frac{\partial u}{\partial x}.$$

Hence by (1.4.2) (and Problem 4), $u + iv$ is analytic on D.

Problems

1. Prove the parallelogram law $|z_1 + z_2|^2 + |z_1 - z_2|^2 = 2[|z_1|^2 + |z_2|^2]$ and give a geometric interpretation.
2. Show that $|z_1 + z_2| = |z_1| + |z_2|$ iff z_1 and z_2 lie on a common ray from 0 iff one of z_1 or z_2 is a nonnegative multiple of the other.
3. Let z_1 and z_2 be nonzero complex numbers, and let $\theta, 0 \le \theta \le \pi$, be the angle between them. Show that
(a) $\operatorname{Re} z_1\overline{z}_2 = |z_1||z_2|\cos\theta$, $\operatorname{Im} z_1\overline{z}_2 = \pm|z_1||z_2|\sin\theta$, and consequently
(b) The area of the triangle formed by z_1, z_2 and $z_2 - z_1$ is $|\operatorname{Im} z_1\overline{z}_2|/2$.
4. Let $g : \Omega \to \mathbb{R}$ be such that $\frac{\partial g}{\partial x}$ and $\frac{\partial g}{\partial y}$ exist at $(x_0, y_0) \in \Omega$, and suppose that one of these partials exists in a neighborhood of (x_0, y_0) and is continuous at (x_0, y_0). Show that g is real-differentiable at (x_0, y_0).
5. Let $f(x) = \overline{z}, z \in \mathbb{C}$. Show that although f is continuous everywhere, it is nowhere differentiable.
6. Let $f(z) = |z|^2, z \in \mathbb{C}$. Show that f is complex-differentiable at $z = 0$, but nowhere else.
7. Let $u(x, y) = \sqrt{|xy|}, (x, y) \in \mathbb{C}$. Show that $\frac{\partial u}{\partial x}$ and $\frac{\partial u}{\partial y}$ both exist at (0,0), but u is not real-differentiable at (0,0).
8. Show that the field of complex numbers is isomorphic to the set of matrices of the form

$$\begin{bmatrix} a & b \\ -b & a \end{bmatrix}$$

with $a, b \in \mathbb{R}$.
9. Show that the complex field cannot be ordered. That is, there is no subset $P \subseteq \mathbb{C}$ of "positive elements" such that
(a) P is closed under addition and multiplication.
(b) If $z \in P$, then exactly one of the relations $z \in P$, $z = 0$, $-z \in P$ holds.

10. (A characterization of absolute value) Show that there is a unique function $\alpha : \mathbb{C} \to \mathbb{R}$ such that
(i) $\alpha(x) = x$ for all real $x \geq 0$;
(ii) $\alpha(zw) = \alpha(z)\alpha(w), z, w \in \mathbb{C}$;
(iii) α is bounded on the unit circle $C(0,1)$.

Hint: First show that $\alpha(z) = 1$ for $|z| = 1$.

11. (Another characterization of absolute value) Show that there is a unique function $\alpha : \mathbb{C} \to \mathbb{R}$ such that
(i) $\alpha(x) = x$ for all real $x \geq 0$;
(ii) $\alpha(zw) = \alpha(z)\alpha(w), z, w \in \mathbb{C}$;
(iii) $\alpha(z + w) \leq \alpha(z) + \alpha(w), z, w \in \mathbb{C}$.

12. Let α be a complex number with $|\alpha| < 1$. Prove that

$$\left|\frac{z - \alpha}{1 - \overline{\alpha}z}\right| = 1 \quad \text{iff} \quad |z| = 1.$$

13. Suppose $z \in \mathbb{C}, z \neq 0$. Show that $z + \frac{1}{z}$ is real iff $\operatorname{Im} z = 0$ or $|z| = 1$.

14. In each case show that u is harmonic and find the harmonic conjugate v such that $v(0,0) = 0$.
(i) $u(x,y) = e^y \cos x$;
(ii) $u(x,y) = 2x - x^3 + 3xy^2$.

15. Let $a, b \in \mathbb{C}$ with $a \neq 0$, and let $T(z) = az + b, z \in \mathbb{C}$.
(i) Show that T maps the circle $C(z_0, r)$ onto the circle $C(T(z_0), r|a|)$.
(ii) For which choices of a and b will T map $C(0,1)$ onto $C(1+i,2)$?
(iii) In (ii), is it possible to choose a and b so that $T(1) = -1 + 3i$?

16. Show that $f(z) = e^{\operatorname{Re} z}$ is nowhere complex-differentiable.

17. Let f be a complex-valued function defined on an open set Ω that is symmetric with respect to the real line, that is, $z \in \Omega$ implies $\overline{z} \in \Omega$. (Examples are $\mathbb{C}$ and $D(x,r)$ where $x \in \mathbb{R}$.) Set $g(z) = \overline{f(\overline{z})}$, and show that g is analytic on Ω if and only if f is analytic on Ω.

18. Show that an equation for the circle $C(z_0, r)$ is $z\overline{z} - \overline{z}_0 z - z_0\overline{z} + z_0\overline{z}_0 = r^2$.

19. (Enestrom's theorem) Suppose that $P(z) = a_0 + a_1 z + \cdots + a_n z^n$, where $n \geq 1$ and $a_0 \geq a_1 \geq a_2 \geq \cdots \geq a_n > 0$. Prove that the zeros of the polynomial $P(z)$ all lie outside the open unit disk $D(0,1)$.

Hint: Look at $(1 - z)P(z)$, and show that $(1 - z)P(z) = 0$ implies that $a_0 = (a_0 - a_1)z + (a_1 - a_2)z^2 + \cdots + (a_{n-1} - a_n)z^n + a_n z^{n+1}$, which is impossible for $|z| < 1$.

20. Continuing Problem 19, show that if $a_{j-1} > a_j$ for all j, then all the zeros of $P(z)$ must be outside the *closed* unit disk $\overline{D}(0,1)$.

Hint: If the last equation of Problem 19 holds for some z with $|z| \leq 1$, then $z = 1$.

Chapter 2

The Elementary Theory

2.1 Integration on Paths

The integral of a complex-valued function on a path in the complex plane will be introduced via the integral of a complex-valued function of a real variable, which in turn is expressed in terms of an ordinary Riemann integral.

2.1.1 Definition

Let $\varphi : [a, b] \to \mathbb{C}$ be a piecewise continuous function on the closed interval $[a, b]$ of reals. The *Riemann integral* of φ is defined in terms of the real and imaginary parts of φ by

$$\int_a^b \varphi(t)\, dt = \int_a^b \operatorname{Re}\varphi(t)\, dt + i \int_a^b \operatorname{Im}\varphi(t)\, dt.$$

2.1.2 Basic Properties of the Integral

The following linearity property is immediate from the above definition and the corresponding result for real-valued functions:

$$\int_a^b (\lambda\varphi(t) + \mu\psi(t))\, dt = \lambda \int_a^b \varphi(t)\, dt + \mu \int_a^b \psi(t)\, dt$$

for any complex numbers λ and μ. A slightly more subtle property is

$$\left|\int_a^b \varphi(t)\, dt\right| \le \int_a^b |\varphi(t)|\, dt.$$

This may be proved by approximating the integral on the left by Riemann sums and using the triangle inequality. A somewhat more elegant argument uses a technique called polarization, which occurs quite frequently in analysis. Define $\lambda = \left|\int_a^b \varphi(t)dt\right| / \int_a^b \varphi(t)\, dt$; then $|\lambda| = 1$. (If the denominator is zero, take λ to be any complex number of absolute value 1.) Then $|\int_a^b \varphi(t)\,dt| = \lambda \int_a^b \varphi(t)\,dt = \int_a^b \lambda\varphi(t)dt$ by linearity. Since the absolute value of a complex number is real,

$$\left|\int_a^b \varphi(t)\,dt\right| = \operatorname{Re}\int_a^b \lambda\varphi(t)\,dt = \int_a^b \operatorname{Re}\lambda\varphi(t)\,dt$$

by definition of the integral. But $\operatorname{Re}|z| \le |z|$, so

$$\int_a^b \operatorname{Re}\lambda\varphi(t)\,dt \le \int_a^b |\lambda\varphi(t)|\,dt = \int_a^b |\varphi(t)|\,dt$$

because $|\lambda| = 1$. ♣

The fundamental theorem of calculus carries over to complex-valued functions. Explicitly, if φ has a continuous derivative on $[a, b]$, then

$$\varphi(x) = \varphi(a) + \int_a^x \varphi'(t)\,dt$$

for $a \le x \le b$. If φ is continuous on $[a, b]$ and $F(x) = \int_a^x \varphi(t)\,dt, a \le x \le b$, then $F'(x) = \varphi(x)$ for all x in $[a, b]$. These assertions are proved directly from the corresponding results for real-valued functions.

2.1.3 Definition

A *curve* in $\mathbb{C}$ is a continuous mapping γ of a closed interval $[a, b]$ into $\mathbb{C}$. If in addition, γ is piecewise continuously differentiable, then γ is called a *path*. A curve (or path) with $\gamma(a) = \gamma(b)$ is called a *closed curve* (or path). The range (or image or trace) of γ will be denoted by γ^*. If γ^* is contained in a set S, γ is said to be a *curve* (or *path*) *in* S.

Intuitively, if $z = \gamma(t)$ and t changes by a small amount dt, then z changes by $dz = \gamma'(t)\,dt$. This motivates the definition of the *length* L of a path γ:

$$L = \int_a^b |\gamma'(t)|\,dt$$

and also motivates the following definition of the *path integral* $\int_\gamma f(z)dz$.

2.1.4 Definition

Let $\gamma : [a, b] \to \mathbb{C}$ be a path, and let f be continuous on γ, that is, $f : \gamma^* \to \mathbb{C}$ is continuous. We define the integral of f on (or along) γ by

$$\int_\gamma f(z)\,dz = \int_a^b f(\gamma(t))\gamma'(t)\,dt.$$

It is convenient to define $\int_\gamma f(z)\,dz$ with γ replaced by certain point sets in the plane. Specifically, if $[z_1, z_2]$ is a line segment in $\mathbb{C}$, we define

$$\int_{[z_1,z_2]} f(z)\,dz = \int_\gamma f(z)\,dz$$

where $\gamma(t) = (1-t)z_1 + tz_2, 0 \le t \le 1$. More generally, if $[z_1, \dots, z_{n+1}]$ is a polygon joining z_1 to z_{n+1}, we define

$$\int_{[z_1,z_2,\dots,z_{n+1}]} f(z)\,dz = \sum_{j=1}^{n} \int_{[z_j,z_{j+1}]} f(z)\,dz.$$

The next estimate will be referred to as the *M-L theorem.*

2.1.5 Theorem

Suppose that f is continuous on the path γ and $|f(z)| \le M$ for all $z \in \gamma^*$. If L is the length of the path γ, then

$$\left| \int_\gamma f(z)\,dz \right| \le ML.$$

Proof. Recall from (2.1.2) that the absolute value of an integral is less than or equal to the integral of the absolute value. Then apply the definition of the path integral in (2.1.4) and the definition of length in (2.1.3). ♣

The familiar process of evaluating integrals by anti-differentiation extends to integration on paths.

2.1.6 Fundamental Theorem for Integrals on Paths

Suppose $f : \Omega \to \mathbb{C}$ is continuous and f has a *primitive* F on Ω, that is, $F' = f$ on Ω. Then for any path $\gamma : [a, b] \to \Omega$ we have

$$\int_\gamma f(z)\,dz = F(\gamma(b)) - F(\gamma(a)).$$

In particular, if γ is a closed path in Ω, then $\int_\gamma f(z)\,dz = 0$.

Proof. $\int_\gamma f(z)dz = \int_a^b F'(\gamma(t))\gamma'(t)\,dt = \int_a^b \frac{d}{dt}F(\gamma(t))\,dt = F(\gamma(b)) - F(\gamma(a))$ by the fundamental theorem of calculus [see (2.1.2)]. ♣

2.1.7 Applications

(a) Let $z_1, z_2 \in \mathbb{C}$ and let γ be any *path from z_1 to z_2*, that is, $\gamma : [a, b] \to \mathbb{C}$ is any path such that $\gamma(a) = z_1$ and $\gamma(b) = z_2$. Then for $n = 0, 1, 2, 3, \dots$ we have

$$\int_\gamma z^n\,dz = (z_2^{n+1} - z_1^{n+1})/(n+1).$$

This follows from (2.1.6) and the fact that $z^{n+1}/(n+1)$ is a primitive of z^n. The preceding remains true for $n = -2, -3, -4, \dots$ provided that $0 \notin \gamma^*$ and the proof is the same: $z^{n+1}/(n+1)$ is a primitive for z^n on $\mathbb{C} \setminus \{0\}$. But if $n = -1$, then the conclusion may fail as the following important computation shows. Take $\gamma(t) = e^{it}, 0 \leq t \leq 2\pi$ (the unit circle, traversed once in the positive sense). Then

$$\int_\gamma \frac{1}{z}\,dz = \int_0^{2\pi} \frac{ie^{it}}{e^{it}}\,dt = 2\pi i \neq 0.$$

This also shows that $f(z) = 1/z$, although analytic on $\mathbb{C} \setminus \{0\}$, does not have a primitive on $\mathbb{C} \setminus \{0\}$.

(b) Suppose f is analytic on the open connected set Ω and $f'(z) = 0$ for all $z \in \Omega$. Then f is constant on Ω.

Proof. Let $z_1, z_2 \in \Omega$. Since Ω is polygonally connected, there is a (polygonal) path $\gamma : [a, b] \to \Omega$ such that $\gamma(a) = z_1$ and $\gamma(b) = z_2$. by (2.1.6), $\int_\gamma f'(z)\,dz = f(z_2) - f(z_1)$. But the left side is zero by hypothesis, and the result follows. ♣

Remark

If we do not assume that Ω is connected, we can prove only that f restricted to any component of Ω is constant.

Suppose that a continuous function f on Ω is given. Theorem 2.1.6 and the applications following it suggest that we should attempt to find conditions on f and/or Ω that are sufficient to guarantee that f has a primitive. Let us attempt to imitate the procedure used in calculus when f is a real-valued continuous function on an open interval in $\mathbb{R}$. We begin by assuming Ω is starlike with star center z_0, say. Define F on Ω by

$$F(z) = \int_{[z_0,z]} f(w)\,dw.$$

If $z_1 \in \Omega$, let us try to show that $F'(z_1) = f(z_1)$. If z is near but unequal to z_1, we have

$$\frac{F(z) - F(z_1)}{z - z_1} = \frac{1}{z - z_1}\left(\int_{[z_0,z]} f(w)\,dw - \int_{[z_0,z_1]} f(w)\,dw\right)$$

and we would like to say, as in the real variables case, that

$$\int_{[z_0,z]} f(w)\,dw - \int_{[z_0,z_1]} f(w)\,dw = \int_{[z_1,z]} f(w)\,dw, \tag{1}$$

from which it would follow quickly that $(F(z) - F(z_1))/(z - z_1) \to f(z_1)$ as $z \to z_1$. Now if T is the triangle $[z_0, z_1, z, z_0]$, equation (1) is equivalent to the statement that $\int_T f(w)\,dw = 0$, but as the example at the end of (2.1.7(a)) suggests, this need not be true, even for analytic functions f. However, in the present setting, we can make the key observation that if $\hat{T}$ is the union of T and its interior (the *convex hull* of T), then $\hat{T} \subseteq \Omega$. If f is analytic on Ω, it must be analytic on $\hat{T}$, and in this case, it turns out that $\int_T f(w)\,dw$ *does* equal 0. This is the content of Theorem 2.1.8; a somewhat different version of this result was first proved by Augustin-Louis Cauchy in 1825.

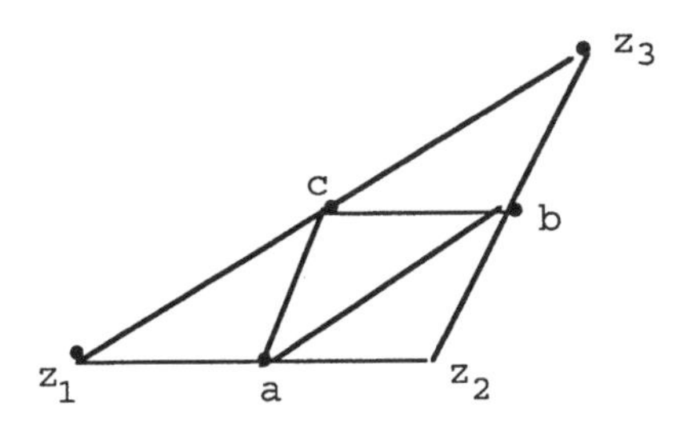

Figure 2.1.1

2.1.8 Cauchy's Theorem for Triangles

Suppose that f is analytic on Ω and $T = [z_1, z_2, z_3, z_1]$ is any triangle such that $\hat{T} \subseteq \Omega$. Then $\int_T f(z)\,dz = 0$.

Proof. Let a, b, c be the midpoints of $[z_1, z_2], [z_2, z_3]$ and $[z_3, z_1]$ respectively. Consider the triangles $[z_1, a, c, z_1], [z_2, b, a, z_2], [z_3, c, b, z_3]$ and $[a, b, c, a]$ (see Figure 2.1.1). Now the integral of f on T is the sum of the integrals on the four triangles, and it follows from the triangle inequality that if T_1 is one of these four triangles chosen so that $|\int_{T_1} f(z)\,dz|$ is as large as possible, then

$$|\int_T f(z)\,dz| \le 4|\int_{T_1} f(z)\,dz|.$$

Also, if L measures length, then $L(T_1) = \frac{1}{2}L(T)$, because a line joining two midpoints of a triangle is half as long as the opposite side. Proceeding inductively, we obtain a sequence $\{T_n : n = 1, 2, \dots\}$ of triangles such that $L(T_n) = 2^{-n}L(T)$, $\hat{T}_{n+1} \subseteq \hat{T}_n$, and

$$|\int_T f(z)\,dz| \le 4^n|\int_{T_n} f(z)\,dz|. \tag{1}$$

Now the $\hat{T}_n$ form a decreasing sequence of nonempty closed and bounded (hence compact) sets in $\mathbb{C}$ whose diameters approach 0 as $n \to \infty$. Thus there is a point $z_0 \in \cap_{n=1}^{\infty}\hat{T}_n$. (If the intersection is empty, then by compactness, some finite collection of $\hat{T}_i$'s would have empty intersection.) Since f is analytic at z_0, there is a continuous function $\epsilon : \Omega \to \mathbb{C}$ with $\epsilon(z_0) = 0$ [see (5) of (1.3.1)] and such that

$$f(z) = f(z_0) + (z - z_0)[f'(z_0) + \epsilon(z)],\ z \in \Omega. \tag{2}$$

By (2) and (2.1.7a), we have

$$\int_{T_n} f(z)\,dz = \int_{T_n} (z - z_0)\epsilon(z)\,dz,\ n = 1, 2, 3, \dots . \tag{3}$$

But by the M-L theorem (2.1.5),

$$\begin{aligned}|\int_{T_n}(z-z_0)\epsilon(z)\,dz| &\leq \sup_{z\in T_n}[|\epsilon(z)|\,|z-z_0|]L(T_n)\\ &\leq \sup_{z\in T_n}|\epsilon(z)|(L(T_n))^2 \text{ since } z\in\hat{T}_n\\ &\leq \sup_{z\in T_n}|\epsilon(z)|4^{-n}(L(T))^2\\ &\to 0 \text{ as } n\to\infty.\end{aligned}$$

Thus by (1) and (3),

$$|\int_T f(z)\,dz| \leq \sup_{z\in T_n}|\epsilon(z)|(L(T))^2 \to 0$$

as $n\to\infty$, because $\epsilon(z_0)=0$. We conclude that $\int_T f(z)\,dz=0$. ♣

We may now state formally the result developed in the discussion preceding Cauchy's theorem.

2.1.9 Cauchy's Theorem for Starlike Regions

Let f be analytic on the starlike region Ω. Then f has a primitive on Ω, and consequently, by (2.1.6), $\int_\gamma f(z)\,dz=0$ for every closed path γ in Ω.

Proof. Let z_0 be a star center for Ω, and define F on Ω by $F(z)=\int_{[z_0,z]}f(w)\,dw$. It follows from (2.1.8) and discussion preceding it that F is a primitive for f. ♣

We may also prove the following converse to Theorem (2.1.6).

2.1.10 Theorem

If $f:\Omega\to\mathbb{C}$ is continuous and $\int_\gamma f(z)\,dz=0$ for every closed path γ in Ω, then f has a primitive on Ω.

Proof. We may assume that Ω is connected (if not we can construct a primitive of f on each component of Ω, and take the union of these to obtain a primitive of f on Ω). So fix $z_0\in\Omega$, and for each $z\in\Omega$, let γ_z be a polygonal path in Ω from z_0 to z. Now define F on Ω by $F(z)=\int_{\gamma_z}f(w)\,dw, z\in\Omega$. Then the discussion preceding (2.1.8) may be repeated without essential change to show that $F'=f$ on Ω. (In Equation (1) in that discussion, $[z_0,z]$ and $[z_0,z_1]$ are replaced by the polygonal paths γ_z and γ_{z_1}, but the line segment $[z_1,z]$ can be retained for all z sufficiently close to z_1.) ♣

2.1.11 Remarks

(a) If $\gamma:[a,b]\to\mathbb{C}$ is a path, we may traverse γ backwards by considering the path λ defined by $\lambda(t)=\gamma(a+b-t)$, $a\leq t\leq b$. Then $\lambda^*=\gamma^*$ and for every continuous function

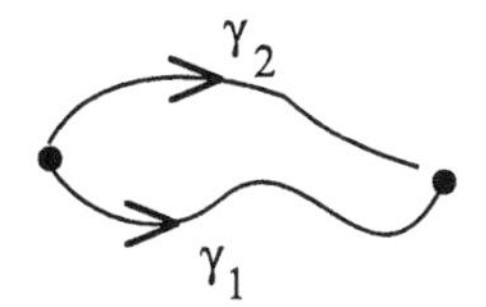

Figure 2.1.2

$f : \gamma^* \to \mathbb{C}$, it follows from the definition of the integral and a brief change of variable argument that

$$\int_{\lambda} f(z)\,dz = -\int_{\gamma} f(z)\,dz.$$

(b) Similarly, if $\gamma_1 : [a,b] \to \mathbb{C}$ and $\gamma_2 : [c,d] \to \mathbb{C}$ are paths with $\gamma_1(b) = \gamma_2(c)$, we may attach γ_2 to γ_1 via the path

$$\gamma(t) = \begin{cases} \gamma_1((1-2t)a + 2tb), & 0 \le t \le 1/2 \\ \gamma_2((2-2t)c + (2t-1)d), & 1/2 \le t \le 1. \end{cases}$$

Then $\gamma^* = \gamma_1^* \cup \gamma_2^*$ and for every continuous function $f : \gamma^* \to \mathbb{C}$,

$$\int_{\gamma} f(z)\,dz = \int_{\gamma_1} f(z)\,dz + \int_{\gamma_2} f(z)\,dz.$$

There is a technical point that should be mentioned. The path $\gamma_1(t), a \le t \le b$, is strictly speaking not the same as the path $\gamma_1((1-2t)a + 2tb), 0 \le t \le 1/2$, since they have different domains of definition. Given the path $\gamma_1 : [a,b] \to \mathbb{C}$, we are forming a new path $\delta = \gamma_1 \circ h$, where $h(t) = (1-2t)a + 2tb, 0 \le t \le 1/2$. It is true then that $\delta^* = \gamma_1^*$ and for every continuous function f on γ_1^*, $\int_{\gamma_1} f(z)\,dz = \int_{\delta} f(z)\,dz$. Problem 4 is a general result of this type.

(c) If γ_1 and γ_2 are paths with the same initial point and the same terminal point, we may form a closed path γ by first traversing γ_1 and then traversing γ_2 backwards. If f is continuous on γ^*, then $\int_{\gamma} f(z)\,dz = 0$ iff $\int_{\gamma_1} f(z)\,dz = \int_{\gamma_2} f(z)\,dz$ (see Figure 2.1.2).

An Application of 2.1.9

Let $\Gamma = [z_1, z_2, z_3, z_4, z_1]$ be a rectangle with center at 0 (see Figure 2.1.3); let us calculate $\int_{\Gamma} \frac{1}{z}\,dz$. Let γ be a circle that circumscribes the rectangle Γ, and let $\gamma_1, \gamma_2, \gamma_3, \gamma_4$ be the arcs of γ joining z_1 to z_2, z_2 to z_3, z_3 to z_4 and z_4 to z_1 respectively. There is an open half plane (a starlike region) excluding 0 but containing both $[z_1, z_2]$ and γ_1^*. By (2.1.9) and Remark (2.1.11c), the integral of $1/z$ on $[z_1, z_2]$ equals the integral of $1/z$ on γ_1. By considering the other segments of Γ and the corresponding arcs of γ, we obtain

$$\int_{\Gamma} \frac{1}{z}\,dz = \int_{\gamma} \frac{1}{z}\,dz = \pm 2\pi i$$

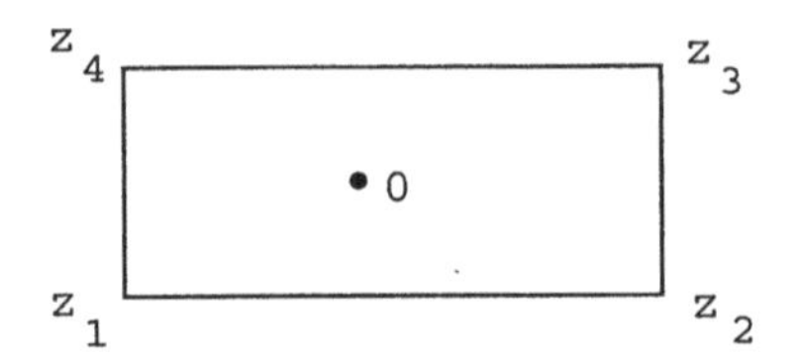

Figure 2.1.3

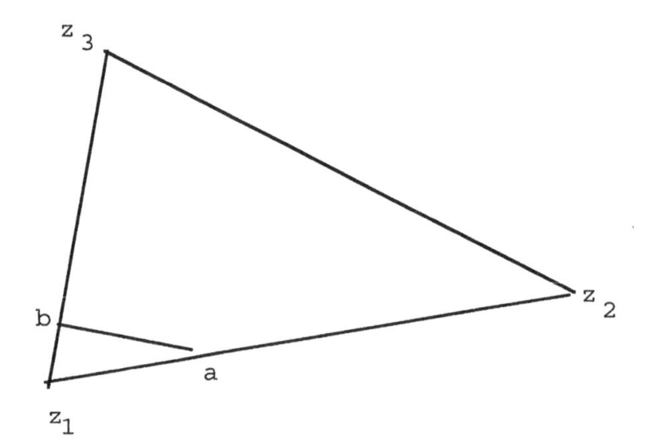

Figure 2.1.4

by a direct calculation, as in (2.1.7a).

The reader who feels that the machinery used to obtain such a simple result is excessive is urged to attempt to compute $\int_\Gamma \frac{1}{z}\,dz$ directly.

The following strengthened form of Cauchy's Theorem for triangles and for starlike regions will be useful in the next section.

2.1.12 Extended Cauchy Theorem for Triangles

Let f be continuous on Ω and analytic on $\Omega \setminus \{z_0\}$. If T is any triangle such that $\hat{T} \subseteq \Omega$, then $\int_T f(z)\,dz = 0$.

Proof. Let $T = [z_1, z_2, z_3, z_1]$. If $z_0 \notin \hat{T}$, the result follows from (2.1.8), Cauchy's theorem for triangles. Also, if z_1, z_2 and z_3 are collinear, then $\int_T f(z)\,dz = 0$ for any continuous (not necessarily analytic) function. Thus assume that z_1, z_2 and z_3 are non-collinear and that $z_0 \in \hat{T}$. Suppose first that z_0 is a vertex, say $z_0 = z_1$. Choose points $a \in [z_1, z_2]$ and $b \in [z_1, z_3]$; see Figure 2.1.4. By (2.1.9),

$$\int_T f(z)\,dz = \int_{[z_1,a]} f(z)\,dz + \int_{[a,b]} f(z)\,dz + \int_{[b,z_1]} f(z).\,dz$$

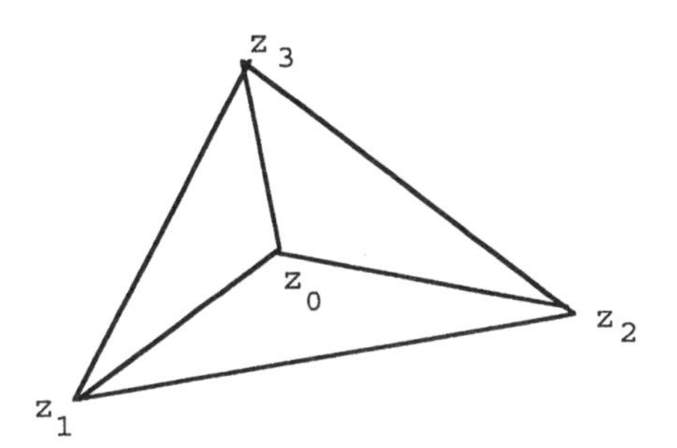

Figure 2.1.5

Since f is continuous at $z_0 = z_1$, each of the integrals on the right approaches zero as $a, b \to z_1$, by the M-L theorem. Therefore $\int_T f(z)\,dz = 0$.

If $z_0 \in \hat{T}$ is not a vertex, join z_0 to each vertex of T by straight line segments (see Figure 2.1.5), and write $\int_T f(z)\,dz$ as a sum of integrals, each of which is zero by the above argument. ♣

2.1.13 Extended Cauchy Theorem for Starlike Regions

Let f be continuous on the starlike region Ω and analytic on $\Omega \setminus \{z_0\}$. Then f has a primitive on Ω, and consequently $\int_\gamma f(z)\,dz = 0$ for every closed path γ in Ω.

Proof. Exactly as in (2.1.9), using (2.1.12) instead of (2.1.8). ♣

Problems

1. Evaluate $\int_{[-i,1+2i]} \operatorname{Im} z\,dz$.
2. Evaluate $\int_\gamma \overline{z}\,dz$ where γ traces the arc of the parabola $y = x^2$ from (1,1) to (2,4).
3. Evaluate $\int_{[z_1,z_2,z_3]} f(z)\,dz$ where $z_1 = -i, z_2 = 2+5i, z_3 = 5i$ and $f(x+iy) = x^2 + iy$.
4. Show that $\int_\gamma f(z)\,dz$ is independent of the parametrization of γ^* in the following sense. Let $h : [c,d] \to [a,b]$ be one-to-one and continuously differentiable, with $h(c) = a$ and $h(d) = b$ (γ is assumed to be defined on $[a,b]$). Let $\gamma_1 = \gamma \circ h$. Show that γ_1 is a path, and prove that if f is continuous on γ^*, then $\int_{\gamma_1} f(z)\,dz = \int_\gamma f(z)\,dz$.
5. In the next section it will be shown that if f is analytic on Ω, then f' is also analytic, in particular continuous, on Ω. Anticipating this result, we can use (2.1.6), the fundamental theorem for integration along paths, to show that $\int_\gamma f'(z)\,dz = f(\gamma(b)) - f(\gamma(a))$. Prove the following.
   (a) If Ω is convex and $\operatorname{Re} f' > 0$ on Ω, then f is one-to-one. (Hint: $z_1, z_2 \in \Omega$ with $z_1 \neq z_2$ implies that $\operatorname{Re}[(f(z_2) - f(z_1))/(z_2 - z_1)] > 0$.)
   (b) Show that (a) does *not* generalize to starlike regions. (Consider $z + 1/z$ on a suitable region.)

(c) Suppose $z_0 \in \Omega$ and $f'(z_0) \neq 0$. Show that there exists $r > 0$ such that f is one-to-one on $D(z_0, r)$. Consequently, if f' has no zeros in Ω, then f is locally one-to-one.

2.2 Power Series

In this section we develop the basic facts about complex series, especially complex power series. The main result is that f is analytic at z_0 iff f can be represented as a convergent power series throughout some neighborhood of z_0. We first recall some elementary facts about complex series in general.

2.2.1 Definition

Given a sequence $w_0, w_1, w_2, \ldots$ of complex numbers, consider the series $\sum_{n=0}^{\infty} w_n$. If $\lim_{n\to\infty} \sum_{k=0}^{n} w_k$ exists and is the complex number w, we say that *the series converges to* w and write $w = \sum_{n=0}^{\infty} w_n$. Otherwise, the series is said to *diverge.*

A useful observation is that a series is convergent iff the partial sums $\sum_{k=0}^{n} w_k$ form a *Cauchy sequence*, that is, $\sum_{k=m}^{n} w_k \to 0$ as $m, n \to \infty$.

The series $\sum_{n=0}^{\infty} w_n$ is said to *converge absolutely* if the series $\sum_{n=0}^{\infty} |w_n|$ is convergent. As in the real variables case, an absolutely convergent series is convergent. A necessary and sufficient condition for absolute convergence is that the sequence of partial sums $\sum_{k=0}^{n} |w_k|$ be bounded. The two most useful tests for absolute convergence of complex series are the ratio and root tests.

2.2.2 The Ratio Test

If $\sum w_n$ is a series of nonzero terms and if $\limsup_{n\to\infty} |\frac{w_{n+1}}{w_n}| < 1$, then the series converges absolutely. If $|\frac{w_{n+1}}{w_n}| \geq 1$ for all sufficiently large n, the series diverges.

2.2.3 The Root Test

Let $\sum w_n$ be any complex series. If $\limsup_{n\to\infty} |w_n|^{1/n} < 1$, the series converges absolutely, while if $\limsup_{n\to\infty} |w_n|^{1/n} > 1$, the series diverges.

The ratio test is usually (but not always) easier to apply in explicit examples, but the root test has a somewhat wider range of applicability and, in fact, is the test that we are going to use to obtain some basic properties of power series. Proofs and a discussion of the relative utility of the tests can be found in most texts on real analysis.

We now consider sequences and series of complex-valued functions.

2.2.4 Theorem

Let $\{f_n\}$ be sequence of complex-valued functions on a set S. Then $\{f_n\}$ *converges pointwise* on S (that is, for each $z \in S$, the sequence $\{f_n(z)\}$ is convergent in $\mathbb{C}$) iff $\{f_n\}$ is *pointwise Cauchy* (that is, for each $z \in S$, the sequence $\{f_n(z)\}$ is a Cauchy sequence in $\mathbb{C}$). Also, $\{f_n\}$ *converges uniformly* iff $\{f_n\}$ is *uniformly Cauchy* on S, in other words, $|f_n(z) - f_m(z)| \to 0$ as $m, n \to \infty$, uniformly for $z \in S$.

(The above result holds just as well if the f_n take their values in an arbitrary complete metric space.)

Proof. As in the real variables case; see Problem 2.2.1. ♣

The next result gives the most useful test for uniform convergence of infinite series of functions.

2.2.5 The Weierstrass M-Test

Let $g_1, g_2, \ldots$ be complex-valued functions on a set S, and assume that $|g_n(z)| \leq M_n$ for all $z \in S$. If $\sum_{n=1}^{\infty} M_n < +\infty$, then the series $\sum_{n=1}^{\infty} g_n(z)$ converges uniformly on S.

Proof. Let $f_n = \sum_{k=1}^{n} g_k$; it follows from the given hypothesis that $\{f_n\}$ is uniformly Cauchy on S. The result now follows from (2.2.4). ♣

We now consider *power series*, which are series of the form $\sum_{n=0}^{\infty} a_n(z-z_0)^n$, where z_0 and the a_n are complex numbers. Thus we are dealing with series of functions $\sum_{n=0}^{\infty} f_n$ of a very special type, namely $f_n(z) = a_n(z-z_0)^n$. Our first task is to describe the sets $S \subseteq \mathbb{C}$ on which such a series will converge.

2.2.6 Theorem

If $\sum_{n=0}^{\infty} a_n(z-z_0)^n$ converges at the point z with $|z-z_0| = r$, then the series converges absolutely on $D(z_0, r)$, uniformly on each closed subdisk of $D(z_0, r)$, hence uniformly on each compact subset of $D(z_0, r)$.

Proof. We have $|a_n(z'-z_0)^n| = |a_n(z-z_0)^n| \left|\frac{z'-z_0}{z-z_0}\right|^n$. The convergence at z implies that $a_n(z-z_0)^n \to 0$, hence the sequence $\{a_n(z-z_0)^n\}$ is bounded. If $|z'-z_0| \leq r' < r$, then

$$\left|\frac{z'-z_0}{z-z_0}\right| \leq \frac{r'}{r} < 1$$

proving absolute convergence at z' (by comparison with a geometric series). The Weierstrass M-test shows that the series converges uniformly on $\overline{D}(z_0, r')$. ♣

We now describe convergence in terms of the coefficients a_n.

2.2.7 Theorem

Let $\sum_{n=0}^{\infty} a_n(z-z_0)^n$ be a power series. Let $r = [\limsup_{n\to\infty}(|a_n|^{1/n})]^{-1}$, the radius of convergence of the series. (Adopt the convention that $1/0 = \infty, 1/\infty = 0$.) The series converges absolutely on $D(z_0, r)$, uniformly on compact subsets. The series diverges for $|z-z_0| > r$.

Proof. We have $\limsup_{n\to\infty} |a_n(z-z_0)^n|^{1/n} = (|z-z_0|)/r$, which will be less than 1 if $|z-z_0| < r$. By (2.2.3), the series converges absolutely on $D(z_0, r)$. Uniform convergence on compact subsets follows from (2.2.6). (We do not necessarily have convergence for $|z-z_0| = r$, but we do have convergence for $|z-z_0| = r'$, where $r' < r$ can be chosen arbitrarily close to r.) If the series converges at some point z with $|z-z_0| > r$, then by (2.2.6) it converges absolutely at points z' such that $r < |z'-z_0| < |z-z_0|$. But then $(|z-z_0|)/r > 1$, contradicting (2.2.3). ♣

2.2.8 Definition

Let $C(z_0, r)$ denote the circle with center z_0 and radius r. then $\int_{C(z_0,r)} f(z)\,dz$ is defined as $\int_\gamma f(z)\,dz$ where $\gamma(t) = z_0 + re^{it}, 0 \le t \le 2\pi$.

The following result provides the essential equipment needed for the theory of power series. In addition, it illustrates the striking difference between the concept of differentiability of complex functions and the analogous idea in the real case. We are going to show that if f is analytic on a closed disk, then the value of f at any interior point is completely determined by its values on the boundary, and furthermore there is an explicit formula describing the dependence.

2.2.9 Cauchy's Integral Formula for a Circle

Let f be analytic on Ω and let $D(z_0, r)$ be a disk such that $\overline{D}(z_0, r) \subseteq \Omega$. Then

$$f(z) = \frac{1}{2\pi i} \int_{C(z_0,r)} \frac{f(w)}{w - z}\,dw, \quad z \in D(z_0, r).$$

Proof. Let $D(z_0, \rho)$ be a disk such that $\overline{D}(z_0, r) \subseteq D(z_0, \rho) \subseteq \Omega$. Fix $z \in D(z_0, r)$ and define a function g on $D(z_0, \rho)$ by

$$g(w) = \begin{cases} \frac{f(w)-f(z)}{w-z} & \text{if } w \neq z \\ f'(z) & \text{if } w = z. \end{cases}$$

Then g is continuous on $D(z_0, \rho)$ and analytic on $D(z_0, \rho) \setminus \{z\}$, so we may apply (2.1.13) to get $\int_{C(z_0,r)} g(w)\,dw = 0$. Therefore

$$\frac{1}{2\pi i} \int_{C(z_0,r)} \frac{f(w)}{w - z}\,dw = \frac{f(z)}{2\pi i} \int_{C(z_0,r)} \frac{1}{w - z}\,dw.$$

Now

$$\int_{C(z_0,r)} \frac{1}{w - z}\,dw = \int_{C(z_0,r)} \frac{1}{(w - z_0) - (z - z_0)}\,dw = \int_{C(z_0,r)} \sum_{n=0}^{\infty} \frac{(z - z_0)^n}{(w - z_0)^{n+1}}\,dw$$

The series converges uniformly on $C(z_0, r)$ by the Weierstrass M-test, and hence we may integrate term by term to obtain

$$\sum_{n=0}^{\infty} (z - z_0)^n \int_{C(z_0,r)} \frac{1}{(w - z_0)^{n+1}}\,dw.$$

But on $C(z_0, r)$ we have $w = z_0 + re^{it}, 0 \le t \le 2\pi$, so the integral on the right is, by (2.2.8),

$$\int_0^{2\pi} r^{-(n+1)} e^{-i(n+1)t} ire^{it}\,dt = \begin{cases} 0 & \text{if } n = 1, 2, \ldots \\ 2\pi i & \text{if } n = 0. \end{cases}$$

We conclude that $\int_{C(z_0,r)} \frac{1}{w-z}\, dw = 2\pi i$, and the result follows. ♣

The integral appearing in Cauchy's formula is an example of what is known as an *integral of the Cauchy type.* The next result, which will be useful later, deals with these integrals.

2.2.10 Theorem

Let γ be a path (not necessarily closed) and let g be a complex-valued continuous function on γ^*. Define a function F on the open set $\Omega = \mathbb{C} \setminus \gamma^*$ by

$$F(z) = \int_\gamma \frac{g(w)}{w-z}\, dw.$$

Then F has derivatives of all orders on Ω, and

$$F^{(n)}(z) = n! \int_\gamma \frac{g(w)}{(w-z)^{n+1}}\, dw$$

for all $z \in \Omega$ and all $n = 0, 1, 2, \ldots$ (take $F^{(0)} = F$). Furthermore, $F^{(n)}(z) \to 0$ as $|z| \to \infty$.

Proof. We use an induction argument. The formula for $F^{(n)}(z)$ is valid for $n = 0$, by hypothesis. Assume that the formula holds for a given n and all $z \in \Omega$; fix $z_1 \in \Omega$ and choose $r > 0$ small enough that $D(z_1, r) \subseteq \Omega$. For any point $z \in D(z_1, r)$ with $z \neq z_1$ we have

$$\frac{F^{(n)}(z) - F^{(n)}(z_1)}{z - z_1} - (n+1)! \int_\gamma \frac{g(w)}{(w-z_1)^{n+2}}\, dw$$

$$= \frac{n!}{z - z_1} \int_\gamma \frac{(w-z_1)^{n+1} - (w-z)^{n+1}}{(w-z)^{n+1}(w-z_1)^{n+1}} g(w)\, dw - (n+1)! \int_\gamma \frac{g(w)}{(w-z_1)^{n+2}} \tag{1}$$

$$= \frac{n!}{z - z_1} \int_\gamma \frac{(z-z_1)\sum_{k=0}^{n}(w-z_1)^{n-k}(w-z)^k}{(w-z)^{n+1}(w-z_1)^{n+1}} g(w)\, dw - (n+1)! \int_\gamma \frac{g(w)}{(w-z_1)^{n+2}}\, dw \tag{2}$$

where the numerator of the first integral in (2) is obtained from that in (1) by applying the algebraic identity $a^{n+1} - b^{n+1} = (a-b)\sum_{k=0}^{n} a^{n-k}b^k$ with $a = w - z_1$ and $b = w - z$. Thus

$$\left| \frac{F^{(n)}(z) - F^{(n)}(z_1)}{z - z_1} - (n+1)! \int_\gamma \frac{g(w)}{(w-z_1)^{n+2}}\, dw \right|$$

$$= n! \left| \int_\gamma \frac{\sum_{k=0}^{n}(w-z_1)^{n-k+1}(w-z)^k - (n+1)(w-z)^{n+1}}{(w-z)^{n+1}(w-z_1)^{n+2}} g(w)\, dw \right|$$

$$\leq n!\left[\max_{w\in\gamma^*}\left|\frac{\sum_{k=0}^{n}(w-z_1)^{n-k+1}(w-z)^k-(n+1)(w-z)^{n+1}}{(w-z)^{n+1}(w-z_1)^{n+2}}g(w)\right|\right]L(\gamma)$$

by the M-L theorem. But the max that appears in brackets approaches 0 as $z \to z_1$, since $\sum_{k=0}^{n}(w-z_1)^{n-k+1}(w-z)^k \to \sum_{k=0}^{n}(w-z_1)^{n+1} = (n+1)(w-z_1)^{n+1}$. Hence

$$\frac{F^{(n)}(z)-F^{(n)}(z_1)}{z-z_1} \to (n+1)!\int_\gamma \frac{g(w)}{(w-z_1)^{n+2}}\,dw$$

as $z \to z_1$, and the statement of the theorem follows by induction. The fact that $|F^{(n)}(z)| \to 0$ as $|z| \to \infty$ is a consequence of the M-L theorem; specifically,

$$|F^{(n)}(z)| \leq n!\left[\max_{w\in\gamma^*}\frac{|g(w)|}{|w-z|^{n+1}}\right]L(\gamma). \;\clubsuit$$

Theorems 2.2.9 and 2.2.10 now yield some useful corollaries.

2.2.11 Corollary

If f is analytic on Ω, then f has derivatives of all orders on Ω. Moreover, if $\overline{D}(z_0, r) \subseteq \Omega$, then

$$f^{(n)}(z) = \frac{n!}{2\pi i}\int_{C(z_0,r)}\frac{f(w)}{(w-z)^{n+1}}\,dw, \quad z \in D(z_0, r).$$

Proof. Apply (2.2.10) to the Cauchy integral formula (2.2.9). ♣

2.2.12 Corollary

If f has a primitive on Ω, then f is analytic on Ω.

Proof. Apply (2.2.11) to any primitive for f. ♣

2.2.13 Corollary

If f is continuous on Ω and analytic on $\Omega \setminus \{z_0\}$, then f is analytic on Ω.

Proof. Choose any disk D such that $D \subseteq \Omega$. By (2.1.13), f has a primitive on D, hence by (2.2.12), f is analytic on D. It follows that f is analytic on Ω. ♣

The next result is a converse to Cauchy's theorem for triangles.

2.2.14 Morera's Theorem

Suppose f is continuous on Ω and $\int_T f(z)\,dz = 0$ for each triangle T such that $\hat{T} \subseteq \Omega$. Then f is analytic on Ω.

Proof. Let D be any disk contained in Ω. The hypothesis implies that f has a primitive on D [see the discussion preceding (2.1.8)]. Thus by (2.2.12), f is analytic on D. Since D is an arbitrary disk in Ω, f is analytic on Ω. ♣

One of many applications of Morera's theorem is the Schwarz reflection principle, which deals with the problem of extending an analytic function to a larger domain.

2.2.15 The Schwarz Reflection Principle

Suppose that f is analytic on the open upper half plane $\mathbb{C}^+ = \{z : \operatorname{Im} z > 0\}$, f is continuous on the closure $\mathbb{C}^+ \cup \mathbb{R}$ of $\mathbb{C}^+$, and $\operatorname{Im} f(z) = 0$ for $z \in \mathbb{R}$. Then f has an analytic extension to all of $\mathbb{C}$.

Proof. We will give an outline of the argument, leaving the details to the problems at the end of the section. Extend f to a function f^* defined on $\mathbb{C}$ by

$$f^*(z) = \begin{cases} f(z), & z \in \mathbb{C}^+ \cup \mathbb{R} \\ \\ \overline{f(\overline{z})}, & z \notin \mathbb{C}^+ \cup \mathbb{R}. \end{cases}$$

Then f^* is analytic on $\mathbb{C} \setminus \mathbb{R}$ and continuous on $\mathbb{C}$ (Problem 10). One can then use Morera's theorem to show that f^* is analytic on $\mathbb{C}$ (Problem 11). ♣

We now complete the discussion of the connection between analytic functions and power series, showing in essence that the two notions are equivalent. We say that a function $f : \Omega \to \mathbb{C}$ is *representable in Ω by power series* if given $D(z_0, r) \subseteq \Omega$, there is a sequence $\{a_n\}$ of complex numbers such that $f(z) = \sum_{n=0}^{\infty} a_n (z - z_0)^n$, $z \in D(z_0, r)$.

2.2.16 Theorem

If f is analytic on Ω, then f is representable in Ω by power series. In fact, if $D(z_0, r) \subseteq \Omega$, then

$$f(z) = \sum_{n=0}^{\infty} \frac{f^{(n)}(z_0)}{n!}(z - z_0)^n, \quad z \in D(z_0, r).$$

As is the usual practice, we will call this series the *Taylor expansion* of f about z_0.

Proof. Let $D(z_0, r) \subseteq \Omega$, fix any $z \in D(z_0, r)$, and choose r_1 such that $|z - z_0| < r_1 < r$. By (2.2.9), Cauchy's formula for a circle,

$$f(z) = \frac{1}{2\pi i} \int_{C(z_0, r_1)} \frac{f(w)}{w - z} \, dw.$$

Now for $w \in C(z_0, r_1)$,

$$\frac{f(w)}{w - z} = \frac{f(w)}{(w - z_0) - (z - z_0)} = \frac{f(w)}{w - z_0} \cdot \frac{1}{1 - \frac{z - z_0}{w - z_0}} = \sum_{n=0}^{\infty} f(w) \frac{(z - z_0)^n}{(w - z_0)^{n+1}}.$$

The n-th term of the series has absolute value at most

$$\max_{w \in C(z_0, r_1)} |f(w)| \cdot \frac{|z - z_0|^n}{r_1^{n+1}} = \frac{1}{r_1} \max_{w \in C(z_0, r_1)} |f(w)| \left[\frac{|z - z_0|}{r_1} \right]^n.$$

Since $\frac{|z - z_0|}{r_1} < 1$, the Weierstrass M-test shows that the series converges uniformly on $C(z_0, r_1)$. Hence we may integrate term by term, obtaining

$$f(z) = \sum_{n=0}^{\infty} \left[\frac{1}{2\pi i} \int_{C(z_0, r_1)} \frac{f(w)}{(w - z_0)^{n+1}} \, dw \right] (z - z_0)^n = \sum_{n=0}^{\infty} \frac{f^{(n)}(z_0)}{n!} (z - z_0)^n$$

by (2.2.11). ♣

In order to prove the converse of (2.2.16), namely that a function representable in Ω by power series is analytic on Ω, we need the following basic result.

2.2.17 Theorem

Let $\{f_n\}$ be a sequence of analytic functions on Ω such that $f_n \to f$ uniformly on compact subsets of Ω. Then f is analytic on Ω, and furthermore, $f_n^{(k)} \to f^{(k)}$ uniformly on compact subsets of Ω for each $k = 1, 2, \dots$.

Proof. First let $\overline{D}(z_0, r)$ be any closed disk contained in Ω. Then we can choose $\rho > r$ such that $\overline{D}(z_0, \rho) \subseteq \Omega$ also. For each $z \in D(z_0, \rho)$ and $n = 1, 2, \dots$, we have, by (2.2.9),

$$f_n(z) = \frac{1}{2\pi i} \int_{C(z_0,\rho)} \frac{f(w)}{w - z}\, dw.$$

By (2.2.10), f is analytic on $D(z_0, \rho)$. It follows that f is analytic on Ω. Now by (2.2.11),

$$f_n^{(k)}(z) - f^{(k)}(z) = \frac{k!}{2\pi i} \int_{C(z_0,\rho)} \frac{f_n(w) - f(w)}{(w - z)^{k+1}}\, dw$$

and if z is restricted to $\overline{D}(z_0, r)$, then by the M-L theorem,

$$|f_n^{(k)}(z) - f^{(k)}(z)| \le \frac{k!}{2\pi} \left[\max_{w \in C(z_0,\rho)} |f_n(w) - f(w)| \right] \frac{2\pi\rho}{(\rho - r)^{k+1}} \to 0 \text{ as } n \to \infty.$$

Thus we have shown that f is analytic on Ω and that $f_n^{(k)} \to f^{(k)}$ uniformly on closed *subdisks* of Ω. Since any compact subset of Ω can be covered by finitely many closed subdisks, the statement of the theorem follows. ♣

The converse of (2.2.16) can now be readily obtained.

2.2.18 Theorem

If f is representable in Ω by power series, then f is analytic on Ω.

Proof. Let $D(z_0, r) \subseteq \Omega$, and let $\{a_n\}$ be such that $f(z) = \sum_{n=0}^{\infty} a_n (z - z_0)^n, z \in D(z_0, r)$. By (2.2.7), the series converges uniformly on compact subsets of $D(z_0, r)$, hence by (2.2.17), f is analytic on Ω. ♣

Remark

Since the above series converges uniformly on compact subsets of $D(z_0, r)$, Theorem 2.2.17 also allows us to derive the power series expansion of $f^{(k)}$ from that of f, and to show that the coefficients $\{a_n\}$ are uniquely determined by z_0 and f. For if $f(z)$ is given by $\sum_{n=0}^{\infty} a_n (z - z_0)^n, z \in D(z_0, r)$, we may differentiate term by term to obtain

$$f^{(k)}(z) = \sum_{n=k}^{\infty} n(n-1) \cdots (n - k + 1) a_n (z - z_0)^{n-k},$$

and if we set $z = z_0$, we find that

$$a_k = \frac{f^{(k)}(z_0)}{k!}.$$

We conclude this section with a result promised in Chapter 1 [see (1.6.1)].

2.2.19 Theorem

If $f = u + iv$ is analytic on Ω, then u and v are harmonic on Ω.

Proof. By (1.4.2), $f' = \frac{\partial u}{\partial x} + i\frac{\partial v}{\partial x} = \frac{\partial v}{\partial y} - i\frac{\partial u}{\partial y}$. But by (2.2.11), f' is also analytic on Ω, and thus the Cauchy-Riemann equations for f' are also satisfied. Consequently,

$$\frac{\partial}{\partial x}\left(\frac{\partial u}{\partial x}\right) = \frac{\partial}{\partial y}\left(-\frac{\partial u}{\partial y}\right), \quad \frac{\partial}{\partial x}\left(\frac{\partial v}{\partial x}\right) = -\frac{\partial}{\partial y}\left(\frac{\partial v}{\partial y}\right).$$

These partials are all continuous because f'' is also analytic on Ω. ♣

Problems

1. Prove Theorem 2.2.4.
2. If $\sum a_n z^n$ has radius of convergence r, show that the differentiated series $\sum n a_n z^{n-1}$ also has radius of convergence r.
3. Let $f(x) = e^{-1/x^2}, x \neq 0; f(0) = 0$. Show that f is infinitely differentiable on $(-\infty, \infty)$ and $f^{(n)}(0) = 0$ for all n. Thus the Taylor series for f is identically 0, hence does not converge to f. Conclude that if $r > 0$, there is no function g analytic on $D(0, r)$ such that $g = f$ on $(-r, r)$.
4. Let $\{a_n : n = 0, 1, 2 \dots\}$ be an arbitrary sequence of complex numbers.
(a) If $\limsup_{n\to\infty} |a_{n+1}/a_n| = \alpha$, what conclusions can be drawn about the radius of convergence of the power series $\sum_{n=0}^{\infty} a_n z^n$?
(b) If $|a_{n+1}/a_n|$ approaches a limit α, what conclusions can be drawn?
5. If f is analytic at z_0, show that it is not possible that $|f^{(n)}(z_0)| > n! b_n$ for all $n = 1, 2, \dots$, where $(b_n)^{1/n} \to \infty$ as $n \to \infty$.
6. Let $R_n(z)$ be the remainder after the term of degree n in the Taylor expansion of a function f about z_0.
(a) Show that
$$R_n(z) = \frac{(z - z_0)^{n+1}}{2\pi i} \int_\Gamma \frac{f(w)}{(w - z)(w - z_0)^{n+1}}\, dw,$$
where $\Gamma = C(z_0, r_1)$ as in (2.2.16).
(b) If $|z - z_0| \leq s < r_1$, show that
$$|R_n(z)| \leq A(s/r_1)^{n+1}, \text{ where } A = M_f(\Gamma) r_1/(r_1 - s)$$
and $M_f(\Gamma) = \max\{|f(w)| : w \in \Gamma\}$.

7. (Summation by parts). Let $\{a_n\}$ and $\{b_n\}$ be sequences of complex numbers. If $\Delta b_k = b_{k+1} - b_k$, show that

$$\sum_{k=r}^{s} a_k \Delta b_k = a_{s+1} b_{s+1} - a_r b_r - \sum_{k=r}^{s} b_{k+1} \Delta a_k.$$

8. (a) If $\{b_n\}$ is bounded and the a_n are real and greater than 0, with $a_1 \geq a_2 \geq \cdots \to 0$, show that $\sum_{n=1}^{\infty} a_n \Delta b_n$ converges.
(b) If $b_n = b_n(z)$, that is, the b_n are *functions* from a set S to $\mathbb{C}$, the b_n are uniformly bounded on S, and the a_n are real and decrease to 0 as in (a), show that $\sum_{n=1}^{\infty} a_n(b_{n+1}(z) - b_n(z))$ converges uniformly on S.
9. (a) Show that $\sum_{n=1}^{\infty} z^n/n$ converges when $|z| = 1$, except at the single point $z = 1$.
(b) Show that $\sum_{n=1}^{\infty} (\sin nx)/n$ converges for real x, uniformly on $\{x : 2k\pi + \delta \leq x \leq 2(k+1)\pi - \delta\}$, $\delta > 0$, k an integer.
(c) Show that $\sum_{n=1}^{\infty} (\sin nz)/n$ diverges if x is not real. (The complex sine function will be discussed in the next chapter. It is defined by $\sin w = (e^{iw} - e^{-iw})/2i$.
10. Show that the function f^* occurring in the proof of the Schwarz reflection principle is analytic on $\mathbb{C} \setminus \mathbb{R}$ and continuous on $\mathbb{C}$.
11. Show that f^* is analytic on $\mathbb{C}$.
12. Use the following outline to give an alternative proof of the Cauchy integral formula for a circle.
(a) Let

$$F(z) = \int_{C(z_0,r)} \frac{1}{w - z}\, dw, \quad z \notin C(z_0, r).$$

Use (2.2.10), (2.1.6) and (2.1.7b) to show that F is constant on $D(z_0, r)$.
(b) $F(z_0) = 2\pi i$ by direct computation.
Theorem 2.2.9 now follows, thus avoiding the series expansion argument that appears in the text.
13. (a) Suppose f is analytic on $D(a, r)$. Prove that for $0 \leq r < R$,

$$|f^{(n)}(a)| \leq \frac{n!}{2\pi r^n} \int_{-\pi}^{\pi} |f(a + re^{it})|\, dt.$$

(b) Prove that if f is an entire function such that for some $M > 0$ and some natural number k, $|f(z)| \leq M|z|^k$ for $|z|$ sufficiently large, then f is a polynomial of degree at most k.
(c) Let f be an entire function such that $|f(z)| \leq 1 + |z|^{3/2}$ for all z. Prove that there are complex numbers a_0, a_1 such that $f(z) = a_0 + a_1 z$.
14. Let $\{a_n : n = 0, 1, \dots\}$ be a sequence of complex numbers such that $\sum_{n=0}^{\infty} |a_n| < \infty$ but $\sum_{n=0}^{\infty} n|a_n| = \infty$. Prove that the radius of convergence of the power series $\sum a_n z^n$ is equal to 1.
15. Let $\{f_n\}$ be a sequence of analytic functions on Ω such that $\{f_n\}$ converges to f uniformly on compact subsets of Ω. Give a proof that f is analytic on Ω, based on

Morera's theorem [rather than (2.2.10), which was the main ingredient in the proof of (2.2.17)]. Note that in the present problem we need not prove that $f_n^{(k)} \to f^{(k)}$ uniformly on compact subsets of Ω.

2.3 The Exponential and Complex Trigonometric Functions

In this section, we use our results on power series to complete the discussion of the exponential function and to introduce some of the other elementary functions.

Recall (Section 1.5) that exp is defined on $\mathbb{C}$ by $\exp(x+iy) = e^x(\cos y + i\sin y)$; thus exp has magnitude e^x and argument y. The function exp satisfies a long list of properties; for the reader's convenience, we give the justification of each item immediately after the statement.

2.3.1 Theorem

(a) exp is an entire function [this was proved in (1.5.2)].

(b) $\exp(z) = \sum_{n=0}^{\infty} z^n/n!$, $z \in \mathbb{C}$.

Apply (a) and (2.2.16), using the fact [see (1.5.2)] that exp is its own derivative.

(c) $\exp(z_1 + z_2) = \exp(z_1)\exp(z_2)$.

Fix $z_0 \in \mathbb{C}$; for each $z \in \mathbb{C}$, we have, by (2.2.16),

$$\exp(z) = \sum_{n=0}^{\infty} \frac{\exp(z_0)}{n!}(z-z_0)^n = \exp(z_0)\sum_{n=0}^{\infty}\frac{(z-z_0)^n}{n!} = \exp(z_0)\exp(z-z_0) \text{ by (b)}.$$

Now set $z_0 = z_1$ and $z = z_1 + z_2$.

(d) exp has no zeros in $\mathbb{C}$.

By (c), $\exp(z-z) = \exp(z)\exp(-z)$. But $\exp(z-z) = \exp(0) = 1$, hence $\exp(z) \neq 0$.

(e) $\exp(-z) = 1/\exp(z)$ (the argument of (d) proves this also).

(f) $\exp(z) = 1$ iff z is an integer multiple of $2\pi i$.

$\exp(x+iy) = 1$ iff $e^x \cos y = 1$ and $e^x \sin y = 0$ iff $e^x \cos y = 1$ and $\sin y = 0$ iff $x = 0$ and $y = 2n\pi$ for some n.

(g) $|\exp(z)| = e^{\operatorname{Re} z}$ (by definition of exp).

(h) exp has $2\pi i$ as a period, and any other period is an integer multiple of $2\pi i$.

$\exp(z+w) = \exp(z)$ iff $\exp(w) = 1$ by (c), and the result follows from (f).

(i) exp maps an arbitrary vertical line $\{z : \operatorname{Re} z = x_0\}$ onto the circle with center 0 and radius e^{x_0}, and exp maps an arbitrary horizontal line $\{z : \operatorname{Im} z = y_0\}$ one-to-one onto the open ray from 0 through $\exp(iy_0)$.

$\{\exp(z) : \operatorname{Re} z = x_0\} = \{e^{x_0}(\cos y + i\sin y) : y \in \mathbb{R}\}$, which is the circle with center 0 and radius e^{x_0} (covered infinitely many times). Similarly, we have $\{\exp(z) : \operatorname{Im} z = y_0\} = \{e^x e^{iy_0} : x \in \mathbb{R}\}$, which is the desired ray.

(j) For each real number α, exp restricted to the horizontal strip $\{x+iy : \alpha \le y < \alpha+2\pi\}$, is a one-to-one map onto $\mathbb{C} \setminus \{0\}$.

This follows from (i) and the observation that as y_0 ranges over $[\alpha, \alpha+2\pi)$, the open rays from 0 through e^{iy_0} sweep out $\mathbb{C} \setminus \{0\}$. ♣

Notation

We will often write e^z for $\exp(z)$. We now define $\sin z$ and $\cos z$ by

$$\sin z = \frac{e^{iz} - e^{-iz}}{2i}, \qquad \cos z = \frac{e^{iz} + e^{-iz}}{2}.$$

These definitions are consistent with, and are motivated by, the fact that $e^{iy} = \cos y + i \sin y, y \in \mathbb{R}$.

Since exp is an entire function, it follows from the chain rule that sin and cos are also entire functions and the usual formulas $\sin' = \cos$ and $\cos' = -\sin$ hold. Also, it follows from property (f) of exp that sin and cos have no additional zeros in the complex plane, other than those on the real line. (Note that $\sin z = 0$ iff $e^{iz} = e^{-iz}$ iff $e^{2iz} = 1$.) However, unlike $\sin z$ and $\cos z$ for real z, sin and cos are *not* bounded functions. This can be deduced directly from the above definitions, or from Liouville's theorem, to be proved in the next section.

The familiar power series representations of sin and cos hold [and may be derived using (2.2.16)]:

$$\sin z = \sum_{n=0}^{\infty} (-1)^n \frac{z^{2n+1}}{(2n+1)!}, \qquad \cos z = \sum_{n=0}^{\infty} (-1)^n \frac{z^{2n}}{(2n)!}.$$

Other standard trigonometric functions can be defined in the usual way; for example, $\tan z = \sin z / \cos z$. Usual trigonometric identities and differentiation formulas hold, for instance, $\sin(z_1 + z_2) = \sin z_1 \cos z_2 + \cos z_1 \sin z_2$, $\frac{d}{dz} \tan z = \sec^2 z$, and so on.

Hyperbolic functions are defined by

$$\cosh z = \frac{e^z + e^{-z}}{2}, \qquad \sinh z = \frac{e^z - e^{-z}}{2}.$$

The following identities can be derived from the definitions:

$$\cos iz = \cosh z, \quad \sin iz = i \sinh z$$

$$\sin(x+iy) = \sin x \cosh y + i \cos x \sinh y, \quad \cos(x+iy) = \cos x \cosh y - i \sin x \sinh y.$$

Also, $\sinh z = 0$ iff $z = in\pi$, n an integer; $\cosh z = 0$ iff $z = i(2n+1)\pi/2$, n an integer.

Problems

1. Show that for any integer k, $\sin z$ maps the strip $\{x+iy : (2k-1)\pi/2 < x < (2k+1)\pi/2\}$ one-to-one onto $\mathbb{C} \setminus \{u+iv : v=0, |u| \geq 1\}$, and maps $\{x+iy : x=(2k+1)\pi/2, y \geq 0\} \cup \{x+iy : x=(2k-1)\pi/2, y \leq 0\}$ one-to-one onto $\{u+iv : v=0, |u| \geq 1\}$.
2. Find all solutions of the equation $\sin z = 3$.
3. Calculate $\int_{C(0,1)} \frac{\sin z}{z^4}\, dz$.
4. Prove that given $r > 0$, there exists n_0 such that if $n \geq n_0$, then $1+z+z^2/2!+\cdots+z^n/n!$ has all its zeros in $|z| > r$.
5. Let f be an entire function such that $f'' + f = 0, f(0) = 0$, and $f'(0) = 1$. Prove that $f(z) = \sin z$ for all $z \in \mathbb{C}$.
6. Let f be an entire function such that $f' = f$ and $f(0) = 1$. What follows and why?

2.4 Further Applications

In this section, we apply the preceding results in a variety of ways. The first two of these are consequences of the Cauchy integral formula for derivatives (2.2.11).

2.4.1 Cauchy's Estimate

Let f be analytic on Ω, and let $\overline{D}(z_0, r) \subseteq \Omega$. Then

$$|f^{(n)}(z_0)| \leq \frac{n!}{r^n} \max_{z \in C(z_0,r)} |f(z)|.$$

Proof. This is immediate from (2.2.11) and the M-L theorem. ♣

Remark

If $f(z) = z^n$ and $z_0 = 0$, we have $f^{(n)}(z_0) = n! = (n!/r^n) \max_{z \in C(z_0,r)} |f(z)|$, so the above inequality is sharp.

2.4.2 Liouville's Theorem

If f is a bounded entire function, then f is constant.

Proof. Assume that $|f(z)| \leq M < \infty$ for all $z \in \mathbb{C}$, and fix $z_0 \in \mathbb{C}$. By (2.4.1), $|f'(z_0)| \leq M/r$ for all $r > 0$. Let $r \to \infty$ to conclude that $f'(z_0) = 0$. Since z_0 is arbitrary, $f' \equiv 0$, hence f is constant on $\mathbb{C}$ by (2.1.7b). ♣

2.4.3 The Fundamental Theorem of Algebra

Suppose $P(z) = a_0 + a_1 z + \cdots + a_n z^n$ is polynomial of degree $n \geq 1$. Then there exists $z_0 \in \mathbb{C}$ such that $P(z_0) = 0$.

Proof. Since

$$|P(z)| = |z|^n \left| a_n + \frac{a_{n-1}}{z} + \cdots + \frac{a_0}{z^n} \right| \geq |z|^n \left| \frac{a_n}{2} \right|$$

for all sufficiently large $|z|$, it follows that $|P(z)| \to \infty$ as $|z| \to \infty$. If $P(z)$ is never 0, then $1/P$ is an entire function. Moreover, $|1/P(z)| \to 0$ as $|z| \to \infty$, and therefore $1/P$ is bounded. By (2.4.2), $1/P$ is constant, contradicting $\deg P \geq 1$. ♣

Recall that if P is a polynomial of degree $n \geq 1$ and $P(z_0) = 0$, we may write $P(z) = (z - z_0)^m Q(z)$ where m is a positive integer and $Q(z)$ is a polynomial (possibly constant) such that $Q(z_0) \neq 0$. In this case P is said to have a zero of order m at z_0. The next definition extends the notion of the order of a zero to analytic functions in general.

2.4.4 Definition

Let f be analytic on Ω and $z_0 \in \Omega$. We say that f *has a zero of order* m *at* z_0 if there is an analytic function g on Ω such that $g(z_0) \neq 0$ and $f(z) = (z - z_0)^m g(z)$ for all $z \in \Omega$.

2.4.5 Remark

In terms of the Taylor expansion $f(z) = \sum_{n=0}^{\infty} a_n (z - z_0)^n$, f has a zero of order m at z_0 iff $a_0 = a_1 = \cdots = a_{m-1} = 0$, while $a_m \neq 0$. Equivalently, $f^{(n)}(z_0) = 0$ for $n = 0, \ldots, m-1$, while $f^{(m)}(z_0) \neq 0$ (see Problem 2).

2.4.6 Definition

If $f : \Omega \to \mathbb{C}$, the *zero set* of f is defined as $Z(f) = \{z \in \Omega : f(z) = 0\}$.

Our next major result, the identity theorem for analytic functions, is a consequence of a topological property of $Z(f)$.

2.4.7 Lemma

Let f be analytic on Ω, and let L be the set of limit points (also called accumulation points or cluster points) of $Z(f)$ in Ω. Then L is both open and closed in Ω.

Proof. First note that $L \subseteq Z(f)$ by continuity of f. Also, L is closed in Ω because the set of limit points of any subset of Ω is closed in Ω. (If $\{z_n\}$ is a sequence in L such that $z_n \to z$, then given $r > 0$, $z_n \in D(z, r)$ for n sufficiently large. Since z_n is a limit point of $Z(f)$, $D(z, r)$ contains infinitely many points of $Z(f)$ different from z_n, and hence infinitely many points of $Z(f)$ different from z. Thus $z \in L$ also.) It remains to show that L is open in Ω. Let $z_0 \in L$, and write $f(z) = \sum_{n=0}^{\infty} a_n (z - z_0)^n, z \in D(z_0, r) \subseteq \Omega$. Now $f(z_0) = 0$, and hence either f has a zero of order m at z_0 (for some m), or else $a_n = 0$ for all n. In the former case, there is a function g analytic on Ω such that $f(z) = (z - z_0)^m g(z), z \in \Omega$, with $g(z_0) \neq 0$. By continuity of g, $g(z) \neq 0$ for all z sufficiently close to z_0, and consequently z_0 is an isolated point of $Z(f)$. But then $z_0 \notin L$, contradicting out assumption. Thus, it must be the case that $a_n = 0$ for all n, so that $f \equiv 0$ on $D(z_0, r)$. Consequently, $D(z_0, r) \subseteq L$, proving that L is open in Ω. ♣

2.4.8 The Identity Theorem

Suppose f is analytic on the open connected set Ω. Then either f is identically zero on Ω or else $Z(f)$ has no limit point in Ω. Equivalently, if $Z(f)$ has a limit point in Ω, then f is identically 0 on Ω.

Proof. By (2.4.7), the set L of limit points of $Z(f)$ is both open and closed in Ω. Since Ω is connected, either $L = \Omega$, in which case $f \equiv 0$ on Ω, or $L = \emptyset$, so that $Z(f)$ has no limit point in Ω. ♣

2.4.9 Corollary

If f and g are analytic on Ω and $\{z \in \Omega : f(z) = g(z)\}$ has a limit point in Ω, then $f \equiv g$.

Proof. Apply the identity theorem to $f - g$. ♣

Our next application will be to show (roughly) that the absolute value of a function analytic on a set S cannot attain a maximum at an interior point of S. As a preliminary we show that the value of an analytic function at the center of a circle is the average of its values on the circumference.

2.4.10 Theorem

Suppose f is analytic on Ω and $\overline{D}(z_0, r) \subseteq \Omega$. Then

$$f(z_0) = \frac{1}{2\pi} \int_0^{2\pi} f(z_0 + re^{it})\, dt.$$

Proof. Use (2.2.9), Cauchy's integral formula for a circle, with $z = z_0$. ♣

The other preliminary to the proof of the maximum principle is the following fact about integrals.

2.4.11 Lemma

Suppose $\varphi : [a, b] \to \mathbb{R}$ is continuous, $\varphi(t) \leq k$ for all t, while the average of φ, namely $\frac{1}{b-a} \int_a^b \varphi(t)\, dt$, is at least k. Then $\varphi(t) = k$ for all t.

Proof. Observe that

$$0 \leq \int_a^b [k - \varphi(t)]\, dt = k(b-a) - \int_a^b \varphi(t)\, dt \leq 0. \quad ♣$$

We now consider the maximum principle, which is actually a collection of closely related results rather than a single theorem. We will prove four versions of the principle, arranged in order of decreasing strength.

2.4.12 Maximum Principle

Let f be analytic on the open connected set Ω.

(a) If $|f|$ assumes a local maximum at some point in Ω, then f is constant on Ω.

(b) If $\lambda = \sup\{|f(z)| : z \in \Omega\}$, then either $|f(z)| < \lambda$ for all $z \in \Omega$ or f is constant on Ω.

(c) If Ω is a bounded region and $M \geq 0$ is such that $\limsup_{n\to\infty} |f(z_n)| \leq M$ for each sequence $\{z_n\}$ in Ω that converges to a boundary point of Ω, then $|f(z)| < M$ for all $z \in \Omega$ or f is constant on Ω.

(d) Let Ω be a bounded region, with f continuous on the closure $\overline{\Omega}$ of Ω. Denote the boundary of Ω by $\partial\Omega$, and let $M_0 = \max\{|f(z)| : z \in \partial\Omega\}$. Then either $|f(z)| < M_0$ for all $z \in \Omega$ or f is constant on Ω. Consequently, $\max\{|f(z)| : z \in \overline{\Omega}\} = \max\{|f(z)| : z \in \partial\Omega\}$.

Proof.

(a) If $|f|$ assumes a local maximum at $z_0 \in \Omega$, then for some $\delta > 0$, $|f(z)| \leq |f(z_0)|$ for $|z - z_0| < \delta$. If $f(z_0) = 0$, then $f(z) = 0$ for all $z \in D(z_0, \delta)$, so $f \equiv 0$ by the identity theorem. So assume that $f(z_0) \neq 0$. If $0 < r < \delta$, then (2.4.10) with both sides divided by $f(z_0)$ yields

$$1 = \frac{1}{2\pi}\int_0^{2\pi} \frac{f(z_0 + re^{it})}{f(z_0)}\,dt.$$

Taking the magnitude of both sides, we obtain

$$1 \leq \frac{1}{2\pi}\int_0^{2\pi} \left|\frac{f(z_0 + re^{it})}{f(z_0)}\right| dt \leq 1$$

because $|f(z_0 + re^{it})| \leq |f(z_0)|$ for all $t \in [0, 2\pi]$. Since this holds for all $r \in (0, \delta)$, the preceding lemma (2.4.11) gives $|f(z)/f(z_0)| = 1, z \in D(z_0, \delta)$. Now take the real part (rather than the magnitude) of both sides of the above integral, and use the fact that for any complex number w, we have $|\operatorname{Re} w| \leq |w|$. We conclude that $\operatorname{Re}(f(z)/f(z_0)) = 1$ on $D(z_0, \delta)$. But if $|w| = \operatorname{Re} w = c$, then $w = c$, hence $f(z) = f(z_0)$ on $D(z_0, \delta)$. By the identity theorem, f is constant on Ω.

(b) If $\lambda = +\infty$ there is nothing to prove, so assume $\lambda < +\infty$. If $|f(z_0)| = \lambda$ for some $z_0 \in \Omega$, then f is constant on Ω by (a).

(c) If λ is defined as in (b), then there is a sequence $\{z_n\}$ in Ω such that $|f(z_n)| \to \lambda$. But since Ω is *bounded*, there is a subsequence $\{z_{n_j}\}$ that converges to a limit z_0. If $z_0 \in \Omega$, then $|f(z_0)| = \lambda$, hence f is constant by (b). On the other hand, if z_0 belongs to the boundary of Ω, then $\lambda \leq M$ by hypothesis. Again by (b), either $|f(z)| < \lambda \leq M$ for all $z \in \Omega$ or f is constant on Ω.

(d) Let $\{z_n\}$ be any sequence in Ω converging to a point $z_0 \in \partial\Omega$. Then $|f(z_n)| \to |f(z_0)| \leq M_0$. By (c), $|f| < M_0$ on Ω or f is constant on Ω. In either case, the maximum of $|f|$ on $\overline{\Omega}$ is equal to the maximum of $|f|$ on $\partial\Omega$. ♣

The absolute value of an analytic function may attain its minimum modulus on an open connected set without being constant (consider $f(z) = z$ on $\mathbb{C}$). However, if the function is never zero, we do have a minimum principle.

2.4.13 Minimum Principle

Let f be analytic and never 0 on the region Ω.

(a) If $|f|$ assumes a local minimum at some point in Ω, then f is constant on Ω.

(b) Let $\mu = \inf\{|f(z)| : z \in \Omega\}$; then either $|f(z)| > \mu$ for all $z \in \Omega$ or f is constant on Ω.
(c) If Ω is a bounded region and $m \geq 0$ is such that $\liminf_{n\to\infty} |f(z_n)| \geq m$ for each sequence $\{z_n\}$ that converges to a boundary point of Ω, then $|f(z)| > m$ for all $z \in \Omega$ or f is constant on Ω.
(d) Let Ω be a bounded region, with f is continuous on $\overline{\Omega}$ and $m_0 = \min\{|f(z)| : z \in \partial\Omega\}$. Then either $|f(z)| > m_0$ for all $z \in \Omega$ or f is constant on Ω. As a consequence, we have $\min\{|f(z)| : z \in \overline{\Omega}\} = \min\{|f(z)| : z \in \partial\Omega\}$.
Proof. Apply the maximum principle to $1/f$. ♣

Suppose f is analytic on the region Ω, and we put $g = e^f$. Then $|g| = e^{\mathrm{Re}\, f}$, and hence $|g|$ assumes a local maximum at $z_0 \in \Omega$ iff $\mathrm{Re}\, f$ has a local maximum at z_0. A similar statement holds for a local minimum. Furthermore, by (2.1.7b), f is constant iff $f' \equiv 0$ iff $f'e^f \equiv 0$ iff $g' \equiv 0$ iff g is constant on Ω. Thus $\mathrm{Re}\, f$ satisfies part (a) of both the maximum and minimum principles (note that $|g|$ is never 0). A similar argument can be given for $\mathrm{Im}\, f$ (put $g = e^{-if}$). Since the real and imaginary parts of an analytic function are, in particular, harmonic functions [see (2.2.19)], the question arises as to whether the maximum and minimum principles are valid for harmonic functions in general. The answer is yes, as we now proceed to show. We will need to establish one preliminary result which is a weak version of the identity theorem (2.4.8) for harmonic functions.

2.4.14 Identity Theorem for Harmonic Functions

If u is harmonic on the region Ω, and u restricted to some subdisk of Ω is constant, then u is constant on Ω.
Proof. Let $A = \{a \in \Omega : u$ is constant on some disk with center at $a\}$. It follows from the definition of A that A is an open subset of Ω. But $\Omega \setminus A$ is also open; to see this, let $z_0 \in \Omega \setminus A$ and $D(z_0, r) \subseteq \Omega$. By (1.6.2), u has a harmonic conjugate v on $D(z_0, r)$, so that u is the real part of an analytic function on $D(z_0, r)$. If u is constant on any subdisk of $D(z_0, r)$, then [since u satisfies (a) of the maximum (or minimum) principle, as indicated in the remarks following (2.4.13)] u is constant on $D(z_0, r)$, contradicting $z_0 \in \Omega \setminus A$. Thus $D(z_0, r) \subseteq \Omega \setminus A$, proving that $\Omega \setminus A$ is also open. Since Ω is connected and $A \neq \emptyset$ by hypothesis, we have $A = \Omega$.

Finally, fix $z_1 \in \Omega$ and let $B = \{z \in \Omega : u(z) = u(z_1)\}$. By continuity of u, B is closed in Ω, and since $A = \Omega$, B is also open in Ω. But B is not empty (it contains z_1), hence $B = \Omega$, proving that u is constant on Ω. ♣

2.4.15 Maximum and Minimum Principle for Harmonic Functions

If u is harmonic on a region Ω and u has either a local maximum or a local minimum at some point of Ω, then u is constant on Ω.
Proof. Say u has a local minimum at $z_0 \in \Omega$ (the argument for a maximum is similar). Then for some $r > 0$ we have $D(z_0, r) \subseteq \Omega$ and $u(z) \geq u(z_0)$ on $D(z_0, r)$. By (1.6.2) again, u is the real part of an analytic function on $D(z_0, r)$, and we may invoke the minimum principle [as we did in proving (2.4.14)] to conclude that u is constant on $D(z_0, r)$ and hence by (2.4.14), constant on Ω. ♣

Remark

The proof of (2.4.12) shows that part (a) of the maximum principle implies part (b), (b) implies (c), and (c) implies (d), and similarly for the minimum principle. Thus harmonic functions satisfy statements (b), (c) and (d) of the maximum and minimum principles.

We conclude this chapter with one of the most important applications of the maximum principle.

2.4.16 Schwarz's Lemma

Let f be analytic on the unit disk $D = D(0,1)$, and assume that $f(0) = 0$ and $|f(z)| \leq 1$ for all $z \in D$. Then (a) $|f(z)| \leq |z|$ on D, and (b) $|f'(0)| \leq 1$. Furthermore, if equality holds in (a) for some $z \neq 0$, or if equality holds in (b), then f is a rotation of D. That is, there is a constant λ with $|\lambda| = 1$ such that $f(z) = \lambda z$ for all $z \in D$.

Proof. Define

$$g(z) = \begin{cases} f(z)/z & \text{if } z \in D \setminus \{0\} \\ f'(0) & \text{if } z = 0. \end{cases}$$

By (2.2.13), g is analytic on D. We claim that $|g(z)| \leq 1$. For if $|z| < r < 1$, part (d) of the maximum principle yields

$$|g(z)| \leq \max\{|g(w)| : |w| = r\} \leq \frac{1}{r} \sup\{|f(w)| : w \in D\} \leq \frac{1}{r}.$$

Since r may be chosen arbitrarily close to 1, we have $|g| \leq 1$ on D, proving both (a) and (b). If equality holds in (a) for some $z \neq 0$, or if equality holds in (b), then g assumes its maximum modulus at a point of D, and hence g is a constant λ on D (necessarily $|\lambda| = 1$). Thus $f(z) = \lambda z$ for all $z \in D$. ♣

Schwarz's lemma will be generalized and applied in Chapter 4 (see also Problem 24).

Problems

1. Give an example of a nonconstant analytic function f on a region Ω such that f has a limit point of zeros at a point outside of Ω.
2. Verify the statements made in (2.4.5).
3. Consider the four forms of the maximum principle (2.4.12), for continuous rather than analytic functions. What can be said about the relative strengths of the statements? The proof in the text shows that (a) implies (b) implies (c) implies (d), but for example, does (b) imply (a)? (The region Ω is assumed to be one particular fixed open connected set, that is, the statement of the theorem does not have "for all Ω" in it.)
4. (L'Hospital's rule). Let f and g be analytic at z_0, and not identically zero in any neighborhood of z_0. If $\lim_{z \to z_0} f(z) = \lim_{z \to z_0} g(z) = 0$, show that $f(z)/g(z)$ approaches a limit (possibly ∞) as $z \to z_0$, and $\lim_{z \to z_0} f(z)/g(z) = \lim_{z \to z_0} f'(z)/g'(z)$.
5. If f is analytic on a region Ω and $|f|$ is constant on Ω, show that f is constant on Ω.

6. Let f be continuous on the closed unit disk $\overline{D}$, analytic on D, and real-valued on ∂D. Prove that f is constant.
7. Let $f(z) = \sin z$. Find $\max\{|f(z)| : z \in K\}$ where $K = \{x + iy : 0 \le x, y \le 2\pi\}$.
8. (A generalization of part (d) of the maximum principle). Suppose K is compact, f is continuous on K, and f is analytic on K°, the interior of K. Show that

$$\max_{z\in K} |f(z)| = \max_{z\in\partial K} |f(z)|.$$

   Moreover, if $|f(z_0)| = \max_{z\in K} |f(z)|$ for some $z_0 \in K^\circ$, then f is constant on the component of K° that contains z_0.
9. Suppose that Ω is a bounded open set (not necessarily connected), f is continuous on $\overline{\Omega}$ and analytic on Ω. Show that $\max\{|f(z)| : z \in \overline{\Omega}\} = \max\{|f(z)| : z \in \partial\Omega\}$.
10. Give an example of a nonconstant harmonic function u on $\mathbb{C}$ such that $u(z) = 0$ for each real z. Thus the disk that appears in the statement of Theorem 2.4.14 cannot be replaced by just any subset of $\mathbb{C}$ having a limit point in $\mathbb{C}$.
11. Prove that an open set Ω is connected iff for all f, g analytic on Ω, the following holds: If $f(z)g(z) = 0$ for every $z \in \Omega$, then either f or g is identically zero on Ω. (This says that the ring of analytic functions on Ω is an integral domain iff Ω is connected.)
12. Suppose that f is analytic on $\mathbb{C}^+ = \{z : \operatorname{Im} z > 0\}$ and continuous on $S = \mathbb{C}^+ \cup (0, 1)$. Assume that $f(x) = x^4 - 2x^2$ for all $x \in (0, 1)$. Show that $f(i) = 3$.
13. Let f be an entire function such that $|f(z)| \ge 1$ for all z. Prove that f is constant.
14. Does there exist an entire function f, not identically zero, for which $f(z) = 0$ for every z in an uncountable set of complex numbers?
15. Explain why knowing that the trigonometric identity $\sin(\alpha + \beta) = \sin\alpha\cos\beta + \cos\alpha\sin\beta$ for all *real* α and β implies that the same identity holds for all *complex* α and β.
16. Suppose f is an entire function and $\operatorname{Im}(f(z)) \ge 0$ for all z. Prove that f is constant. (Consider $\exp(if)$.)
17. Suppose f and g are analytic and nonzero on $D(0, 1)$, and $\frac{f'(1/n)}{f(1/n)} = \frac{g'(1/n)}{g(1/n)}, n = 2, 3, \ldots$. Prove that f/g is constant on $D(0, 1)$.
18. Suppose that f is an entire function, $f(0) = 0$ and $|f(z) - e^z \sin z| < 4$ for all z. Find a formula for $f(z)$.
19. Let f and g be analytic on $D = D(0, 1)$ and continuous on $\overline{D}$. Assume that $\operatorname{Re} f(z) = \operatorname{Re} g(z)$ for all $z \in \partial D$. Prove that $f - g$ is constant.
20. Let f be analytic on $D = D(0, 1)$. Prove that either f has a zero in D, or there is a sequence $\{z_n\}$ in D such that $|z_n| \to 1$ and $\{f(z_n)\}$ is bounded.
21. Let u be a nonnegative harmonic function on $\mathbb{C}$. Prove that f is constant.
22. Suppose f is analytic on $\Omega \supseteq \overline{D}(0, 1)$, $f(0) = i$, and $|f(z)| > 1$ whenever $|z| = 1$. Prove that f has a zero in $D(0, 1)$.
23. Find the maximum value of $\operatorname{Re} z^3$ for z in the unit square $[0, 1] \times [0, 1]$.

24. Suppose that f is analytic on $D(0,1)$, with $f(0) = 0$. Define $f_n(z) = f(z^n)$ for $n = 1, 2, \ldots, z \in D(0,1)$. Prove that $\sum f_n$ is uniformly convergent on compact subsets of $D(0,1)$. (Use Schwarz's lemma.)

25. It follows from (2.4.12c) that if f is analytic on $D(0,1)$ and $f(z_n) \to 0$ for each sequence $\{z_n\}$ in $D(0,1)$ that converges to a point of $C(0,1)$, then $f \equiv 0$. Prove the following strengthened version for *bounded* f. Assume only that $f(z_n) \to 0$ for each sequence $\{z_n\}$ that converges to a point in some given arc $\{e^{it}, \alpha \leq t \leq \beta\}$ where $\alpha < \beta$, and deduce that $f \equiv 0$. [Hint: Assume without loss of generality that $\alpha = 0$. Then for sufficiently large n, the arcs $A_j = \{e^{it} : (j-1)\beta \leq t \leq j\beta\}, j = 1, 2, \ldots, n$ cover $C(0,1)$. Now consider $F(z) = f(z)f(e^{i\beta}z)f(e^{i2\beta}z) \cdots f(e^{in\beta}z)$.]

26. (a) Let Ω be a bounded open set and let $\{f_n\}$ be a sequence of functions that are analytic on Ω and continuous on the closure $\overline{\Omega}$. Suppose that $\{f_n\}$ is uniformly Cauchy on the boundary of Ω. Prove that $\{f_n\}$ converges uniformly on $\overline{\Omega}$. If f is the limit function, what are some properties of f?

    (b) What complex-valued functions on the unit circle $C(0,1)$ can be uniformly approximated by polynomials in z?

Chapter 3

The General Cauchy Theorem

In this chapter, we consider two basic questions. First, for a given open set Ω, we try to determine which closed paths γ in Ω have the property that $\int_\gamma f(z)\,dz = 0$ for *every* analytic function f on Ω. Then second, we try to characterize those open sets Ω having the property that $\int_\gamma f(z)\,dz = 0$ for all closed paths γ in Ω and all analytic functions f on Ω. The results, which may be grouped under the name "Cauchy's theorem", form the cornerstone of analytic function theory.

A basic concept in the general Cauchy theory is that of *winding number* or *index* of a point with respect to a closed curve not containing the point. In order to make this precise, we need several preliminary results on logarithm and argument functions.

3.1 Logarithms and Arguments

In (2.3.1), property (j), we saw that given a real number α, the exponential function when restricted to the strip $\{x + iy : \alpha \le y < \alpha + 2\pi\}$ is a one-to-one analytic map of this strip onto the nonzero complex numbers. With this in mind, we make the following definition.

3.1.1 Definition

We take $\log_\alpha$ to be the inverse of the exponential function restricted to the strip $S_\alpha = \{x + iy : \alpha \le y < \alpha + 2\pi\}$. We define $\arg_\alpha$ to the the imaginary part of $\log_\alpha$.

Consequently, $\log_\alpha(\exp z) = z$ for each $z \in S_\alpha$, and $\exp(\log_\alpha z) = z$ for all $z \in \mathbb{C}\setminus\{0\}$.

Several important properties of $\log_\alpha$ and $\arg_\alpha$ follow readily from Definition 3.1.1 and the basic properties of exp.

3.1.2 Theorem

(a) If $z \neq 0$, then $\log_\alpha(z) = \ln|z| + i\arg_\alpha(z)$, and $\arg_\alpha(z)$ is the unique number in $[\alpha, \alpha + 2\pi)$ such that $z/|z| = e^{i\arg_\alpha(z)}$, in other words, the unique argument of z in $[\alpha, \alpha + 2\pi)$.

(b) Let R_α be the ray $[0, e^{i\alpha}, \infty) = \{re^{i\alpha} : r \geq 0\}$. The functions $\log_\alpha$ and $\arg_\alpha$ are continuous at each point of the "slit" complex plane $\mathbb{C} \setminus R_\alpha$, and discontinuous at each point of R_α.

(c) The function $\log_\alpha$ is analytic on $\mathbb{C} \setminus R_\alpha$, and its derivative is given by $\log'_\alpha(z) = 1/z$.

Proof.

(a) If $w = \log_\alpha(z), z \neq 0$, then $e^w = z$, hence $|z| = e^{\operatorname{Re} w}$ and $z/|z| = e^{i \operatorname{Im} w}$. Thus $\operatorname{Re} w = \ln |z|$, and $\operatorname{Im} w$ is an argument of $z/|z|$. Since $\operatorname{Im} w$ is restricted to $[\alpha, \alpha + 2\pi)$ by definition of $\log_\alpha$, it follows that $\operatorname{Im} w$ is the unique argument for z that lies in the interval $[\alpha, \alpha + 2\pi)$.

(b) By (a), it suffices to consider $\arg_\alpha$. If $z_0 \in \mathbb{C} \setminus R_\alpha$ and $\{z_n\}$ is a sequence converging to z_0, then $\arg_\alpha(z_n)$ must converge to $\arg_\alpha(z_0)$. (Draw a picture.) On the other hand, if $z_0 \in R_\alpha \setminus \{0\}$, there is a sequence $\{z_n\}$ converging to z_0 such that $\arg_\alpha(z_n) \to \alpha + 2\pi \neq \arg_\alpha(z_0) = \alpha$.

(c) This follows from Theorem 1.3.2 (with $g = \exp, \Omega_1 = \mathbb{C}, f = \log_\alpha$, and $\Omega = \mathbb{C} \setminus R_\alpha$) and the fact that exp is its own derivative. ♣

3.1.3 Definition

The *principal branches* of the logarithm and argument functions, to be denoted by Log and Arg, are obtained by taking $\alpha = -\pi$. Thus, $\operatorname{Log} = \log_{-\pi}$ and $\operatorname{Arg} = \arg_{-\pi}$.

Remark

The definition of principal branch is not standardized; an equally common choice for α is $\alpha = 0$. Also, having made a choice of principal branch, one can define $w^z = \exp(z \operatorname{Log} w)$ for $z \in \mathbb{C}$ and $w \in \mathbb{C} \setminus \{0\}$. We will not need this concept, however.

3.1.4 Definition

Let S be a subset of $\mathbb{C}$ (or more generally any metric space), and let $f : S \to \mathbb{C} \setminus \{0\}$ be continuous. A function $g : S \to \mathbb{C}$ is a *continuous logarithm* of f if g is continuous on S and $f(s) = e^{g(s)}$ for all $s \in S$. A function $\theta : S \to \mathbb{R}$ is a *continuous argument* of f if θ is continuous on S and $f(s) = |f(s)|e^{i\theta(s)}$ for all $s \in S$.

3.1.5 Examples

(a) If $S = [0, 2\pi]$ and $f(s) = e^{is}$, then f has a continuous argument on S, namely $\theta(s) = s + 2k\pi$ for any fixed integer k.

(b) If for some α, f is a continuous mapping of S into $\mathbb{C} \setminus R_\alpha$, then f has a continuous argument, namely $\theta(s) = \arg_\alpha(f(s))$.

(c) If $S = \{z : |z| = 1\}$ and $f(z) = z$, then f does not have a continuous argument on S.

Part (a) is a consequence of Definition 3.1.4, and (b) follows from (3.1.4) and (3.1.2b). The intuition underlying (c) is that if we walk entirely around the unit circle, a continuous argument of z must change by 2π. Thus the argument of z must abruptly jump by 2π

at the end of the trip, which contradicts continuity. A formal proof will be easier after further properties of continuous arguments are developed (see Problem 3.2.5).

Continuous logarithms and continuous arguments are closely related, as follows.

3.1.6 Theorem

Let $f : S \to \mathbb{C}$ be continuous.

(a) If g is a continuous logarithm of f, then $\operatorname{Im} g$ is a continuous argument of f.

(b) If θ is a continuous argument of f, then $\ln|f| + i\theta$ is a continuous logarithm of f.

Thus f has a continuous logarithm iff f has a continuous argument.

(c) Assume that S is connected, and f has continuous logarithms g_1 and g_2, and continuous arguments θ_1 and θ_2. Then there are integers k and l such that $g_1(s) - g_2(s) = 2\pi ik$ and $\theta_1(s) - \theta_2(s) = 2\pi l$ for all $s \in S$. Thus $g_1 - g_2$ and $\theta_1 - \theta_2$ are constant on S.

(d) If S is connected and $s, t \in S$, then $g(s) - g(t) = \ln|f(s)| - \ln|f(t)| + i(\theta(s) - \theta(t))$ for all continuous logarithms g and all continuous arguments θ of f.

Proof.

(a) If $f(s) = e^{g(s)}$, then $|f(s)| = e^{\operatorname{Re} g(s)}$, hence $f(s)/|f(s)| = e^{i \operatorname{Im} g(s)}$ as required.

(b) If $f(s) = |f(s)|e^{i\theta(s)}$, then $f(s) = e^{\ln|f(s)|+i\theta(s)}$, so $\ln|f| + i\theta$ is a continuous logarithm.

(c) We have $f(s) = e^{g_1(s)} = e^{g_2(s)}$, hence $e^{g_1(s)-g_2(s)} = 1$, for all $s \in S$. By (2.3.1f), $g_1(s) - g_2(s) = 2\pi ik(s)$ for some integer-valued function k. Since g_1 and g_2 are continuous on S, so is k. But S is connected, so k is a constant function. A similar proof applies to any pair of continuous arguments of f.

(d) If θ is a continuous argument of f, then $\ln|f| + i\theta$ is a continuous logarithm of f by part (b). Thus if g is any continuous logarithm of f, then $g = \ln|f| + i\theta + 2\pi ik$ by (c). The result follows. ♣

As Example 3.1.5(c) indicates, a given zero-free continuous function on a set S need not have a continuous argument. However, a continuous argument must exist when S is an *interval*, as we now show.

3.1.7 Theorem

Let $\gamma : [a, b] \to \mathbb{C} \setminus \{0\}$ be continuous, that is, γ is a curve and $0 \notin \gamma^*$. Then γ has a continuous argument, hence by (3.1.6), a continuous logarithm.

Proof. Let ϵ be the distance from 0 to γ^*, that is, $\epsilon = \min\{|\gamma(t)| : t \in [a, b]\}$. Then $\epsilon > 0$ because $0 \notin \gamma^*$ and γ^* is a closed set. By the uniform continuity of γ on $[a, b]$, there is a partition $a = t_0 < t_1 < \cdots < t_n = b$ of $[a, b]$ such that if $1 \leq j \leq n$ and $t \in [t_{j-1}, t_j]$, then $\gamma(t) \in D(\gamma(t_j), \epsilon)$. By (3.1.5b), the function γ, restricted to the interval $[t_0, t_1]$, has a continuous argument θ_1, and γ restricted to $[t_1, t_2]$ has a continuous argument θ_2. Since $\theta_1(t_1)$ and $\theta_2(t_1)$ differ by an integer multiple of 2π, we may (if necessary) redefine θ_2 on $[t_1, t_2]$ so that the relation $\theta_1 \cup \theta_2$ is a continuous argument of γ on $[t_0, t_2]$. Proceeding in this manner, we obtain a continuous argument of γ on the entire interval $[a, b]$. ♣

For a generalization to other subsets S, see Problem 3.2.6.

3.1.8 Definition

Let f be analytic on Ω. We say that g is an *analytic logarithm* of f if g is analytic on Ω and $e^g = f$.

Our next goal is to show that if Ω satisfies certain conditions, in particular, if Ω is a starlike region, then every zero-free analytic function f on Ω has an analytic logarithm on Ω. First, we give necessary and sufficient conditions for f to have an analytic logarithm.

3.1.9 Theorem

Let f be analytic and never zero on the open set Ω. Then f has an analytic logarithm on Ω iff the "logarithmic derivative" f'/f has a primitive on Ω. Equivalently, by (2.1.6) and (2.1.10), $\int_\gamma \frac{f'(z)}{f(z)}\,dz = 0$ for every closed path γ in Ω.

Proof. If g is an analytic logarithm of f, then $e^g = f$, hence $f'/f = g'$. Conversely, if f'/f has a primitive g, then $f'/f = g'$, and therefore

$$(fe^{-g})' = -fe^{-g}g' + f'e^{-g} = e^{-g}(f' - fg')$$

which is identically zero on Ω. Thus fe^{-g} is constant on each component of Ω. If $fe^{-g} = k_A$ on the component A, then k_A cannot be zero, so we can write $k_A = e^{l_A}$ for some constant l_A. We then have $f = e^{g+l_A}$, so that $g + l_A$ is an analytic logarithm of f on A. Finally, $\cup_A(g + l_A)$ is an analytic logarithm of f on Ω. ♣

We may now give a basic sufficient condition on Ω under which every zero-free analytic function on Ω has an analytic logarithm.

3.1.10 Theorem

If Ω is an open set such that $\int_\gamma h(z)\,dz = 0$ for every analytic function h on Ω and every closed path γ in Ω, in particular if Ω is a starlike region, then every zero-free analytic function f on Ω has an analytic logarithm.

Proof. The result is a consequence of (3.1.9). If Ω is starlike, then $\int_\gamma h(z)\,dz = 0$ by Cauchy's theorem for starlike regions (2.1.9). ♣

3.1.11 Remark

If g is an analytic logarithm of f on Ω, then f has an analytic n-th root, namely $f^{1/n} = \exp(g/n)$. If $f(z) = z$ and $g = \log_\alpha$, we obtain

$$z^{1/n} = \exp\left(\frac{1}{n}\ln|z| + i\frac{1}{n}\arg_\alpha z\right) = |z|^{1/n}\exp\left(\frac{i}{n}\arg_\alpha z\right).$$

More generally, we may define an analytic version of f^w for any complex number w, via $f^w = e^{wg}$.

3.2 The Index of a Point with Respect to a Closed Curve

In the introduction to this chapter we raised the question of which closed paths γ in an open set Ω have the property that $\int_\gamma f(z)\,dz = 0$ for every analytic function f on Ω. As we will see later, a necessary and sufficient condition on γ is that "γ not wind around any points outside of Ω." That is to say, if $z_0 \notin \Omega$ and γ is defined on $[a,b]$, there is "no net change in the argument of $\gamma(t) - z_0$" as t increases from a to b. To make this precise, we define the notion of the *index* (or *winding number*) of a point with respect to a closed curve. The following observation will be crucial in showing that the index is well-defined.

3.2.1 Theorem

Let $\gamma : [a,b] \to \mathbb{C}$ be a closed curve. Fix $z_0 \notin \gamma^*$, and let θ be a continuous argument of $\gamma - z_0$ [θ exists by (3.1.7)]. Then $\theta(b) - \theta(a)$ is an integer multiple of 2π. Furthermore, if θ_1 is another continuous argument of $\gamma - z_0$, then $\theta_1(b) - \theta_1(a) = \theta(b) - \theta(a)$.

Proof. By (3.1.4), we have $(\gamma(t) - z_0)/|\gamma(t) - z_0| = e^{i\theta(t)}, a \le t \le b$. Since γ is a closed curve, $\gamma(a) = \gamma(b)$, hence

$$1 = \frac{\gamma(b) - z_0}{|\gamma(b) - z_0|} \cdot \frac{|\gamma(a) - z_0|}{\gamma(a) - z_0} = e^{i(\theta(b) - \theta(a))}.$$

Consequently, $\theta(b) - \theta(a)$ is an integer multiple of 2π. If θ_1 is another continuous argument of $\gamma - z_0$, then by (3.1.6c), $\theta_1 - \theta = 2\pi l$ for some integer l. Thus $\theta_1(b) = \theta(b) + 2\pi l$ and $\theta_1(a) = \theta(a) + 2\pi l$, so $\theta_1(b) - \theta_1(a) = \theta(b) - \theta(a)$. ♣

It is now possible to define the index of a point with respect to a closed curve.

3.2.2 Definition

Let $\gamma : [a,b] \to \mathbb{C}$ be a closed curve. If $z_0 \notin \gamma^*$, let θ_{z_0} be a continuous argument of $\gamma - z_0$. The *index of* z_0 *with respect to* γ, denoted by $n(\gamma, z_0)$, is

$$n(\gamma, z_0) = \frac{\theta_{z_0}(b) - \theta_{z_0}(a)}{2\pi}.$$

By (3.2.1), $n(\gamma, z_0)$ is well-defined, that is, $n(\gamma, z_0)$ does not depend on the particular continuous argument chosen. Intuitively, $n(\gamma, z_0)$ is the net number of revolutions of $\gamma(t), a \le t \le b$, about the point z_0. This is why the term *winding number* is often used for the index. Note that by the above definition, for any complex number w we have $n(\gamma, z_0) = n(\gamma + w, z_0 + w)$.

If γ is sufficiently smooth, an integral representation of the index is available.

3.2.3 Theorem

Let γ be a closed *path*, and z_0 a point not belonging to γ^*. Then

$$n(\gamma, z_0) = \frac{1}{2\pi i} \int_\gamma \frac{1}{z - z_0}\,dz.$$

More generally, if f is analytic on an open set Ω containing γ^*, and $z_0 \notin (f \circ \gamma)^*$, then

$$n(f \circ \gamma, z_0) = \frac{1}{2\pi i} \int_\gamma \frac{f'(z)}{f(z) - z_0}\, dz.$$

Proof. Let ϵ be the distance from z_0 to γ^*. As in the proof of (3.1.7), there is a partition $a = t_0 < t_1 < \cdots < t_n = b$ such that $t_{j-1} \le t \le t_j$ implies $\gamma(t) \in D(\gamma(t_j), \epsilon)$. For each j, $z_0 \notin D(\gamma(t_j), \epsilon)$ by definition of ϵ. Consequently, by (3.1.10), the analytic function $z \to z - z_0$, when restricted to $D(\gamma(t_j), \epsilon)$ has an analytic logarithm g_j. Now if g is an analytic logarithm of f, then $g' = f'/f$ [see (3.1.9)]. Therefore $g_j'(z) = 1/(z - z_0)$ for all $z \in D(\gamma(t_j), \epsilon)$. The path γ restricted to $[t_{j-1}, t_j]$ lies in the disk $D(\gamma(t_j), \epsilon)$, and hence by (2.1.6),

$$\int_{\gamma|_{[t_{j-1}, t_j]}} \frac{1}{z - z_0}\, dz = g_j(\gamma(t_j)) - g_j(\gamma(t_{j-1})).$$

Thus

$$\int_\gamma \frac{1}{z - z_0}\, dz = \sum_{j=1}^{n} [g_j(\gamma(t_j)) - g_j(\gamma(t_{j-1}))].$$

If $\theta_j = \operatorname{Im} g_j$, then by (3.1.6a), θ_j is a continuous argument of $z \to z - z_0$ on $D(\gamma(t_j), \epsilon)$. By (3.1.6d), then,

$$\int_\gamma \frac{1}{z - z_0}\, dz = \sum_{j=1}^{n} [\theta_j(\gamma(t_j)) - \theta_j(\gamma(t_{j-1}))].$$

If θ is any continuous argument of $\gamma - z_0$, then $\theta|_{[t_{j-1}, t_j]}$ is a continuous argument of $(\gamma - z_0)|_{[t_{j-1}, t_j]}$. But so is $\theta_j \circ \gamma|_{[t_{j-1}, t_j]}$, hence by (3.1.6c),

$$\theta_j(\gamma(t_j)) - \theta_j(\gamma(t_{j-1})) = \theta(t_j) - \theta(t_{j-1}).$$

Therefore,

$$\int_\gamma \frac{1}{z - z_0}\, dz = \sum_{j=1}^{n} [\theta(t_j) - \theta(t_{j-1})] = \theta(b) - \theta(a) = 2\pi n(\gamma, z_0)$$

completing the proof of the first part of the theorem. Applying this result to the path $f \circ \gamma$, we get the second statement. Specifically, if $z_0 \notin (f \circ \gamma)^*$, then

$$n(f \circ \gamma, z_0) = \frac{1}{2\pi i} \int_{f \circ \gamma} \frac{1}{z - z_0}\, dz = \frac{1}{2\pi i} \int_\gamma \frac{f'(z)}{f(z) - z_0}\, dz. \quad ♣$$

The next result contains additional properties of winding numbers that will be useful later, and which are also interesting (and amusing, in the case of (d)) in their own right.

3.2.4 Theorem

Let $\gamma, \gamma_1, \gamma_2 : [a, b] \to \mathbb{C}$ be closed curves.
(a) If $z \notin \gamma^*$, then $n(\gamma, z) = n(\gamma - z, 0)$.
(b) If $0 \notin \gamma_1^* \cup \gamma_2^*$, then $n(\gamma_1\gamma_2, 0) = n(\gamma_1, 0) + n(\gamma_2, 0)$ and $n(\gamma_1/\gamma_2, 0) = n(\gamma_1, 0) - n(\gamma_2, 0)$.
(c) If $\gamma^* \subseteq D(z_0, r)$ and $z \notin D(z_0, r)$, then $n(\gamma, z) = 0$.
(d) If $|\gamma_1(t) - \gamma_2(t)| < |\gamma_1(t)|, a \le t \le b$, then $0 \notin \gamma_1^* \cup \gamma_2^*$ and $n(\gamma_1, 0) = n(\gamma_2, 0)$.
Proof.
(a) This follows from Definition 3.2.2.
(b) Since $0 \notin \gamma_1^* \cup \gamma_2^*$, both $n(\gamma_1, 0)$ and $\gamma_2, 0)$ are defined. If θ_1 and θ_2 are continuous arguments of γ_1 and γ_2 respectively, then $\gamma_j(t) = |\gamma_j(t)|e^{i\theta_j(t)}, j = 1, 2$, so

$$\gamma_1(t)\gamma_2(t) = |\gamma_1(t)\gamma_2(t)|e^{i(\theta_1(t)+\theta_2(t))}, \quad \gamma_1(t)/\gamma_2(t) = |\gamma_1(t)/\gamma_2(t)|e^{i(\theta_1(t)-\theta_2(t))}.$$

Thus

$$\begin{aligned} n(\gamma_1\gamma_2, 0) &= (\theta_1(b) + \theta_2(b)) - (\theta_1(a) + \theta_2(a)) = (\theta_1(b) - \theta_1(a)) + (\theta_2(b) - \theta_2(a)) \\ &= n(\gamma_1, 0) + n(\gamma_2, 0). \end{aligned}$$

Similarly, $n(\gamma_1/\gamma_2, 0) = n(\gamma_1, 0) - n(\gamma_2, 0)$.
(c) If $z \notin D(z_0, r)$, then by (3.1.10), the function f defined by $f(w) = w - z, w \in D(z_0, r)$, has an analytic logarithm g. If θ is the imaginary part of g, then by (3.1.6a), $\theta \circ \gamma$ is a continuous argument of $\gamma - z$. Consequently, $n(\gamma, z) = (2\pi)^{-1}[\theta(\gamma(b)) - \theta(\gamma(a))] = 0$ since $\gamma(b) = \gamma(a)$.
(d) First note that if $\gamma_1(t) = 0$ or $\gamma_2(t) = 0$, then $|\gamma_1(t) - \gamma_2(t)| < |\gamma_1(t)|$ is false; therefore, $0 \notin \gamma_1^* \cup \gamma_2^*$. Let γ be the closed curve defined by $\gamma(t) = \gamma_2(t)/\gamma_1(t)$. By the hypothesis, we have $|1 - \gamma(t)| < 1$ on $[a, b]$, hence $\gamma^* \subseteq D(1, 1)$. But by (c) and (b), $0 = n(\gamma, 0) = n(\gamma_2, 0) - n(\gamma_1, 0)$. ♣

Part (d) of (3.2.4) is sometimes called the "dog-walking theorem". (See the text by W. Veech, *A Second Course in Complex Analysis*, page 30.) For if $\gamma_1(t)$ and $\gamma_2(t)$ are respectively the positions of a man and a dog on a variable length leash, and a tree is located at the origin, then the hypothesis states that the length of the leash is always less than the distance from the man to the tree. The conclusion states that the man and the dog walk around the tree exactly the same number of times. See Problem 4 for a generalization of (d).

The final theorem of this section deals with $n(\gamma, z_0)$ when viewed as a function of z_0.

3.2.5 Theorem

If γ is a closed curve, then the function $z \to n(\gamma, z), z \notin \gamma^*$, is constant on each component of $\mathbb{C} \setminus \gamma^*$, and is 0 on the unbounded component of $\mathbb{C} \setminus \gamma^*$.

Proof. Let $z_0 \in \mathbb{C} \setminus \gamma^*$, and choose $r > 0$ such that $D(z_0, r) \subseteq \mathbb{C} \setminus \gamma^*$. If $z \in D(z_0, r)$, then by parts (a) and (b) of (3.2.4),

$$n(\gamma, z) - n(\gamma, z_0) = n(\gamma - z, 0) - n(\gamma - z_0, 0) = n\left(\frac{\gamma - z}{\gamma - z_0}, 0\right) = n\left(1 + \frac{z_0 - z}{\gamma - z_0}, 0\right).$$

But for each t,

$$\left|\frac{z_0 - z}{\gamma(t) - z_0}\right| < \frac{r}{|\gamma(t) - z_0|} \leq 1$$

since $D(z_0, r) \subseteq \mathbb{C}\setminus\gamma^*$. Therefore the curve $1+(z_0-z)/(\gamma-z_0)$ lies in $D(1,1)$. By part (c) of (3.2.4), $n(1+(z_0-z)/(\gamma-z_0), 0) = 0$, so $n(\gamma, z) = n(\gamma, z_0)$. This proves that the function $z \to n(\gamma, z)$ is continuous on the open set $\mathbb{C}\setminus\gamma^*$ and locally constant. By an argument that we have seen several times, the function is constant on components of $\mathbb{C}\setminus\gamma^*$. (If $z_0 \in \mathbb{C}\setminus\gamma^*$ and Ω is that component of $\mathbb{C} \setminus \gamma^*$ containing z_0, let $A = \{z \in \Omega : n(\gamma, z) = n(\gamma, z_0)\}$. Then A is a nonempty subset of Ω and A is both open and closed in Ω, so $A = \Omega$.) To see that $n(\gamma, z) = 0$ on the unbounded component of $\mathbb{C} \setminus \gamma^*$, note that $\gamma^* \subseteq D(0, R)$ for R sufficiently large. By (3.2.4c), $n(\gamma, z) = 0$ for $z \notin D(0, R)$. Since all z outside of $D(0, R)$ belong to the unbounded component of $\mathbb{C} \setminus \gamma^*$, we are finished. ♣

Problems

1. Suppose Ω is a region in $\mathbb{C} \setminus \{0\}$ such that every ray from 0 meets Ω.
   (a) Show that for any $\alpha \in \mathbb{R}$, $\log_\alpha$ is not analytic on Ω.
   (b) Show, on the other hand, that there exist regions of this type such that z *does* have an analytic logarithm on Ω.
2. Let $f(z) = (z - a)(z - b)$ for z in the region $\Omega = \mathbb{C} \setminus [a, b]$, where a and b are distinct complex numbers. Show that f has an analytic square root, but not an analytic logarithm, on Ω.
3. Let f be an analytic zero-free function on Ω. Show that the following are equivalent.
   (a) f has an analytic logarithm on Ω.
   (b) f has an analytic k-th root on Ω (that is, an analytic function h such that $h^k = f$) for every positive integer k.
   (c) f has an analytic k-th root on Ω for infinitely many positive integers k.
4. Prove the following extension of (3.2.4d), the "generalized dog-walking theorem". Let $\gamma_1, \gamma_2 : [a, b] \to \mathbb{C}$ be closed curves such that $|\gamma_1(t) - \gamma_2(t)| < |\gamma_1(t)| + |\gamma_2(t)|$ for all $t \in [a, b]$. Prove that $n(\gamma_1, 0) = n(\gamma_2, 0)$. (Hint: Define γ as in the proof of (3.2.4d), and investigate the location of γ^*.) Also, what does the hypothesis imply about the dog and the man in this case?
5. Prove the result given in Example 3.1.5(c).
6. Let f be a continuous mapping of the rectangle $S = \{x + iy : a \leq x \leq b, c \leq y \leq d\}$ into $\mathbb{C} \setminus \{0\}$. Show that f has a continuous logarithm. This can be viewed as a generalization of Theorem 3.1.7; to obtain (3.1.7) (essentially), take $c = d$.
7. Let f be analytic and zero-free on Ω, and suppose that g is a continuous logarithm of f on Ω. Show that g is actually analytic on Ω.
8. Characterize the entire functions f, g such that $f^2 + g^2 = 1$. (Hint: $1 = f^2 + g^2 = (f + ig)(f - ig)$, so $f + ig$ is never 0.)
9. Let f and g be continuous mappings of the connected set S into $\mathbb{C} \setminus \{0\}$.
   (a) If $f^n = g^n$ for some positive integer n, show that $f = g\exp(i2\pi k/n)$ for some

$k = 0, 1, \dots, n-1$.Hence if $f(s_0) = g(s_0)$ for some $s_0 \in S$, then $f \equiv g$.
(b) Show that $\mathbb{C} \setminus \{0\}$ cannot be replaced by $\mathbb{C}$ in the hypothesis.

3.3 Cauchy's Theorem

This section is devoted to a discussion of the global (or homology) version of Cauchy's theorem. The elementary proof to be presented below is due to John Dixon, and appeared in Proc. Amer. Math. Soc. 29 (1971), pp. 625-626, but the theorem as stated is originally due to E.Artin.

3.3.1 Cauchy's Theorem

Let γ be closed path in Ω such that $n(\gamma, z) = 0$ for all $z \in \mathbb{C} \setminus \Omega$.

(i) For all analytic functions f on Ω, $\int_\gamma f(w)\,dw = 0$;

(ii) If $z \in \Omega \setminus \gamma^*$, then

$$n(\gamma, z) f(z) = \frac{1}{2\pi i} \int_\gamma \frac{f(w)}{w - z}\,dw.$$

A path γ in Ω with $n(\gamma, z) = 0$ for all $z \in \mathbb{C} \setminus \Omega$ is said to be *Ω-homologous to zero.* Dixon's proof requires two preliminary lemmas.

3.3.2 Lemma

Let f be analytic on Ω, and define g on $\Omega \times \Omega$ by

$$g(w, z) = \begin{cases} \frac{f(w)-f(z)}{w-z}, & w \neq z \\ f'(z), & w = z. \end{cases}$$

Then g is continuous, and for each fixed $w \in \Omega$, the function given by $z \to g(w, z)$ is analytic on Ω.

Proof. Let $\{(w_n, z_n), n = 1, 2, \dots\}$ be any sequence in $\Omega \times \Omega$ converging to $(w, z) \in \Omega \times \Omega$. If $w \neq z$, then eventually $w_n \neq z_n$, and by continuity of f, $g(w_n, z_n) = \frac{f(w_n)-f(z_n)}{w_n - z_n} \to \frac{f(w)-f(z)}{w-z} = g(w, z)$. However, if $w = z$, then

$$g(w_n, z_n) = \begin{cases} \frac{1}{w_n - z_n} \int_{[z_n, w_n]} f'(\tau)\,d\tau & \text{if } w_n \neq z_n \\ \\ f'(z_n) & \text{if } w_n = z_n. \end{cases}$$

In either case, the continuity of f' at z implies that $g(w_n, z_n) \to f'(z) = g(z, z)$.

Finally, the function $z \to g(w, z)$ is continuous on Ω and analytic on $\Omega \setminus \{w\}$ (because f is analytic on Ω). Consequently, $z \to g(w, z)$ is analytic on Ω by (2.2.13). ♣

3.3.3 Lemma

Suppose $[a,b] \subseteq \mathbb{R}$, and let φ be a continuous complex-valued function on the product space $\Omega \times [a,b]$. Assume that for each $t \in [a,b]$, the function $z \to \varphi(z,t)$ is analytic on Ω. Define F on Ω by $F(z) = \int_a^b \varphi(z,t)\,dt$, $z \in \Omega$. Then F is analytic on Ω and

$$F'(z) = \int_a^b \frac{\partial \varphi}{\partial z}(z,t)\,dt, z \in \Omega.$$

Note that Theorem 2.2.10 on integrals of the Cauchy type is special case of this result. However, (2.2.10) will itself play a part in the proof of (3.3.3).

Proof. Fix any disk $D(z_0,r)$ such that $\overline{D}(z_0,r) \subseteq \Omega$. Then for each $z \in D(z_0,r)$,

$$\begin{aligned} F(z) &= \int_a^b \varphi(z,t)\,dt \\ &= \frac{1}{2\pi i}\int_a^b \left(\int_{C(z_0,r)} \frac{\varphi(w,t)}{w-z}\,dw\right) dt \qquad \text{(by 2.2.9)} \\ &= \frac{1}{2\pi i}\int_{C(z_0,r)} \left(\int_a^b \varphi(w,t)\,dt\right) \frac{1}{w-z}\,dw \end{aligned}$$

(Write the path integral as an ordinary definite integral and observe that the interchange in the order of integration is justified by the result that applies to continuous functions on rectangles.) Now $\int_a^b \varphi(w,t)\,dt$ is a continuous function of w (to see this use the continuity of φ on $\Omega \times [a,b]$), hence by (2.2.10), F is analytic on $D(z_0,r)$ and for each $z \in D(z_0,r)$,

$$\begin{aligned} F'(z) &= \frac{1}{2\pi i}\int_{C(z_0,r)} \left[\int_a^b \varphi(w,t)\,dt\right] \frac{1}{(w-z)^2}\,dw \\ &= \int_a^b \left[\frac{1}{2\pi i}\int_{C(z_0,r)} \frac{\varphi(w,t)}{(w-z)^2}\,dw\right] dt \\ &= \int_a^b \frac{\partial \varphi}{\partial z}(z,t)\,dt \end{aligned}$$

by (2.2.10) again. ♣

Proof of Cauchy's Theorem.

Let γ be a closed path in the open set Ω such that $n(\gamma,z) = 0$ for all $z \in \mathbb{C} \setminus \Omega$, and let f be an analytic function on Ω. Define $\Omega' = \{z \in \mathbb{C} \setminus \gamma^* : n(\gamma,z) = 0\}$. Then $\mathbb{C} \setminus \Omega \subseteq \Omega'$, so $\Omega \cup \Omega' = \mathbb{C}$; furthermore, Ω' is open by (3.2.5). If $z \in \Omega \cap \Omega'$ and g is defined as in (3.3.2), then $g(w,z) = (f(w) - f(z))/(w-z)$ since $z \notin \gamma^*$. Thus

$$\int_\gamma g(w,z)\,dw = \int_\gamma \frac{f(w)}{w-z}\,dw - 2\pi i n(\gamma,z) f(z) = \int_\gamma \frac{f(w)}{w-z}\,dw$$

since $n(\gamma, z) = 0$ for $z \in \Omega'$. The above computation shows that we can define a function h on $\mathbb{C}$ by

$$h(z) = \begin{cases} \int_\gamma g(w,z)\,dw & \text{if } z \in \Omega \\ \int_\gamma \frac{f(w)}{w-z}\,dw & \text{if } z \in \Omega'. \end{cases}$$

By (2.2.10), h is analytic on Ω', and by (3.3.2) and (3.3.3), h is analytic on Ω. Thus h is an entire function. But for $|z|$ sufficiently large, $n(\gamma, z) = 0$ by (3.2.5), hence $z \in \Omega'$. Consequently, $h(z) = \int_\gamma \frac{f(w)}{w-z}\,dw \to 0$ as $|z| \to \infty$. By Liouville's theorem (2.4.2), $h \equiv 0$. Now if $z \in \Omega \setminus \gamma^*$ we have, as at the beginning of the proof,

$$0 = h(z) = \int_\gamma g(w,z)\,dw = \int_\gamma \frac{f(w)}{w-z}\,dw - 2\pi i n(\gamma, z) f(z)$$

proving (ii). To obtain (i), choose any $z \in \Omega \setminus \gamma^*$ and apply (ii) to the function $w \to (w - z)f(w), w \in \Omega$. ♣

3.3.4 Remarks

Part (i) of (3.3.1) is usually referred to as Cauchy's *theorem*, and part (ii) as Cauchy's *integral formula*. In the above proof we derived (i) from (ii); see Problem 1 for the reverse implication.

Also, there is a converse to part (i): If γ is a closed path in Ω such that $\int_\gamma f(w)\,dw = 0$ for every f analytic on Ω, then $n(\gamma, z) = 0$ for every $z \notin \Omega$. To prove this, take $f(w) = 1/(w - z)$ and apply (3.2.3).

It is sometimes convenient to integrate over objects slightly more general than closed paths.

3.3.5 Definitions

Let $\gamma_1, \gamma_2, \ldots, \gamma_m$ be closed paths. If $k_1, k_2, \ldots, k_m$ are integers, then the formal sum $\gamma = k_1\gamma_1 + \cdots + k_m\gamma_m$ is called a *cycle*. We define $\gamma^* = \cup_{j=1}^m \gamma_j^*$, and for any continuous function f on γ^*,

$$\int_\gamma f(w)\,dw = \sum_{j=1}^m k_j \int_{\gamma_j} f(w)\,dw.$$

Finally, for $z \notin \gamma^*$, define

$$n(\gamma, z) = \sum_{j=1}^m k_j n(\gamma_j, z).$$

It follows directly from the above definitions that the integral representation (3.2.3) for winding numbers extends to cover cycles as well. Also, the proof of Cauchy's theorem (3.3.1) may be repeated almost verbatim for cycles (Problem 2).

Cauchy's theorem, along with the remarks and definitions following it combine to yield the following equivalence.

3.3.6 Theorem

Let γ be a closed path (or cycle) in the open set Ω. Then $\int_\gamma f(z)\,dz = 0$ for every analytic function f on Ω iff $n(\gamma, z) = 0$ for every $z \notin \Omega$.

Proof. Apply (3.3.1), (3.3.4) and (3.3.5). Note that the proof of the converse of (i) of (3.3.1) given in (3.3.4) works for cycles, because the integral representation (3.2.3) still holds. ♣

3.3.7 Corollary

Let γ_1 and γ_2 be closed paths (or cycles) in the open set Ω. Then $\int_{\gamma_1} f(w)\,dw = \int_{\gamma_2} f(w)\,dw$ for every analytic function f on Ω iff $n(\gamma_1, z) = n(\gamma_2, z)$ for every $z \notin \Omega$.

Proof. Apply (3.3.6) to the cycle $\gamma_1 - \gamma_2$. ♣

Note that Theorem 3.3.6 now provides a solution of the first problem posed at the beginning of the chapter, namely, a characterization of those closed paths γ in Ω such that $\int_\gamma f(z)\,dz = 0$ for every analytic function f on Ω.

Problems

1. Show that (i) implies (ii) in (3.3.1).
2. Explain briefly how the proof of (3.3.1) is carried out for cycles.
3. Let Ω, γ and f be as in (3.3.1). Show that for each $k = 0, 1, 2, \dots$ and $z \in \Omega \setminus \gamma^*$, we have

$$n(\gamma, z) f^{(k)}(z) = \frac{k!}{2\pi i} \int_\gamma \frac{f(w)}{(w-z)^{k+1}}\,dw.$$

4. Compute $\int_{C(0,2)} \frac{1}{z^2-1}\,dz$.
5. Use Problem 3 to calculate each of the integrals $\int_{\gamma_j} \frac{e^z + \cos z}{z^4}\,dz$, $j = 1, 2$, where the γ_j are the paths indicated in Figure 3.3.1.
6. Consider $\gamma : [0.2\pi] \to \mathbb{C}$ given by $\gamma(t) = a\cos t + ib\sin t$, where a and b are nonzero real numbers. Evaluate $\int_\gamma dz/z$, and using this result, deduce that

$$\int_0^{2\pi} \frac{dt}{a^2\cos^2 t + b^2 \sin^2 t} = \frac{2\pi}{ab}.$$

3.4 Another Version of Cauchy's Theorem

In this section we consider the second question formulated at the beginning of the chapter: Which open sets Ω have the property that $\int_\gamma f(z)\,dz = 0$ for all analytic functions f on Ω and all closed paths (or cycles) γ in Ω? A concise answer is given by Theorem 3.4.6, but several preliminaries are needed.

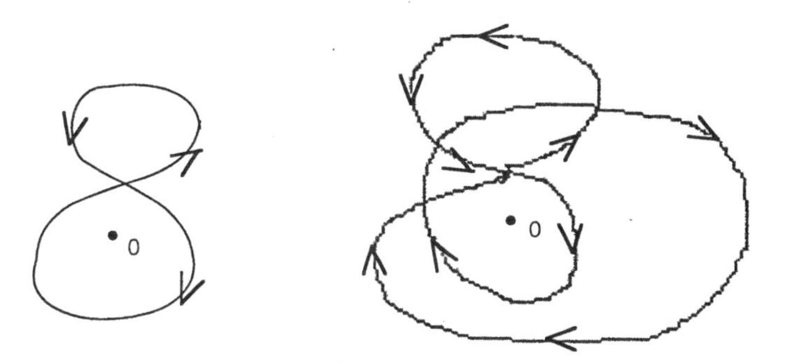

Figure 3.3.1

3.4.1 The Extended Complex Plane

Let $S = \{(x_1, x_2, x_3) \in \mathbb{R}^3 : x_1^2 + x_2^2 + (x_3 - 1/2)^2 = 1/4\}$. Thus S is the sphere in $\mathbb{R}^3$ (called the *Riemann sphere*) with center at (0,0,1/2) and radius 1/2 (Figure 3.4.1). The line segment joining (0,0,1), the north pole of S, to a point $(x, y, 0)$ is $\{(tx, ty, 1-t) : 0 \le t \le 1\}$, and this segment meets S when and only when

$$t^2(x^2 + y^2) + (\frac{1}{2} - t)^2 = \frac{1}{4}, \quad \text{or} \quad t = \frac{1}{1 + x^2 + y^2}.$$

Therefore the intersection point is (x_1, x_2, x_3), where

$$x_1 = \frac{x}{1 + x^2 + y^2}, \; x_2 = \frac{y}{1 + x^2 + y^2}, \; x_3 = \frac{x^2 + y^2}{1 + x^2 + y^2}. \tag{1}$$

Since $1 - x_3 = 1/(1 + x^2 + y^2)$, it follows from (1) that

$$x = \frac{x_1}{1 - x_3}, \; y = \frac{x_2}{1 - x_3}. \tag{2}$$

Let h be the mapping that takes $(x, y, 0)$ to the point (x_1, x_2, x_3) of S. Then h maps $\mathbb{R}^2 \times \{0\}$, which can be identified with $\mathbb{C}$, one-to-one onto $S \setminus \{(0, 0, 1)\}$. Also, by (2), $h^{-1}(x_1, x_2, x_3) = (\frac{x_1}{1-x_3}, \frac{x_2}{1-x_3}, 0)$. Consequently, h is a homeomorphism, that is, h and h^{-1} are continuous.

We can identify $\mathbb{C}$ and $S \setminus \{(0, 0, 1)\}$ formally as follows. Define $k : \mathbb{C} \to \mathbb{R}^2 \times \{0\}$. by $k(x+iy) = (x, y.0)$. Then k is an isometry (a one-to-one, onto, distance-preserving map), hence $h \circ k$ is a homeomorphism of $\mathbb{C}$ onto $S \setminus \{(0, 0, 1)\}$. Next let ∞ denote a point not belonging to $\mathbb{C}$, and take $\hat{\mathbb{C}}$ to be $\mathbb{C} \cup \{\infty\}$. Define $g : \hat{\mathbb{C}} \to S$ by

$$g(z) = \begin{cases} h(k(z)), & z \in \mathbb{C} \\ (0, 0, 1), & z = \infty. \end{cases}$$

Then g maps $\hat{\mathbb{C}}$ one-to-one onto S. If ρ is the usual Euclidean metric of $\mathbb{R}^3$ and $\hat{d}$ is defined on $\hat{\mathbb{C}} \times \hat{\mathbb{C}}$ by

$$\hat{d}(z, w) = \rho(g(z), g(w)),$$

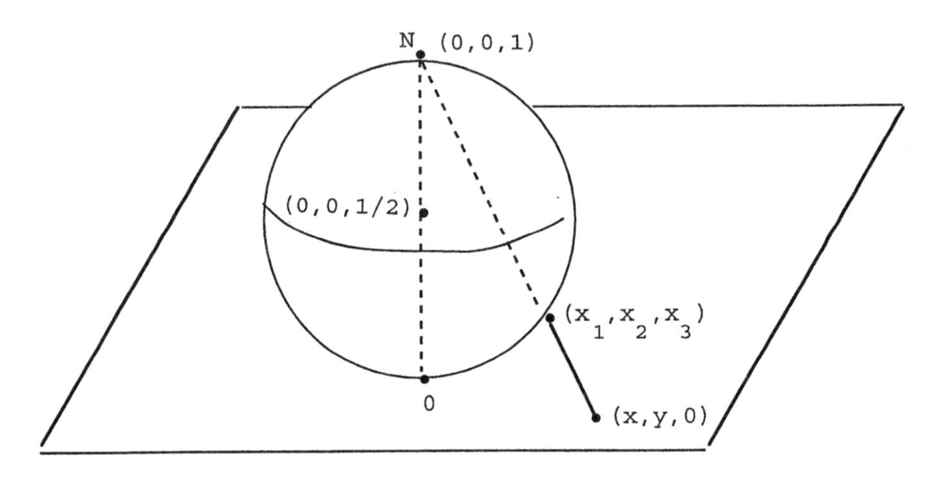

Figure 3.4.1

then $\hat{d}$ is a metric on $\hat{\mathbb{C}}$. (The $\hat{d}$-distance between points of $\hat{\mathbb{C}}$ is the Euclidean distance between the corresponding points on the Riemann sphere.) The metric space $(\hat{\mathbb{C}}, \hat{d})$ is called the *extended plane*, and $\hat{d}$ is called the *chordal metric* on $\hat{\mathbb{C}}$. It is a consequence of the definition of $\hat{d}$ that $(\hat{\mathbb{C}}, \hat{d})$ and (S, ρ) are isometric spaces. The following formulas for $\hat{d}$ hold.

3.4.2 Lemma

$$\hat{d}(z,w) = \begin{cases} \frac{|z-w|}{(1+|z|^2)^{1/2}(1+|w|^2)^{1/2}}, & z, w \in \mathbb{C} \\ \frac{1}{(1+|z|^2)^{1/2}}, & z \in \mathbb{C}, w = \infty. \end{cases}$$

Proof. Suppose $z = x + iy, w = u + iv$. Then by (1) of (3.4.1),

$$\begin{aligned} [\hat{d}(z,w)]^2 &= \left[\frac{x}{1+|z|^2} - \frac{u}{1+|w|^2}\right]^2 + \left[\frac{y}{1+|z|^2} - \frac{v}{1+|w|^2}\right]^2 + \left[\frac{|z|^2}{1+|z|^2} - \frac{|w|^2}{1+|w|^2}\right]^2 \\ &= \frac{x^2+y^2+|z|^4}{(1+|z|^2)^2} + \frac{u^2+v^2+|w|^4}{(1+|w|^2)^2} - 2\frac{xu+yv+|z|^2|w|^2}{(1+|z|^2)(1+|w|^2)} \\ &= \frac{|z|^2}{1+|z|^2} + \frac{|w|^2}{1+|w|^2} - \frac{|z|^2+|w|^2-|z-w|^2+2|z|^2|w|^2}{(1+|z|^2)(1+|w|^2)} \\ &= \frac{|z-w|^2}{(1+|z|^2)(1+|w|^2)} \end{aligned}$$

as desired. Also,

$$\begin{aligned}
[\hat{d}(z,\infty)]^2 &= [\rho(g(z),(0,0,1))]^2 \\
&= \frac{x^2+y^2+1}{(1+x^2+y^2)^2} \text{ by (1) of (3.4.1)} \\
&= \frac{1}{1+|z|^2}. \quad ♣
\end{aligned}$$

Here is a list of the most basic properties of $\hat{\mathbb{C}}$.

3.4.3 Theorem

(a) The metric space $(\hat{\mathbb{C}}, \hat{d})$ is compact, and the identity function on $\mathbb{C}$ is a homeomorphism of $\mathbb{C}$ (with the usual metric) onto $(\mathbb{C}, \hat{d})$.

(b) The complex plane is a dense subspace of $\hat{\mathbb{C}}$. In fact, a sequence $\{z_n\}$ in $\mathbb{C}$ converges to ∞ iff $\{|z_n|\}$ converges to $+\infty$.

(c) The metric space $(\hat{\mathbb{C}}, \hat{d})$ is connected and complete.

(d) Let γ be a closed curve in $\mathbb{C}$, and define $n(\gamma,\infty) = 0$. Then the function $n(\gamma,\cdot)$ is continuous on $\hat{\mathbb{C}} \setminus \gamma^*$.

(e) The identity map on $\hat{\mathbb{C}}$ is a homeomorphism of $(\hat{\mathbb{C}}, \hat{d})$ with the one-point compactification $(\mathbb{C}_\infty, \mathcal{T})$ of $\mathbb{C}$. (Readers unfamiliar with the one-point compactification of a locally compact space may simply ignore this part of the theorem, as it will not be used later.)

Proof. Since the Riemann sphere is compact, connected and complete, so is $(\hat{\mathbb{C}}, \hat{d})$. The formula for $\hat{d}$ in (3.4.2) shows that the identity map on $\mathbb{C}$ is a homeomorphism of $\mathbb{C}$ into $\hat{\mathbb{C}}$, and that $z_n \to \infty$ iff $|z_n| \to +\infty$. This proves (a), (b) and (c). Part (d) follows from (3.2.5). For (e), see Problem 4. ♣

We are now going to make precise, in two equivalent ways, the notion that an open set has no holes.

3.4.4 Theorem

Let Ω be open in $\mathbb{C}$. Then $\hat{\mathbb{C}} \setminus \Omega$ is connected iff each closed curve (and each cycle) γ in Ω is Ω-homologous to 0, that is, $n(\gamma, z) = 0$ for all $z \notin \Omega$.

Proof. Suppose first that $\hat{\mathbb{C}} \setminus \Omega$ is connected, and let γ be a closed curve in Ω. Since $z \to n(\gamma, z)$ is a continuous integer-valued function on $\mathbb{C} \setminus \gamma^*$ [by (3.2.5) and (3.4.3d)], it must be constant on the connected set $\hat{\mathbb{C}} \setminus \Omega$. But $n(\gamma,\infty) = 0$, hence $n(\gamma, z) = 0$ for all $z \in \hat{\mathbb{C}} \setminus \Omega$. The statement for cycles now follows from the result for closed curves.

The converse is considerably more difficult, and is a consequence of what we will call the *hexagon lemma*. As we will see, this lemma has several applications in addition to its use in the proof of the converse.

3.4.5 The Hexagon Lemma

Let Ω be an open subset of $\mathbb{C}$, and let K be a nonempty compact subset of Ω. Then there are closed polygonal paths $\gamma_1, \gamma_2, \dots, \gamma_m$ in $\Omega \setminus K$ such that

$$\sum_{j=1}^{m} n(\gamma_j, z) = \begin{cases} 1 & \text{if } z \in K \\ 0 & \text{if } z \notin \Omega. \end{cases}$$

The lemma may be expressed by saying there is a (polygonal) cycle in $\Omega \setminus K$ which winds around each point of K exactly once, but does not wind around any point of $\mathbb{C} \setminus \Omega$.

Proof. For each positive integer n, let $\mathcal{P}_n$ be the hexagonal partition of $\mathbb{C}$ determined by the hexagon with base $[0, 1/n]$; see Figure 3.4.2. Since K is a compact subset of the open

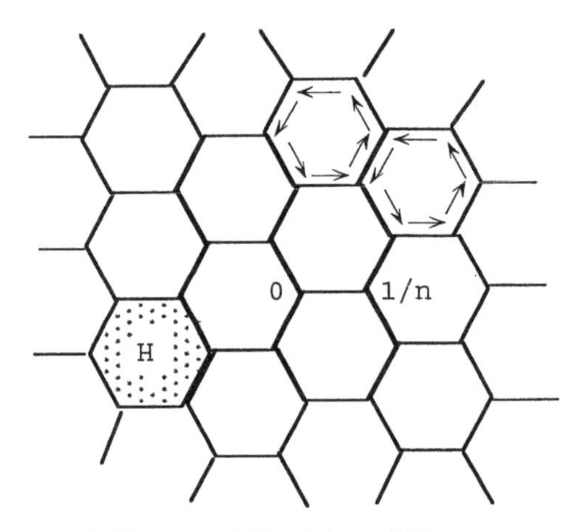

Figure 3.4.2. A Hexagonal Partition of $\mathbb{C}$.

set Ω, we have $\operatorname{dist}(K, \mathbb{C} \setminus \Omega) > 0$, and therefore we can choose n large enough so that if $H \in \mathcal{P}_n$ and $H \cap K \neq \emptyset$, then $H \subseteq \Omega$. Define $\mathcal{K} = \{H \in \mathcal{P}_n : H \cap K \neq \emptyset\}$. Since K is nonempty and bounded, $\mathcal{K}$ is a nonempty finite collection and

$$K \subseteq \cup\{H : H \in \mathcal{K}\} \subseteq \Omega.$$

Now assign a positive (that is, counterclockwise) orientation to the sides of each hexagon (see Figure 3.4.2). Let S denote the collection of all oriented sides of hexagons in $\mathcal{K}$ that are sides of exactly one member of $\mathcal{K}$. Observe that given an oriented side $\vec{ab} \in S$, there are *unique* oriented sides $\vec{ca}$ and $\vec{bd}$ in S. (This uniqueness property is the motivation for tiling with hexagons instead of squares. If we used squares instead, as in Figure 3.4.3, we have $\vec{ab}, \vec{bc}$ and $\vec{bd} \in S$, so $\vec{ab}$ does not have a unique successor, thus complicating the argument that follows.)

By the above observations, and the fact that S is a finite collection, it follows that given $\vec{a_1a_2} \in S$, there is a *uniquely* defined closed polygonal path $\gamma_1 = [a_1, a_2, \dots, a_k, a_1]$ with all sides in S. If S_1 consists of the edges of γ_1 and $S \setminus S_1 \neq \emptyset$, repeat the above

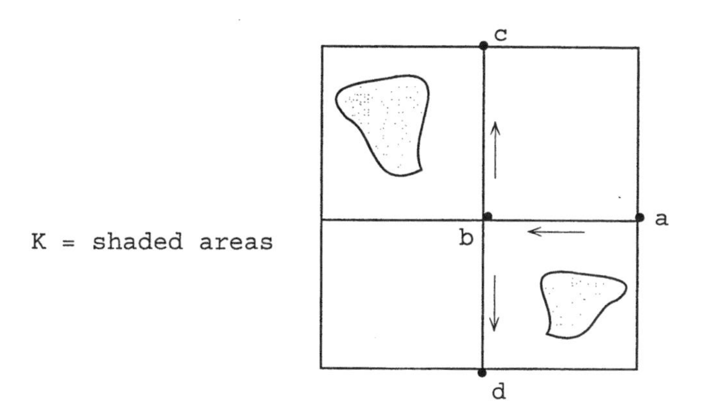

Figure 3.4.3 Nonuniqueness when squares are used.

construction with S replaced by $S \setminus S_1$. Continuing in this manner, we obtain pairwise disjoint collections $S_1, S_2, \ldots, S_n$ such that $S = \cup_{j=1}^{m} S_j$, and corresponding closed polygonal paths $\gamma_1, \gamma_2, \ldots, \gamma_m$ (Figure 3.4.4).

Suppose now that the hexagons in $\mathcal{K}$ are $H_1, H_2, \ldots, H_p$, and let σ_j denote the boundary of H_j, oriented positively. If z belongs to the interior of some H_r, then $n(\sigma_r, z) = 1$ and $n(\sigma_j, z) = 0, j \neq r$. Consequently, $n(\sigma_1 + \sigma_2 + \cdots + \sigma_p, z) = 1$ by (3.3.5). But by construction, $n(\gamma_1 + \cdots + \gamma_m, z) = n(\sigma_1 + \cdots + \sigma_p, z)$. (The key point is that if both hexagons containing a particular side $[a, b]$ belong to $\mathcal{K}$, then $\vec{ab} \notin S$ and $\vec{ba} \notin S$. Thus $[a, b]$ will not contribute to either $n(\gamma_1 + \cdots + \gamma_m, z)$ or to $n(\sigma_1 + \cdots + \sigma_p, z)$. If only one hexagon containing $[a, b]$ belongs to $\mathcal{K}$, then $\vec{ab}$ (or $\vec{ba}$) appears in both cycles.) Therefore $n(\gamma_1 + \cdots + \gamma_m, z) = 1$. Similarly, if $z \notin \Omega$, then $n(\gamma_1 + \cdots + \gamma_m, z) = n(\sigma_1 + \cdots + \sigma_p, z) = 0$.

Finally, assume $z \in K$ and z belongs to a side s of some H_r. Then s cannot be in S, so $z \notin (\gamma_1 + \cdots + \gamma_m)^*$. Let $\{w_k\}$ be a sequence of interior points of H_r with w_k converging to z. We have shown that $n(\gamma_1 + \cdots + \gamma_m, w_k) = 1$ for all k, so by (3.2.5), $n(\gamma_1 + \cdots + \gamma_m, z) = 1$. ♣

Completion of the Proof of (3.4.4)

If $\hat{\mathbb{C}} \setminus \Omega$ is not connected, we must exhibit a cycle in Ω that is not Ω-homologous to 0. Now since $\hat{\mathbb{C}}$ is closed and not connected, it can be expressed as the union of two nonempty disjoint closed sets K and L. One of these two sets must contain ∞; assume that $\infty \in L$. Then K must be a compact subset of the complex plane $\mathbb{C}$, and K is contained in the plane open set $\Omega_1 = \mathbb{C} \setminus L$. apply the hexagon lemma (3.4.5) to Ω_1 and K to obtain a cycle σ in $\Omega_1 \setminus K = \mathbb{C} \setminus (K \cup L) = \Omega$ such that $n(\sigma, z) = 1$ for each $z \in K$ (and $n(\sigma, z) = 0$ for $z \notin \Omega_1$). Pick any point z in the nonempty set $K \subseteq \mathbb{C} \setminus \Omega$. Then $z \notin \Omega$ and $n(\sigma, z) = 1 \neq 0$. ♣

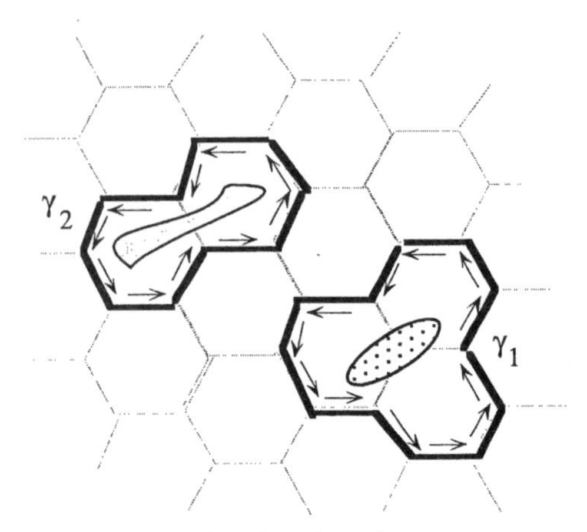

Figure 3.4.4. Construction of the closed paths.

Remark

As a consequence of the definition (3.3.5) of the index of a cycle, if $\hat{\mathbb{C}} \setminus \Omega$ is not connected, there must actually be a closed *path* γ in Ω such that $n(\gamma, z) \neq 0$ for some $z \notin \Omega$.

The list of equivalences below is essentially a compilation of results that have already been established.

3.4.6 Second Cauchy Theorem

Let Ω be an open subset of $\mathbb{C}$. The following are equivalent.

(1) $\hat{\mathbb{C}} \setminus \Omega$ is connected.

(2) $n(\gamma, z) = 0$ for each closed path (or cycle) γ in Ω and each point $z \in \mathbb{C} \setminus \Omega$.

(3) $\int_\gamma f(z)\, dz = 0$ for every closed path (or cycle) γ in Ω and every analytic function f on Ω.

(4) Every analytic function on Ω has a primitive on Ω.

(5) Every zero-free analytic function on Ω has an analytic logarithm.

(6) Every zero-free analytic function on Ω has an analytic n-th root for $n = 1, 2, \dots$.

Proof.

(1) is equivalent to (2) by Theorem 3.4.4.

(2) is equivalent to (3) by Theorem 3.3.6.

(3) is equivalent to (4) by Theorems 2.1.6 and 2.1.10.

(3) implies (5) by Theorem 3.1.10.

(5) is equivalent to (6) by Problem 3.2.3.

(5) implies (2): If $z_0 \notin \Omega$, let $f(z) = z - z_0, z \in \Omega$. Then f has an analytic logarithm on Ω, and hence for each closed path (or cycle) γ in Ω we have, by (3.2.3) and (3.1.9),

$$n(\gamma, z_0) = \frac{1}{2\pi i} \int_\gamma \frac{1}{z - z_0}\, dz = \frac{1}{2\pi i} \int_\gamma \frac{f'(z)}{f(z)}\, dz = 0. \clubsuit$$

We will be adding to the above list in later chapters. An open subset of $\mathbb{C}$ satisfying any (and hence all) of the conditions of (3.4.6) is said to be *(homologically) simply connected.*

It is true that in complex analysis, the implications (1) $\Rightarrow$ (2) $\Rightarrow$ (3) are used almost exclusively. The rather tedious hexagon lemma was required to establish the reverse implication (2) $\Rightarrow$ (1). Thus one might wonder why we have gone to the trouble of obtaining the hexagon lemma at all. One answer is that it has other applications, including the following global integral representation formula. This formula should be compared with Cauchy's integral formula for a circle (2.2.9). It will also be used later in the proof of Runge's theorem on rational approximation.

3.4.7 Theorem

Let K be a compact subset of the open set Ω. Then there is a cycle γ in $\Omega \setminus K$ such that γ is a formal sum of closed polygonal paths, and for every analytic function f on Ω,

$$f(z) = \frac{1}{2\pi i}\int_\gamma \frac{f(w)}{w-z}\,dw = 0 \quad \text{for all } z \in K.$$

Proof. Apply the hexagon lemma and part (ii) of (3.3.1). ♣

Problems

1. (a) Give an example of an open connected set that is not simply connected. For this set, describe explicitly an analytic function f and a closed path γ such that $\int_\gamma f(z)\,dz \neq 0$.
   (b) Give an example of an open, simply connected set that is not connected.
2. Suppose that in the hexagon lemma, Ω is assumed to be connected. Can a *cycle* that satisfies the conclusion be taken to be a closed *path*?
3. Let Γ_1 be the ray $[1, i/2, \infty) = \{1 - t + ti/2 : 0 \le t < \infty\}$ and let Γ_2 be the ray $[1, 2, \infty)$.
   (a) Show that $1 - z$ has analytic square roots f and g on $\mathbb{C}\setminus\Gamma_1$ and $\mathbb{C}\setminus\Gamma_2$ respectively, such that $f(0) = g(0) = 1$.
   (b) Show that $f = g$ below $\Gamma = \Gamma_1 \cup \Gamma_2$ and $f = -g$ above Γ. (Compare Problem 3.2.9.)
   (c) Let $h(z)$ be given by the binomial expansion of $(1-z)^{1/2}$, that is,

$$h(z) = \sum_{n=0}^{\infty} \binom{1/2}{n}(-z)^n, \quad |z| < 1,$$

   where $\binom{w}{n} = \frac{w(w-1)\cdots(w-n+1)}{n!}$. What is the relationship between h and f?
4. Prove Theorem 3.4.3(e).

Chapter 4

Applications Of The Cauchy Theory

This chapter contains several applications of the material developed in Chapter 3. In the first section, we will describe the possible behavior of an analytic function near a singularity of that function.

4.1 Singularities

We will say that f has an *isolated singularity* at z_0 if f is analytic on $D(z_0, r) \setminus \{z_0\}$ for some r. What, if anything, can be said about the behavior of f near z_0? The basic tool needed to answer this question is the *Laurent series*, an expansion of $f(z)$ in powers of $z - z_0$ in which negative as well as positive powers of $z - z_0$ may appear. In fact, the number of negative powers in this expansion is the key to determining how f behaves near z_0.

From now on, the punctured disk $D(z_0, r) \setminus \{z_0\}$ will be denoted by $D'(z_0, r)$. We will need a consequence of Cauchy's integral formula.

4.1.1 Theorem

Let f be analytic on an open set Ω containing the *annulus* $\{z : r_1 \le |z - z_0| \le r_2\}$, $0 < r_1 < r_2 < \infty$, and let γ_1 and γ_2 denote the positively oriented inner and outer boundaries of the annulus. Then for $r_1 < |z - z_0| < r_2$, we have

$$f(z) = \frac{1}{2\pi i} \int_{\gamma_2} \frac{f(w)}{w - z} \, dw - \frac{1}{2\pi i} \int_{\gamma_1} \frac{f(w)}{w - z} \, dw.$$

Proof. Apply Cauchy's integral formula [part (ii) of (3.3.1)] to the cycle $\gamma_2 - \gamma_1$. ♣

4.1.2 Definition

For $0 \leq s_1 < s_2 \leq +\infty$ and $z_0 \in \mathbb{C}$, we will denote the open annulus $\{z : s_1 < |z-z_0| < s_2\}$ by $A(z_0, s_1, s_2)$.

4.1.3 Laurent Series Representation

If f is analytic on $\Omega = A(z_0, s_1, s_2)$, then there is a unique two-tailed sequence $\{a_n\}_{n=-\infty}^{\infty}$ such that

$$f(z) = \sum_{n=-\infty}^{\infty} a_n(z - z_0)^n, z \in \Omega.$$

In fact, if r is such that $s_1 < r < s_2$, then the coefficients a_n are given by

$$a_n = \frac{1}{2\pi i}\int_{C(z_0,r)} \frac{f(w)}{(w - z_0)^{n+1}}\, dw,\ n = 0, \pm 1, \pm 2, \dots .$$

Also, the above series converges absolutely on Ω and uniformly on compact subsets of Ω.

Proof. Choose r_1 and r_2 such that $s_1 < r_1 < r_2 < s_2$ and consider the Cauchy type integral

$$\frac{1}{2\pi i}\int_{C(z_0,r_2)} \frac{f(w)}{w - z}\, dw,\ z \in D(z_0, r_2).$$

Then proceeding just as we did in the proof of Theorem 2.2.16, we obtain

$$\frac{1}{2\pi i}\int_{C(z_0,r_2)} \frac{f(w)}{w - z}\, dw = \sum_{n=0}^{\infty} a_n(z - z_0)^n$$

where

$$a_n = \frac{1}{2\pi i}\int_{C(z_0,r_2)} \frac{f(w)}{(w - z_0)^{n+1}}\, dw.$$

The series converges absolutely on $D(z_0, r_2)$, and uniformly on compact subsets of $D(z_0, r)$. Next, consider the Cauchy type integral

$$-\frac{1}{2\pi i}\int_{C(z_0,r_1)} \frac{f(w)}{w - z}\, dw,\ |z - z_0| > r_1.$$

This can be written as

$$\frac{1}{2\pi i}\int_{C(z_0,r_1)} \frac{f(w)}{(z - z_0)[1 - \frac{w-z_0}{z-z_0}]}\, dw = \frac{1}{2\pi i}\int_{C(z_0,r_1)} \left[\sum_{n=1}^{\infty} f(w)\frac{(w - z_0)^{n-1}}{(z - z_0)^n}\right] dw.$$

By the Weierstrass M-test, the series converges absolutely and uniformly for $w \in C(z_0, r_1)$. Consequently, we may integrate term by term to obtain the series

$$\sum_{n=1}^{\infty} b_n(z - z_0)^{-n}, \text{ where } b_n = \frac{1}{2\pi i}\int_{C(z_0,r_1)} \frac{f(w)}{(w - z_0)^{-n+1}}\, dw.$$

This is a power series in $1/(z-z_0)$, and it converges for $|z-z_0| > r_1$, and hence uniformly on sets of the form $\{z : |z-z_0| \geq 1/\rho\}$ where $(1/\rho) > r_1$. It follows that the convergence is uniform on compact (indeed on closed) subsets of $\{z : z - z_0| > r_1$.

The existence part of the theorem now follows from (4.1.1) and the above computations, if we note two facts. First, if $s_1 < r < s_2$ and $k = 0, \pm 1, \pm 2, \ldots,$

$$\int_{C(z_0,r)} \frac{f(w)}{(w-z_0)^{k+1}}\,dw = \int_{C(z_0,r_1)} \frac{f(w)}{(w-z_0)^{k+1}}\,dw = \int_{C(z_0,r_2)} \frac{f(w)}{(w-z_0)^{k+1}}\,dw.$$

Second, any compact subset of $A(z_0, s_1, s_2)$ is contained in $\{z : \rho_1 \leq |z-z_0| \leq \rho_2\}$ for some ρ_1 and ρ_2 with $s_1 < \rho_1 < \rho_2 < s_2$.

We turn now to the question of uniqueness. Let $\{b_n\}$ be a sequence such that $f(z) = \sum_{n=-\infty}^{\infty} b_n(z-z_0)^n$ for $z \in A(z_0, s_1, s_2)$. As in the above argument, this series must converge uniformly on compact subsets of $A(z_0, s_1, s_2)$. Therefore if k is any integer and $s_1 < r < s_2$, then

$$\begin{aligned}\frac{1}{2\pi i}\int_{C(z_0,r)} \frac{f(w)}{(w-z_0)^{k+1}}\,dw &= \frac{1}{2\pi i}\int_{C(z_0,r)} \left[\sum_{n=-\infty}^{\infty} b_n(w-z_0)^{n-k-1}\right] dw \\ &= \sum_{n=-\infty}^{\infty} b_n \frac{1}{2\pi i}\int_{C(z_0,r)} (w-z_0)^{n-k-1}\,dw \\ &= b_k,\end{aligned}$$

because

$$\frac{1}{2\pi i}\int_{C(z_0,r)} (w-z_0)^{n-k-1}\,dw = \begin{cases} 1 & \text{if } n-k-1 = -1 \\ 0 & \text{otherwise.}\end{cases}$$

The theorem is completely proved. ♣

We are now in a position to analyze the behavior of f near an isolated singularity. As the preceding discussion shows, if f has an isolated singularity at z_0, then f can be represented uniquely by

$$f(z) = \sum_{n=-\infty}^{\infty} a_n(z-z_0)^n$$

in some deleted neighborhood of z_0.

4.1.4 Definition

Suppose f has an isolated singularity at z_0, and let $\sum_{n=-\infty}^{\infty} a_n(z-z_0)^n$ be the *Laurent expansion of f about z_0*, that is, the series given in (4.1.3). We say that f has a *removable singularity* at z_0 if $a_n = 0$ for all $n < 0$; f has a *pole of order m* at z_0 if m is the largest positive integer such that $a_{-m} \neq 0$. (A pole of order 1 is called a *simple pole*.) Finally, if $a_n \neq 0$ for infinitely many $n < 0$, we say that f has an *essential singularity* at z_0.

The next theorem relates the behavior of $f(z)$ for z near z_0 to the type of singularity that f has at z_0.

4.1.5 Theorem

Suppose that f has an isolated singularity at z_0. Then

(a) f has a removable singularity at z_0 iff $f(z)$ approaches a finite limit as $z \to z_0$ iff $f(z)$ is bounded on the punctured disk $D'(z_0, \delta)$ for some $\delta > 0$.

(b) For a given positive integer m, f has a pole of order m at z_0 iff $(z-z_0)^m f(z)$ approaches a finite nonzero limit as $z \to z_0$. Also, f has a pole at z_0 iff $|f(z)| \to +\infty$ as $z \to z_0$.

(c) f has an essential singularity at z_0 iff $f(z)$ does not approach a finite or infinite limit as $z \to z_0$, that is, $f(z)$ has no limit in $\hat{\mathbb{C}}$ as $z \to z_0$.

Proof. Let $\{a_n\}_{n=-\infty}^{\infty}$ and $r > 0$ be such that $f(z) = \sum_{-\infty}^{\infty} a_n(z-z_0)^n$ for $0 < |z-z_0| < r$.

(a) If $a_n = 0$ for all $n < 0$, then $\lim_{z\to z_0} f(z) = a_0$. Conversely, if $\lim_{z\to z_0} f(z)$ exists (in $\mathbb{C}$), then f can be defined (or redefined) at z_0 so that f is analytic on $D(z_0, r)$. It follows that there is a sequence $\{b_n\}_{n=0}^{\infty}$ such that $f(z) = \sum_{n=0}^{\infty} b_n(z - z_0)^n$ for $z \in D'(z_0, r)$. By uniqueness of the Laurent expansion, we conclude that $a_n = 0$ for $n < 0$ and $a_n = b_n$ for $n \geq 0$. (Thus in this case, the Laurent and Taylor expansions coincide.) The remaining equivalence stated in (a) is left as an exercise (Problem 1).

(b) If f has a pole of order m at z_0, then for $0 < |z - z_0| < r$,

$$f(z) = a_{-m}(z - z_0)^{-m} + \cdots + a_{-1}(z - z_0)^{-1} + \sum_{n=0}^{\infty} a_n(z - z_0)^n$$

where $a_{-m} \neq 0$. Consequently, $(z - z_0)^m f(z) \to a_{-m} \neq 0$ as $z \to z_0$. Conversely, if $\lim_{z\to z_0}(z - z_0)^m f(z) \neq 0$, then by (a) applied to $(z - z_0)^m f(z)$, there is a sequence $\{b_n\}_{n=0}^{\infty}$ such that

$$(z - z_0)^m f(z) = \sum_{n=0}^{\infty} b_n(z - z_0)^n, z \in D'(z_0, r).$$

Let $z \to z_0$ to obtain $b_0 = \lim_{z\to z_0}(z - z_0)^m f(z) \neq 0$. Thus $f(z)$ can be written as $b_0(z - z_0)^{-m} + b_1(z - z_0)^{-m+1} + \cdots$, showing that f has a pole of order m at z_0. The remaining equivalence in (b) is also left as an exercise (Problem 1).

(c) If $f(z)$ does not have a limit in $\hat{\mathbb{C}}$ as $z \to z_0$, then by (a) and (b), f must have an essential singularity at z_0. Conversely, if f has an essential singularity at z_0, then (a) and (b) again imply that $\lim_{z\to z_0} f(z)$ cannot exist in $\hat{\mathbb{C}}$. ♣

The behavior of a function near an essential singularity is much more pathological even than (4.1.5c) suggests, as the next theorem shows.

4.1.6 Casorati-Weierstrass Theorem

Let f have an isolated essential singularity at z_0. Then for any complex number w, $f(z)$ comes arbitrarily close to w in every deleted neighborhood of z_0. That is, for any $\delta > 0$, $f(D'(z_0, \delta))$ is a dense subset of $\mathbb{C}$.

Proof. Suppose that for some $\delta > 0$, $f(D'(z_0, \delta))$ is not dense in $\mathbb{C}$. Then for some $w \in \mathbb{C}$, there exists $\epsilon > 0$ such that $D(w, \epsilon)$ does not meet $f(D'(z_0, \delta))$. For $z \in D'(z_0, \delta)$, put

$g(z) = 1/(f(z)-w))$. Then g is bounded and analytic on $D'(z_0, \delta)$, and hence by (4.1.5a), g has a removable singularity at z_0. Let m be the order of the zero of g at z_0 (set $m = 0$ if $g(z_0) \neq 0$) and write $g(z) = (z-z_0)^m g_1(z)$ where g_1 is analytic on $D(z_0, \delta)$ and $g_1(z_0) \neq 0$ [see (2.4.4)]. Then $(z-z_0)^m g_1(z) = 1/(f(z) - w)$, so as z approaches z_0,

$$(z-z_0)^m f(z) = (z-z_0)^m w + \frac{1}{g_1(z)} \longrightarrow \begin{cases} w + 1/g_1(z_0) & \text{if } m = 0 \\ 1/g_1(z_0) & \text{if } m \neq 0. \end{cases}$$

Thus f has a removable singularity or a pole at z_0. ♣

4.1.7 Remark

The Casorati-Weierstrass theorem is actually a weak version of a much deeper result called the "big Picard theorem", which asserts that if f has an isolated essential singularity at z_0, then for any $\delta > 0$, $f(D'(z_0, \delta))$ is either the complex plane $\mathbb{C}$ or $\mathbb{C}$ minus one point. We will not prove this result.

The behavior of a complex function f at ∞ may be studied by considering $g(z) = f(1/z)$ for z near 0. This allows us to talk about isolated singularities at ∞. Here are the formal statements.

4.1.8 Definition

We say that f has an *isolated singularity at* ∞ if f is analytic on $\{z : |z| > r\}$ for some r; thus the function $g(z) = f(1/z)$ has an isolated singularity at 0. The type of singularity of f at ∞ is then defined as that of g at 0.

4.1.9 Remark

Liouville's theorem implies that if an entire function f has a removable singularity at ∞, then f is constant. (By (4.1.5a), f is bounded on $\mathbb{C}$.)

Problems

1. Complete the proofs of (a) and (b) of (4.1.5). (Hint for (a): If f is bounded on $D'(z_0, \delta)$, consider $g(z) = (z - z_0)f(z)$.)
2. Classify the singularities of each of the following functions (include the point at ∞). (a) $z/\sin z$ (b) $\exp(1/z)$ (c) $z \cos 1/z$ (d) $1/[z(e^z - 1)]$ (e) $\cot z$
3. Obtain three different Laurent expansions of $(7z-2)/z(z+1)(z-2)$ about $z = -1$. (Use partial fractions.)
4. Obtain all Laurent expansions of $f(z) = z^{-1} + (z-1)^{-2} + (z+2)^{-1}$ about $z = 0$, and indicate where each is valid.
5. Find the first few terms in the Laurent expansion of $\frac{1}{z^2(e^z - e^{-z})}$ valid for $0 < |z| < \pi$.
6. Without carrying out the computation in detail, indicate a relatively easy procedure for finding the Laurent expansion of $1/\sin z$ valid for $\pi < |z| < 2\pi$.

7. (Partial Fraction Expansion). Let $R(z) = P(z)/Q(z)$, where P and Q are polynomials and $\deg P < \deg Q$. (If this is not the case, then by long division we may write $P(z)/Q(z) = a_n z^n + \cdots + a_1 z + a_0 + P_1(z)/Q(z)$ where $\deg P_1 < \deg Q$.) Suppose that the zeros of Q are at $z_1, \ldots, z_k$ with respective orders $n_1, \ldots, n_k$. Show that $R(z) = \sum_{j=1}^{k} B_j(z)$, where $B_j(z)$ is of the form

$$\frac{A_{j,0}}{(z-z_j)^{n_j}} + \cdots + \frac{A_{j,(n_j-1)}}{(z-z_j)},$$

with

$$A_{j,r} = \lim_{z \to z_j} \frac{1}{r!} \frac{d^r}{dz^r}[(z-z_j)^{n_j} R(z)]$$

($\frac{d^r}{dz^r} f(z)$ is interpreted as $f(z)$ when $r = 0$).

Apply this result to $R(z) = 1/[z(z+i)^3]$.

8. Find the sum of the series $\sum_{n=0}^{\infty} e^{-n} \sin nz$ (in closed form), and indicate where the series converges. Make an appropriate statement about uniform convergence. (Suggestion: Consider $\sum_{n=0}^{\infty} e^{-n} e^{inz}$ and $\sum_{n=0}^{\infty} e^{-n} e^{-inz}$.)

9. (a) Show that if f is analytic on $\hat{\mathbb{C}}$, then f is constant.
(b) Suppose f is entire and there exists $M > 0$ and $k > 0$ such that $|f(z)| \le M|z|^k$ for $|z|$ sufficiently large. Show that $f(z)$ is a polynomial of degree at most k. (This can also be done without series; see Problem 2.2.13.)
(c) Prove that if f is entire and has a nonessential singularity at ∞, then f is a polynomial.
(d) Prove that if f is meromorphic on $\hat{\mathbb{C}}$ (that is, any singularity of f in $\hat{\mathbb{C}}$ is a pole), then f is a rational function.

10. Classify the singularities of the following functions (include the point at ∞).

$$\text{(a)} \frac{\sin^2 z}{z^4} \quad \text{(b)} \frac{1}{z^2(z+1)} + \sin\frac{1}{z} \quad \text{(c)} \csc z - \frac{k}{z} \quad \text{(d)} \exp(\tan\frac{1}{z}) \quad \text{(e)} \frac{1}{\sin(\sin z)}.$$

11. Suppose that a and b are distinct complex numbers. Show that $(z-a)/(z-b)$ has an analytic logarithm on $\mathbb{C} \setminus [a,b]$, call it g. Then find the possible Laurent expansions of $g(z)$ about $z = 0$.

12. Suppose f is entire and $f(\mathbb{C})$ is not dense in $\mathbb{C}$. Show that f is constant.

13. Assume f has a pole of order m at α, and P is a polynomial of degree n. Prove that the composition $P \circ f$ has a pole of order mn at α.

4.2 Residue Theory

We now develop a technique that often allows for the rapid evaluation of integrals of the form $\int_\gamma f(z)\,dz$, where γ is a closed path (or cycle) in Ω and f is analytic on Ω except possibly for isolated singularities.

4.2.1 Definition

Let f have an isolated singularity at z_0, and let the Laurent expansion of f about z_0 be $\sum_{n=-\infty}^{\infty} a_n(z-z_0)^n$. The *residue* of f at z_0, denoted by $\text{Res}(f, z_0)$, is defined to be a_{-1}.

4.2.2 Remarks

In many cases, the evaluation of an integral can be accomplished by the computation of residues. This is illustrated by (a) and (b) below.

(a) Suppose f has an isolated singularity at z_0, so that f is analytic on $D'(z_0, \rho)$ for some $\rho > 0$. Then for any r such that $0 < r < \rho$, we have

$$\int_{C(z_0,r)} f(w)\, dw = 2\pi i \,\text{Res}(f, z_0).$$

Proof. Apply the integral formula (4.1.3) for a_{-1}. ♣

(b) More generally, if γ is a closed path or cycle in $D'(z_0, \rho)$ such that $n(\gamma, z_0) = 1$ and $n(\gamma, z) = 0$ for every $z \notin D(z_0, \rho)$, then

$$\int_{\gamma} f(w)\, dw = 2\pi i \,\text{Res}(f, z_0).$$

Proof. This follows from (3.3.7). ♣

(c) $\text{Res}(f, z_0)$ is that number k such that $f(z) - [k/(z - z_0)]$ has a primitive on $D'(z_0, \rho)$.

Proof. Note that if $0 < r < \rho$, then by (a),

$$\int_{C(z_0,r)} \left(f(w) - \frac{k}{w - z_0} \right) dw = 2\pi i[\text{Res}(f, z_0) - k].$$

Thus if $f(z) - [k/(z - z_0)]$ has a primitive on $D'(z_0, \rho)$, then the integral is zero, and hence $\text{Res}(f, z_0) = k$. Conversely, if $\text{Res}(f, z_0) = k$, then

$$f(z) - \frac{k}{z - z_0} = \sum_{\substack{n=-\infty \\ n \neq -1}}^{\infty} a_n(z - z_0)^n,$$

which has a primitive on $D'(z_0, \rho)$, namely

$$\sum_{\substack{n=-\infty \\ n \neq -1}}^{\infty} \frac{a_n}{n+1}(z - z_0)^{n+1}. \text{ ♣}$$

(d) If f has a pole of order m at z_0, then

$$\text{Res}(f, z_0) = \frac{1}{(m-1)!} \lim_{z \to z_0} \left\{ \frac{d^{m-1}}{dz^{m-1}} [(z - z_0)^m f(z)] \right\}.$$

In particular, if f has a simple pole at z_0, then

$$\operatorname{Res}(f, z_0) = \lim_{z \to z_0} [(z - z_0) f(z)].$$

Proof. Let $\{a_n\}$ be the Laurent coefficient sequence for f about z_0, so that $a_n = 0$ for $n < -m$ and $a_{-m} \neq 0$. Then for $z \in D'(z_0, \rho)$,

$$(z - z_0)^m f(z) = a_{-m} + a_{-m+1}(z - z_0) + \cdots + a_{-1}(z - z_0)^{m-1} + a_0(z - z_0)^m + \cdots,$$

hence

$$\frac{d^{m-1}}{dz^{m-1}}[(z - z_0)^m f(z)] = (m-1)! a_{-1} + (z - z_0) g(z)$$

where g has a removable singularity at z_0. The result follows. ♣

(e) Suppose f is analytic at z_0 and has a zero of order k at z_0. Then f'/f has a simple pole at z_0 and $\operatorname{Res}(f'/f, z_0) = k$.

Proof. There exists $\rho > 0$ and a zero-free analytic function g on $D(z_0, \rho)$ such that $f(z) = (z - z_0)^k g(z)$ for $z \in D(z_0, \rho)$. Then $f'(z) = k(z - z_0)^{k-1} g(z) + (z - z_0)^k g'(z)$, and hence for $z \in D'(z_0, \rho)$,

$$\frac{f'(z)}{f(z)} = \frac{k}{z - z_0} + \frac{g'(z)}{g(z)}.$$

Since g'/g is analytic on $D(z_0, \rho)$, it follows that f'/f has a simple pole at z_0 and $\operatorname{Res}(f'/f, z_0) = k$. ♣

We are now ready for the main result of this section.

4.2.3 Residue Theorem

Let f be analytic on $\Omega \setminus S$, where S is a subset of Ω with no limit point in Ω. In other words, f is analytic on Ω except for isolated singularities. Then for any closed path (or cycle) γ in $\Omega \setminus S$ such that γ is Ω-homologous to 0, we have

$$\int_\gamma f(w)\, dw = 2\pi i \sum_{w \in S} n(\gamma, w) \operatorname{Res}(f, w).$$

Proof. Let $S_1 = \{w \in S : n(\gamma, w) \neq 0\}$. Then $S_1 \subseteq Q = \mathbb{C} \setminus \{z \notin \gamma^* : n(\gamma, z) = 0\}$. Since γ is Ω-homologous to 0, Q is a subset of Ω. Furthermore, by (3.2.5), Q is closed and bounded. Since S has no limit point in Ω, S_1 has no limit points at all. Thus S_1 is a finite set. Consequently, the sum that appears in the conclusion of the theorem is the finite sum obtained by summing over S_1. Let $w_1, w_2, \ldots, w_k$ denote the distinct points of S_1. [If S_1 is empty, we are finished by Cauchy's theorem (3.3.1).] Choose positive numbers $r_1, r_2, \ldots, r_k$ so small that

$$D'(w_j, r_j) \subseteq \Omega \setminus S, \quad j = 1, 2, \ldots, k.$$

Let σ be the cycle $\sum_{j=1}^k n(\gamma, w_j)\gamma_j$, where γ_j is the positively oriented boundary of $D(w_j, r_j)$. Then σ is cycle in the open set $\Omega \setminus S$, and you can check that if $z \notin \Omega \setminus S$, then

$n(\gamma, z) = n(\sigma, z)$. Since f is analytic on $\Omega \setminus S$, it follows from (3.3.7) that $\int_\gamma f(w)\,dw = \int_\sigma f(w)\,dw$. But by definition of σ,

$$\int_\gamma f(w)\,dw = \sum_{j=1}^{k} n(\gamma, w_j) \int_{\gamma_j} f(w)\,dw = 2\pi i \sum_{j=1}^{k} n(\gamma, w_j)\,\mathrm{Res}(f, w_j)$$

by part (a) of (4.2.2). ♣

In many applications of the residue theorem, the integral $\int_\gamma f(w)\,dw$ is computed by evaluating the sum $2\pi i \sum_{w\in S} n(\gamma, w)\,\mathrm{Res}(f, w)$. Thus it is important to have methods available for calculating residues. For example, (4.2.2d) is useful when f is a rational function, since the only singularities of f are poles. The residue theorem can also be applied to obtain a basic geometric property of analytic functions called the argument principle. Before discussing the general result, let's look at a simple special case. Suppose z traverses the unit circle once in the positive sense, that is, $z = e^{it}, 0 \le t \le 2\pi$. Then the argument of z^2, namely $2t$, changes by 4π, so that z^2 makes two revolutions around the origin. Thus the number of times that z^2 winds around the origin as z traverses the unit circle is the number of zeros of z^2 inside the circle, counting multiplicity.

The index of a point with respect to a closed path allows us to formalize the notion of the number of times that $f(z)$ winds around the origin as z traverses a path γ. For we are looking at the net number of revolutions about 0 of $f(\gamma(t)), a \le t \le b$, and this, as we have seen, is $n(f \circ \gamma, 0)$. We may now state the general result.

4.2.4 Argument Principle

Let f be analytic on Ω, and assume that f is not identically zero on any component of Ω. If $Z(f) = \{z : f(z) = 0\}$ and γ is any closed path in $\Omega \setminus Z(f)$ such that γ is Ω-homologous to 0,then

$$n(f \circ \gamma, 0) = \sum_{z\in Z(f)} n(\gamma, z) m(f, z)$$

where $m(f, z)$ is the order of the zero of f at z.

Proof. The set $S = Z(f)$ and the function f'/f satisfy the hypothesis of the residue theorem. Applying it, we get

$$\frac{1}{2\pi i} \int_\gamma \frac{f'(z)}{f(z)}\,dz = \sum_{z\in Z(f)} n(\gamma, z)\,\mathrm{Res}(f'/f, z).$$

But the left side equals $n(f\circ\gamma, 0)$ by (3.2.3), and the right side equals $\sum_{z\in Z(f)} n(\gamma, z) m(f, z)$ by (4.2.2e). ♣

4.2.5 Remarks

Assuming that for each $z \in Z(f)$, $n(\gamma, z) = 1$ or 0, the argument principle says that the net increase in the argument of $f(z)$ as z traverses γ^* in the positive direction is equal to the number of zeros of f "inside γ" ($n(\gamma, z) = 1$) with multiplicities taken into account.

There is a useful generalization of (4.2.4) to meromorphic functions. A function f is *meromorphic* on Ω if f is analytic on Ω except possibly for poles. That is, there is a subset $S \subseteq \Omega$ with no limit points in Ω such that f is analytic on $\Omega \setminus S$ and f has a pole at each point of S. For example, any rational function is meromorphic on $\mathbb{C}$. More generally, the quotient f/g of two analytic functions is meromorphic, provided g is not identically zero on any component of Ω. (This follows from (2.4.8) and (4.1.5).) Conversely, every meromorphic function is a quotient of two analytic functions. (This is a much deeper result, which will be proved in a later chapter.)

4.2.6 Definition

For f meromorphic on Ω, let $Z(f)$ denote the set of zeros of f, and $P(f)$ the set of poles of f. If $z \in Z(f) \cup P(f)$, let $m(f, z)$ be the order of the zero or pole of f at z.

4.2.7 Argument Principle for Meromorphic Functions

Suppose f is meromorphic on Ω. Then for any closed path (or cycle) γ in $\Omega\setminus(Z(f)\cup P(f))$ such that γ is Ω-homologous to 0, we have

$$n(f \circ \gamma, 0) = \sum_{z\in Z(f)} n(\gamma, z)m(f, z) - \sum_{z\in P(f)} n(\gamma, z)m(f, z).$$

Proof. Take $S = Z(f) \cup P(f)$, and apply the residue theorem to f'/f. The analysis is the same as in the proof of (4.2.4), if we note that if $z_0 \in P(f)$, then $\text{Res}(f'/f, z_0) = -m(f, z_0)$. To see this, write $f(z) = g(z)/(z - z_0)^k$ where $k = m(f, z_0)$ and g is analytic at z_0, with $g(z_0) \neq 0$. Then $f'(z)/f(z) = [g'(z)/g(z)] - [k/(z - z_0)]$. ♣

Under certain conditions, the argument principle allows a very useful comparison of the number of zeros of two functions.

4.2.8 Rouché's Theorem

Suppose f and g are analytic on Ω, with neither f nor g identically zero on any component of Ω. Let γ be a closed path in Ω such that γ is Ω-homologous to 0. If

$$|f(z) + g(z)| < |f(z)| + |g(z)| \text{ for each } z \in \gamma^*, \tag{1}$$

then

$$\sum_{z\in Z(f)} n(\gamma, z)m(f, z) = \sum_{z\in Z(g)} n(\gamma, z)m(g, z).$$

Thus f and g have the same number of zeros, counting multiplicity and index.

Proof. The inequality (1) implies that $\gamma^* \subseteq \Omega\setminus[Z(f) \cup P(f)]$, and hence by the argument principle, applied to each of f and g, we obtain

$$n(f \circ \gamma, 0) = \sum_{z\in Z(f)} n(\gamma, z)m(f, z) \quad \text{and} \quad n(g \circ \gamma, 0) = \sum_{z\in Z(g)} n(\gamma, z)m(g, z).$$

But again by (1), $|f(\gamma(t)) + g(\gamma(t))| < |f(\gamma(t))| + |g(\gamma(t))|$ for all t in the domain of the closed path γ. Therefore by the generalized dog-walking theorem (Problem 3.2.4), $n(f \circ \gamma, 0) = n(g \circ \gamma, 0)$. The result follows. ♣

4.2.9 Remarks

Rouché's theorem is true for cycles as well. To see this, suppose that γ is the formal sum $k_1\gamma_1 + \cdots + k_r\gamma_r$. Then just as in the proof of (4.2.8), we have $n(f \circ \gamma, 0) = \sum_{z \in Z(f)} n(\gamma, z)m(f, z)$ and $n(g \circ \gamma, 0) = \sum_{z \in Z(g)} n(\gamma, z)m(g, z)$. But now $|f(z) + g(z)| < |f(z)| + |g(z)|$ for each $z \in \gamma^* = \cup_{j=1}^r \gamma_j^*$ implies, as before, that $n(f \circ \gamma_j, 0) = n(g \circ \gamma_j, 0)$ for $j = 1, \dots, r$, hence $n(f \circ \gamma, 0) = n(g \circ \gamma, 0)$ and the proof is complete. ♣

In the hypothesis of (4.2.8), (1) is often replaced by

$$|f(z) - g(z)| < |f(z)| \text{ for each } z \in \gamma^*. \tag{2}$$

But now if (2) holds, then $|f(z) + (-g(z))| < |f(z)| \le |f(z)| + |-g(z)|$ on γ^*, so f and $-g$, hence f and g, have the same number of zeros.

Problems

1. Let $f(z) = (z-1)(z-3+4i)/(z+2i)^2$, and let γ be as shown in Figure 4.2.1. Find $n(f \circ \gamma, 0)$, and interpret the result geometrically.

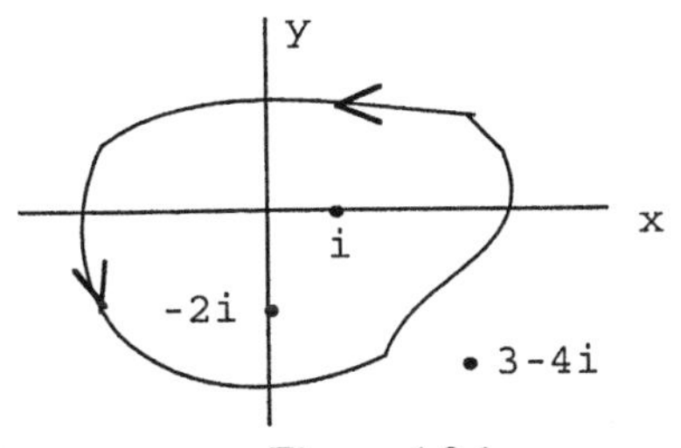

Figure 4.2.1

2. Use the argument principle to find (geometrically) the number of zeros of $z^3 - z^2 + 3z + 5$ in the right half plane.

3. Use Rouché's theorem to prove that any polynomial of degree $n \ge 1$ has exactly n zeros, counting multiplicity.

4. Evaluate the following integrals using residue theory or Cauchy's theorem.

   (a) $\int_{-\infty}^{\infty} \frac{x \sin ax}{x^4+4}\,dx,\ a > 0$ (b) $\int_{-\infty}^{\infty} \frac{x}{(x^2+1)(x^2+2x+2)}\,dx$

   (c) $\int_{-\infty}^{\infty} \frac{1}{(x^2-4x+5)^2}\,dx$ (d) $\int_0^{2\pi} \frac{\cos\theta}{5+4\cos\theta}\,d\theta$ (e) $\int_0^{\infty} \frac{1}{x^4+a^4}\,dx,\ a > 0$

   (f) $\int_0^{\infty} \frac{\cos x}{x^2+1}\,dx$ (g) $\int_0^{2\pi} (\sin\theta)^{2n}\,d\theta$

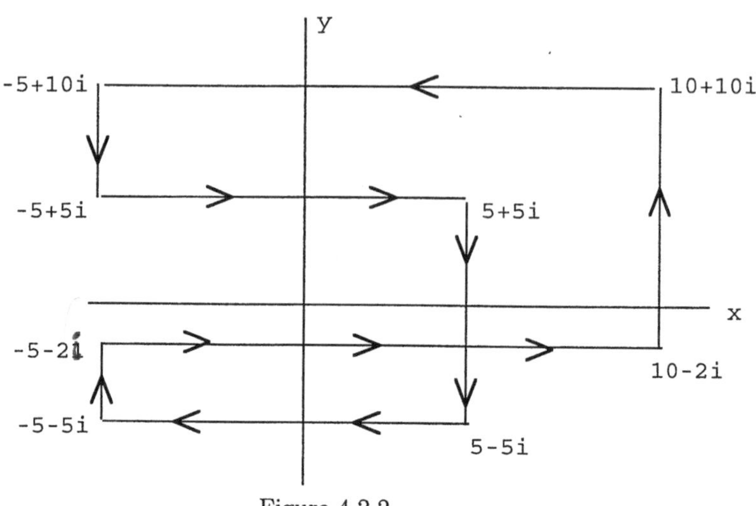

Figure 4.2.2

5. Evaluate $\int_\gamma \frac{\operatorname{Log} z}{1+e^z}\,dz$ along the path γ indicated in Figure 4.2.2.
6. Find the residue at $z=0$ of (a) $\csc^2 z$, (b) $z^{-3}\csc(z^2)$, (c) $z\cos(1/z)$.
7. Find the residue of $\sin(e^z/z)$ at $z=0$. (Leave the answer in the form of an infinite series.)
8. The results of this exercise are necessary for the calculations that are to be done in Problem 9.

   (a) Show that for any $r>0$,

$$\int_0^{\pi/2} e^{-r\sin\theta}\,d\theta \le \frac{\pi}{2r}(1-e^{-r}).$$

   (Hint: $\sin\theta \ge 2\theta/\pi$ for $0\le\theta\le\pi/2$.)

   (b) Suppose f has a simple pole at z_0, and let γ_ϵ be a circular arc with center z_0 and radius ϵ which subtends an angle α at $z_0, 0<\alpha\le 2\pi$ (see Figure 4.2.3). Prove that

$$\lim_{\epsilon\to 0}\int_{\gamma_\epsilon} f(z)\,dz = \alpha i \operatorname{Res}(f,z_0).$$

   In particular, if the γ_ϵ are semicircular arcs ($\alpha=\pi$), then

$$\lim_{\epsilon\to 0}\int_{\gamma_\epsilon} f(z)\,dz = \pi i \operatorname{Res}(f,z_0) = (1/2)2\pi i \operatorname{Res}(f,z_0).$$

   (Hint: $f(z)-[\operatorname{Res}(f,z_0)/(z-z_0)]$ has a removable singularity at z_0.)
9. (a) Show that $\int_{-\infty}^{\infty}\frac{\sin x}{x}\,dx=\pi$ by integrating e^{iz}/z on the closed path $\gamma_{R,\epsilon}$ indicated in Figure 4.2.4.

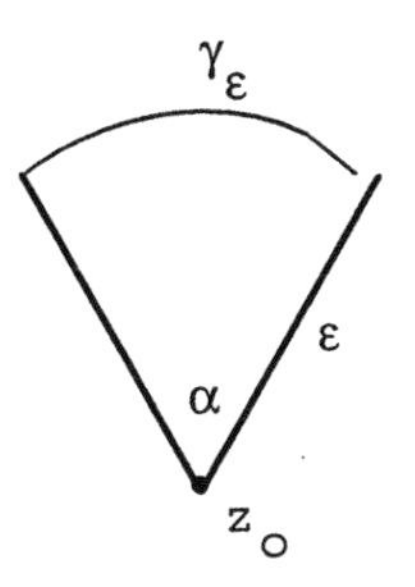

Figure 4.2.3

(b) Show that $\int_0^\infty \cos x^2\,dx = \int_0^\infty \sin x^2\,dx = \frac{1}{2}\sqrt{\pi/2}$. (Integrate e^{iz^2} around the closed path indicated in Figure 4.2.5; assume as known the result that $\int_0^\infty e^{-x^2}\,dx = \frac{1}{2}\sqrt{\pi}$.)
(c) Compute $\int_0^\infty \frac{\ln(x^2+1)}{x^2+1}\,dx$ by integrating $\frac{\text{Log}(z+i)}{z^2+1}$ around the closed path of Figure 4.2.6.
(d) Derive formulas for $\int_0^{\pi/2} \ln\cos\theta\,d\theta$ and $\int_0^{\pi/2} \ln\sin\theta\,d\theta$ by making the change of variable $x = \tan\theta$ in (c).

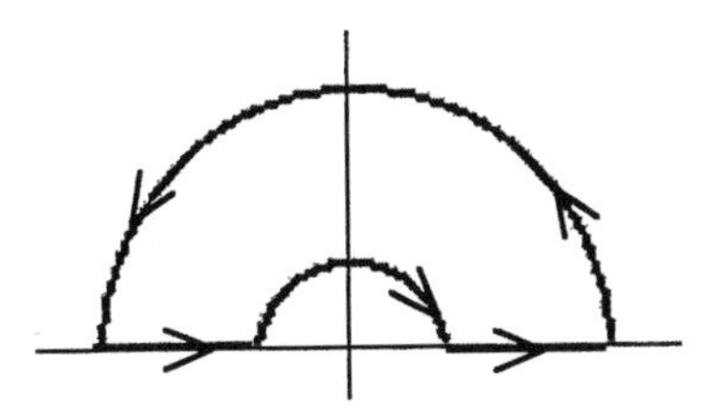

Figure 4.2.4

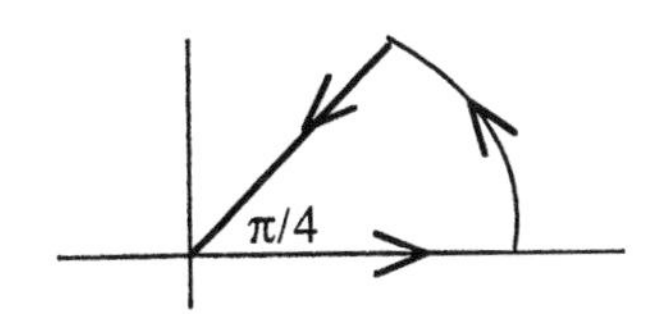

Figure 4.2.5

10. Use Rouché's theorem to show that all the zeros of $z^4 + 6z + 3$ are in $|z| < 2$, and three of them are in $1 < |z| < 2$.
11. Suppose f is analytic on an open set $\Omega \supset \overline{D}(0,1)$, and $|f(z)| < 1$ for $|z| = 1$. Show that for each n, the function $f(z) - z^n$ has exactly n zeros in $D(0,1)$, counting multiplicity. In particular, f has exactly one fixed point in $D(0,1)$.
12. Prove the following version of Rouché's theorem. Suppose K is compact, Ω is an open subset of K, f and g are continuous on K and analytic on Ω, and we have the

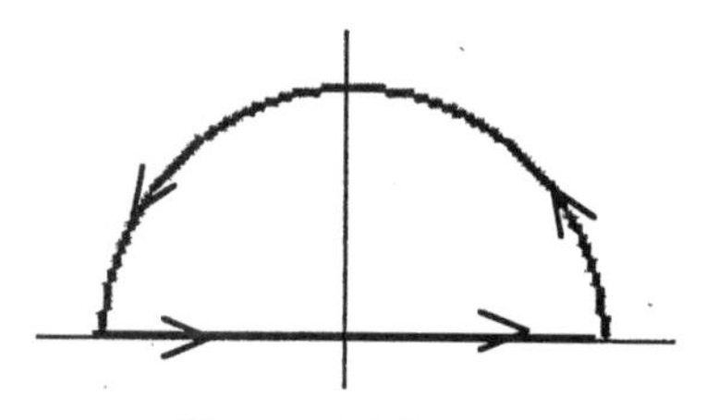

Figure 4.2.6

inequality $|f(z)+g(z)| < |f(z)|+|g(z)|$ for every $z \in K \setminus \Omega$. Show that f and g have the same number of zeros in Ω, that is,

$$\sum_{z\in Z(f)} m(f,z) = \sum_{z \in Z(g)} m(g,z).$$

[Hint: Note that $Z(f) \cup Z(g) \subseteq \{z : |f(z)+g(z)| = |f(z)|+|g(z)|$, and the latter set is a compact subset of Ω. Now apply the hexagon lemma and (4.2.9).]

13. Show that $\frac{1}{\pi}\int_{-\infty}^{\infty} \frac{e^{iux}}{1+x^2}\,dx = e^{-|u|}$ for real u.
14. Evaluate the integral of $\exp[\sin(1/z)]$ around the unit circle $|z|=1$.
15. Suppose f and g are analytic at z_0. Establish the following:
(a) If f has a zero of order k and g has a zero of order $k+1$ at z_0, then f/g has a simple pole at z_0 and

$$\operatorname{Res}(f/g, z_0) = (k+1)f^{(k)}(z_0)/g^{(k+1)}(z_0).$$

(The case $k=0$ is allowed.)
(b) If $f(z_0) \neq 0$ and g has a zero of order 2 at z_0, then f/g has a pole of order 2 at z_0 and

$$\operatorname{Res}(f/g, z_0) = 2\frac{f'(z_0)}{g''(z_0)} - \frac{2}{3}\frac{f(z_0)g'''(z_0)}{[g''(z_0)]^2}.$$

16. Show that the equation $3z = e^{-z}$ has exactly one root in $|z|<1$.
17. Let f be analytic on $D(0,1)$ with $f(0)=0$. Suppose $\epsilon > 0, 0 < r < 1$, and $\min_{|z|=r}|f(z)| \geq \epsilon$. Prove that $D(0,\epsilon) \subseteq f(D(0,r))$.
18. Evaluate

$$\int_{C(1+i,2)} \left[\frac{e^{\pi z}}{z^2+1} + \cos\frac{1}{z} + \frac{1}{e^z}\right] dz.$$

19. Suppose that P and Q are polynomials, the degree of Q exceeds that of P by at least 2, and the rational function P/Q has no poles on the real axis. Prove that $\int_{-\infty}^{\infty}[P(x)/Q(x)]\,dx$ is $2\pi i$ times the sum of the residues of P/Q at its poles in the upper half plane. Then compute this integral with $P(x)=x^2$ and $Q(x)=1+x^4$.

20. Prove that the equation $e^z - 3z^7 = 0$ has seven roots in the unit disk $|z| < 1$. More generally, if $|a| > e$ and n is a positive integer, prove that $e^z - az^n$ has exactly n roots in $|z| < 1$.
21. Prove that $e^z = 2z + 1$ for exactly one $z \in D(0,1)$.
22. Show that $f(z) = z^7 - 5z^4 + z^2 - 2$ has exactly 4 zeros inside the unit circle.
23. If $f(z) = z^5 + 15z + 1$, prove that all zeros of f are in $\{z : |z| < 2\}$, but only one zero of f is in $\{z : |z| < 1/2\}$.
24. Show that all the roots of $z^5 + z + 1 = 0$ satisfy $|z| < 5/4$.
25. Let $\{f_n\}$ be a sequence of analytic functions on an open connected set Ω such that $f_n \to f$ uniformly on compact subsets of Ω. Assume that f is not identically zero, and let $z_0 \in \Omega$. Prove that $f(z_0) = 0$ iff there is a subsequence $\{f_{n_k}\}$ and a sequence $\{z_k\}$ such that $z_k \to z_0$ and $f_{n_k}(z_k) = 0$ for all k. (Suggestion: Rouché's theorem.)
26. Let $p(z) = a_n z^n + \cdots + a_1 z + a_0, a_n \neq 0$, define $q(z) = \overline{a}_0 z^n + \cdots + \overline{a}_{n-1} z + \overline{a}_n$, and put

$$f(z) = \overline{a}_0 p(z) - a_n q(z).$$

Assume that p has $k \geq 0$ zeros in $|z| < 1$, but no zeros on $|z| = 1$. Establish the following.
(a) For $z \neq 0, q(z) = z^n \overline{p}(1/\overline{z})$.
(b) q has $n - k$ zeros in $|z| < 1$.
(c) $|p(z)| = |q(z)|$ for $|z| = 1$.
(d) If $|a_0| > |a_n|$, then f also has k zeros in $|z| < 1$, while if $|a_0| < |a_n|$, then f has $n - k$ zeros in $|z| < 1$.
(e) If $|a_0| > |a_n|$, then p has at least one zero in $|z| > 1$, while if $|a_0| < |a_n|$, then p has at least one zero in $|z| < 1$.

4.3 The Open Mapping Theorem for Analytic Functions

Our aim in this section is to show that a non-constant analytic function on a region Ω maps Ω to a region, and that a one-to-one analytic function has an analytic inverse. These facts, among others, are contained in the following theorem.

4.3.1 Open Mapping Theorem

Let f be a non-constant analytic function on an open connected set Ω. Let $z_0 \in \Omega$ and $w_0 = f(z_0)$, and let $k = m(f - w_0, z_0)$ be the order of the zero which $f - w_0$ has at z_0.

(a) There exists $\epsilon > 0$ such that $\overline{D}(z_0, \epsilon) \subseteq \Omega$ and such that neither $f - w_0$ nor f' has a zero in $\overline{D}(z_0, \epsilon) \setminus \{z_0\}$.

(b) Let γ be the positively oriented boundary of $\overline{D}(z_0, \epsilon)$, let W_0 be the component of $\mathbb{C} \setminus (f \circ \gamma)^*$ that contains w_0, and let $\Omega_1 = D(z_0, \epsilon) \cap f^{-1}(W_0)$. Then f is a k-to-one map of $\Omega_1 \setminus \{z_0\}$ onto $W_0 \setminus \{w_0\}$.

(c) f is a one-to-one map of Ω_1 onto W_0 iff $f'(z_0) \neq 0$.

(d) $f(\Omega)$ is open.

(e) $f : \Omega \to \mathbb{C}$ maps open subsets of Ω onto open sets.

(f) If f is one-to-one, then f^{-1} is analytic.

Proof.

(a) This follows from the identity theorem; the zeros of a non-constant analytic function and its derivative have no limit point in Ω.

(b) If $w \in W_0$, then by the argument principle, $n(f \circ \gamma)$ is the number of zeros of $f - w$ in $D(z_0, \epsilon)$. But $n(f \circ \gamma, w) = n(f \circ \gamma, w_0)$, because the index is constant on components of the complement of $(f \circ \gamma)^*$. Since $n(f \circ \gamma, w_0) = k$, and f' has no zeros in $D'(z_0, \epsilon)$, it follows that for $w \neq w_0, f - w$ has exactly k zeros in $D(z_0, \epsilon)$, all simple. This proves (b).

(c) If $f'(z_0) \neq 0$, then $k = 1$. Conversely, if $f'(z_0) = 0$, then $k > 1$.

(d) This is a consequence of (a) and (b), as they show that $f(z_0)$ is an interior point of the range of f.

(e) This is a consequence of (d) as applied to an arbitrary open subdisk of Ω.

(f) Assume that f is one-to-one from Ω onto $f(\Omega)$. Since f maps open subsets of Ω onto open subsets of $f(\Omega)$, f^{-1} is continuous on $f(\Omega)$. By (c), f' has no zeros in Ω, and Theorem 1.3.2 then implies that f^{-1} is analytic. ♣

4.3.2 Remarks

If Ω is not assumed to be connected, but f is non-constant on each component of Ω, then the conclusions of (4.3.1) are again true. In particular, if f is one-to-one, then surely f is non-constant on components of Ω and hence f^{-1} is analytic on $f(\Omega)$. Finally, note that the maximum principle is an immediate consequence of the open mapping theorem. (Use (4.3.1d), along with the observation that given any disk $D(w_0, r)$, there exists $w \in D(w_0, r)$ with $|w| > |w_0|$.)

The last result of this section is an integral representation theorem for f^{-1} in terms of the given function f. It can also be used to give an alternative proof that f^{-1} is analytic.

4.3.3 Theorem

Let f and g be analytic on Ω and assume that f is one-to-one. Then for each $z_0 \in \Omega$ and each r such that $\overline{D}(z_0, r) \subseteq \Omega$, we have

$$g(f^{-1}(w)) = \frac{1}{2\pi i} \int_{C(z_0,r)} g(z) \frac{f'(z)}{f(z) - w} \, dw$$

for every $w \in f(D(z_0, r))$. In particular, with $g(z) = z$, we have

$$f^{-1}(w) = \frac{1}{2\pi i} \int_{C(z_0,r)} z \frac{f'(z)}{f(z) - w} \, dw.$$

Proof. Let $w \in f(D(z_0, r))$. The function $h(z) = g(z)\frac{f'(z)}{f(z)-w}$ is analytic on $\Omega \setminus \{f^{-1}(w)\}$, and hence by the residue theorem [or even (4.2.2a)],

$$\frac{1}{2\pi i}\int_{C(z_0,r)} g(z)\frac{f'(z)}{f(z)-w}\,dz = \operatorname{Res}(h, w).$$

But g is analytic at $f^{-1}(w)$ and $f - w$ has a simple zero at $f^{-1}(w)$ (because f is one-to-one), hence (Problem 1)

$$\operatorname{Res}(h, w) = g(f^{-1}(w))\operatorname{Res}(\frac{f'}{f-w}, w) = g(f^{-1}(w)) \text{ by (4.2.2e).} \quad ♣$$

In Problem 2, the reader is asked to use the above formula to give another proof that f^{-1} is analytic on $f(\Omega)$.

Problems

1. Suppose g is analytic at z_0 and f has a simple pole at z_0. show that $\operatorname{Res}(gf, z_0) = g(z_0)\operatorname{Res}(f, z_0)$. Show also that the result is false if the word "simple" is omitted.
2. Let f be as in Theorem 4.3.3. Use the formula for f^{-1} derived therein to show that f^{-1} is analytic on $f(\Omega)$. (Show that f^{-1} is representable in $f(\Omega)$ by power series.)
3. The goal of this problem is an open mapping theorem for meromorphic functions. Recall from (4.2.5) that f is meromorphic on Ω if f is analytic on $\Omega \setminus P$ where P is a subset of Ω with no limit point in Ω such that f has a pole at each point of P. Define $f(z) = \infty$ if $z \in P$, so that by (4.1.5b), f is a continuous map of Ω into the extended plane $\hat{\mathbb{C}}$. Prove that if f is non-constant on each component of Ω, then $f(\Omega)$ is open in $\hat{\mathbb{C}}$.
4. Suppose f is analytic on Ω, $\overline{D}(z_0, r) \subseteq \Omega$, and f has no zeros on $C(z_0, r)$. Let $a_1, a_2, \dots, a_n$ be the zeros of f in $D(z_0, r)$. Prove that for any g that is analytic on Ω,

   $$\frac{1}{2\pi i}\int_{C(z_0,r)} \frac{f'(z)}{f(z)} g(z)\,dz = \sum_{j=1}^{n} m(f, a_j)g(a_j)$$

   where (as before) $m(f, a_j)$ is the order of the zero of f at a_j.
5. Let f be a non-constant analytic function on an open connected set Ω. How does the open mapping theorem imply that neither $|f|$ nor $\operatorname{Re} f$ nor $\operatorname{Im} f$ takes on a local maximum in Ω?

4.4 Linear Fractional Transformations

In this section we will study the mapping properties of a very special class of functions on $\mathbb{C}$, the *linear fractional transformations* (also known as *Möbius transformations*).

4.4.1 Definition

If a, b, c, d are complex numbers such that $ad-bc \neq 0$, the *linear fractional transformation* $T : \hat{\mathbb{C}} \to \hat{\mathbb{C}}$ associated with a, b, c, d is defined by

$$T(z) = \begin{cases} \frac{az+b}{cz+d}, & z \neq \infty, z \neq -d/c \\ a/c, & z = \infty \\ \infty, & z = -d/c. \end{cases}$$

Note that the condition $ad - bc \neq 0$ guarantees that T is not constant. Also, if $c = 0$, then $a \neq 0$ and $d \neq 0$, so that the usual agreements regarding ∞ can be made, that is,

$$T(\infty) = \begin{cases} a/c & \text{if } c \neq 0, \\ \infty & \text{if } c = 0 \end{cases} \qquad \text{and} \qquad T(-d/c) = \infty \text{ if } c \neq 0.$$

It follows from the definition that T is a one-to-one continuous map of $\hat{\mathbb{C}}$ onto $\hat{\mathbb{C}}$. Moreover, T is analytic on $\hat{\mathbb{C}} \setminus \{-d/c\}$ with a simple pole at the point $-d/c$. Also, each such T is a composition of maps of the form

(i) $z \to z + B$ (translation)
(ii) $z \to \lambda z$, where $|\lambda| = 1$ (rotation)
(iii) $z \to \rho z$, $\rho > 0$ (dilation)
(iv) $z \to 1/z$ (inversion).

To see that T is always such a composition, recall that if $c = 0$, then $a \neq 0 \neq d$, so

$$T(z) = |a/d|\frac{a/d}{|a/d|}z + \frac{b}{d},$$

and if $c \neq 0$, then

$$T(z) = \frac{(bc - ad)/c^2}{z + (d/c)} + \frac{a}{c}.$$

Linear fractional transformations have the important property of mapping the family of lines and circles in $\mathbb{C}$ onto itself. This is most easily seen by using complex forms of equations for lines and circles.

4.4.2 Theorem

Let $L = \{z : \alpha z\overline{z} + \overline{\beta}z + \beta\overline{z} + \gamma = 0\}$ where α and γ are real numbers, β is complex, and $s^2 = \beta\overline{\beta} - \alpha\gamma > 0$. If $\alpha \neq 0$, then L is a circle, while if $\alpha = 0$, then L is a line. Conversely, each line or circle can be expressed as one of the sets L for appropriate α, γ, β.

Proof. First let us suppose that $\alpha \neq 0$. Then the equation defining L is equivalent to $|z + (\beta/\alpha)|^2 = (\beta\overline{\beta} - \alpha\gamma)/\alpha^2$, which is the equation of a circle with center at $-\beta/\alpha$ and radius $s/|\alpha|$. Conversely, the circle with center z_0 and radius $r > 0$ has equation $|z - z_0|^2 = r^2$, which is equivalent to $z\overline{z} - \overline{z}_0 z - z_0\overline{z} + |z_0|^2 - r^2 = 0$. This has the required form with $\alpha = 1$, $\beta = -z_0$, $\gamma = |z_0|^2 - r^2$. On the other hand, if $\alpha = 0$, then $\beta \neq 0$, and the equation defining L becomes $\overline{\beta}z+\beta\overline{z}+\gamma=0$, which is equivalent to $\text{Re}(\overline{\beta}z)+\gamma/2=0$. This has the form $Ax+By+\gamma/2=0$ where $z=x+iy$ and $\beta=A+iB$, showing that L is a line in this case. Conversely, an equation of the form $Ax+By+C=0$, where A and B are not both zero, can be written in complex form as $\text{Re}(\overline{\beta}z)+\gamma/2=0$, where $\beta=A+iB$ and $\gamma=2C$. ♣

4.4.3 Theorem

Suppose L is a line or circle, and T is a linear fractional transformation. Then $T(L)$ is a line or circle.

Proof. Since T is a composition of maps of the types (i)-(iv) of (4.4.1), it is sufficient to show that $T(L)$ is a line or circle if T is any one of these four types. Now translations, dilations, and rotations surely map lines to lines and circles to circles, so it is only necessary to look at the case where $T(z)=1/z$. But if z satisfies $\alpha z\overline{z}+\overline{\beta}z+\beta\overline{z}+\gamma=0$, then $w=1/z$ satisfies $\gamma w\overline{w}+\beta w+\overline{\beta}\overline{w}+\alpha=0$, which is also an equation of a line or circle. ♣

Note, for example, that if $T(z)=1/z$, $\gamma=0$ and $\alpha\neq 0$, then L is a circle through the origin, but $T(L)$, with equation $\beta w+\overline{\beta}\overline{w}+\alpha=0$, is a line not through the origin. This is to be expected because inversion interchanges 0 and ∞.

Linear fractional transformations also have an angle-preserving property that is possessed, more generally, by all analytic functions with non-vanishing derivatives. This will be discussed in the next section. Problems on linear fractional transformations will be postponed until the end of Section 4.5.

4.5 Conformal Mapping

We saw in the open mapping theorem that if $f'(z_0)\neq 0$, then f maps small neighborhoods of z_0 onto neighborhoods of $f(z_0)$ in a one-to-one fashion. In particular, f maps smooth arcs (that is, continuously differentiable arcs) through z_0 onto smooth arcs through $f(z_0)$. Our objective now is to show that f preserves angles between any two such arcs. This is made precise as follows.

4.5.1 Definition

Suppose f is a complex function defined on a neighborhood of z_0, with $f(z)\neq f(z_0)$ for all z near z_0 but not equal to z_0. If there exists a unimodular complex number $e^{i\varphi}$ such that for all θ,

$$\frac{f(z_0+re^{i\theta})-f(z_0)}{|f(z_0+re^{i\theta})-f(z_0)|}\to e^{i\varphi}e^{i\theta}$$

as $r\to 0^+$, then we say that f *preserves angles* at z_0.

To gain some insight and intuitive feeling for the meaning of the above condition, note that for any θ and small $r_0>0$, $\frac{f(z_0+re^{i\theta})-f(z_0)}{|f(z_0+r_0e^{i\theta})-f(z_0)|}$ is a unit vector from $f(z_0)$ to $f(z_0+r_0e^{i\theta})$. The vectors from z_0 to $z_0+re^{i\theta}, 0<r\leq r_0$, have argument θ, so the

condition states that f maps these vectors onto an arc from $f(z_0)$ whose unit tangent vector at $f(z_0)$ has argument $\varphi + \theta$. Since φ is to be the same for all θ, f rotates all short vectors from z_0 through the fixed angle φ. Thus we see that f preserves angles between tangent vectors to smooth arcs through z_0.

4.5.2 Theorem

Suppose f is analytic at z_0. Then f preserves angles at z_0 iff $f'(z_0) \neq 0$.

Proof. If $f'(z_0) \neq 0$, then for any θ,

$$\lim_{r\to 0+} \frac{f(z_0 + re^{i\theta}) - f(z_0)}{|f(z_0 + re^{i\theta}) - f(z_0)|} = e^{i\theta} \lim_{r\to 0+} \frac{[f(z_0 + re^{i\theta}) - f(z_0)]/re^{i\theta}}{|[f(z_0 + re^{i\theta}) - f(z_0)]|/r} = e^{i\theta} \frac{f'(z_0)}{|f'(z_0)|}.$$

Thus the required unimodular complex number of Definition 4.5.1 is $f'(z_0)/|f'(z_0)|$. Conversely, suppose that $f'(z_0) = 0$. Assuming that f is not constant, $f - f(z_0)$ has a zero of some order $m > 1$ at z_0, hence we may write $f(z) - f(z_0) = (z - z_0)^m g(z)$ where g is analytic at z_0 and $g(z_0) \neq 0$. For any θ and small $r > 0$,

$$\frac{f(z_0 + re^{i\theta}) - f(z_0)}{|f(z_0 + re^{i\theta}) - f(z_0)|} = e^{im\theta} \frac{g(z_0 + re^{i\theta})}{|g(z_0 + re^{i\theta})|} = e^{i\theta} e^{i(m-1)\theta} \frac{g(z_0 + re^{i\theta})}{|g(z_0 + re^{i\theta})|}$$

and the expression on the right side approaches $e^{i\theta}e^{i(m-1)\theta}g(z_0)/|g(z_0)|$ as $r \to 0^+$. Since the factor $e^{i(m-1)\theta}g(z_0)/|g'(z_0)|$ depends on θ, f does *not* preserve angles at z_0. Indeed, the preceding shows that angles are increased by a factor of m, the order of the zero of $f - f(z_0)$ at z_0. ♣

A function f on Ω that is analytic and has a nonvanishing derivative will be called a *conformal map*; it is locally one-to-one and preserves angles. Examples are the exponential function and the linear fractional transformation (on their domains of analyticity). The angle-preserving property of the exponential function was illustrated in part (i) of (2.3.1), where it was shown that exp maps any pair of vertical and horizontal lines onto, respectively, a circle with center 0 and an open ray emanating from 0. Thus the exponential function preserves the orthogonality of vertical and horizontal lines.

Problems

1. Show that the inverse of a linear fractional transformation and the composition of two linear fractional transformations is again a linear fractional transformation.
2. Consider the linear fractional transformation $T(z) = (1 + z)/(1 - z)$.
   (a) Find a formula for the inverse of T.
   (b) Show that T maps $|z| < 1$ onto $\operatorname{Re} z > 0$, $|z| = 1$ onto $\{z : \operatorname{Re} z = 0\} \cup \{\infty\}$, and $|z| > 1$ onto $\operatorname{Re} z < 0$.
3. Find linear fractional transformations that map
   (a) $1, i, -1$ to $1, 0, -1$ respectively.
   (b) $1, i, -1$ to $-1, i, 1$ respectively.

4. Let (z_1, z_2, z_3) be a triple of distinct complex numbers.
   (a) Prove that there is a unique linear fractional transformation T with the property that $T(z_1) = 0,\ T(z_2) = 1,\ T(z_3) = \infty$.
   (b) Prove that if one of z_1, z_2, z_3 is ∞, then the statement of (a) remains true.
   (c) Let each of (z_1, z_2, z_3) and (w_1, w_2, w_3) be triples of distinct complex numbers (or extended complex numbers in $\hat{\mathbb{C}}$). Show that there is a unique linear fractional transformation such that $T(z_j) = w_j, j = 1, 2, 3$.
5. Let f be meromorphic on $\mathbb{C}$ and assume that f is one-to-one. Show that f is a linear fractional transformation. In particular, if f is entire, then f is linear, that is, a first degree polynomial in z. Here is a suggested outline:
   (a) f has at most one pole in $\mathbb{C}$, consequently ∞ is an isolated singularity of f.
   (b) $f(D(0,1))$ and $f(\mathbb{C} \setminus \overline{D}(0,1))$ are disjoint open sets in $\hat{\mathbb{C}}$.
   (c) f has a pole or removable singularity at ∞, so f is meromorphic on $\hat{\mathbb{C}}$.
   (d) f has exactly one pole in $\hat{\mathbb{C}}$.
   (e) Let the pole of f be at z_0. If $z_0 = \infty$, then f is a polynomial, which must be of degree 1. If $z_0 \in \mathbb{C}$, consider $g(z) = 1/f(z), z \neq z_0;\ g(z_0) = 0$. Then g is analytic at z_0 and $g'(z_0) \neq 0$.
   (f) f has a simple pole at z_0.
   (g) $f(z) - [\text{Res}(f, z_0))/(z - z_0)]$ is constant, hence f is a linear fractional transformation.

4.6 Analytic Mappings of One Disk to Another

In this section we will investigate the behavior of analytic functions that map one disk into another. The linear fractional transformations are examples which are, in addition, one-to-one. Schwarz's lemma (2.4.16) is an important illustration of the type of conclusion that can be drawn about such functions, and will be generalized in this section. We will concentrate on the special case of maps of the unit disk $D = D(0,1)$ into itself. The following lemma supplies us with an important class of examples.

4.6.1 Lemma

Fix $a \in D$, and define a function φ_a on $\hat{\mathbb{C}}$ by

$$\varphi_a(z) = \frac{z - a}{1 - \overline{a}z},$$

where the usual conventions regarding ∞ are made: $\varphi_a(\infty) = -1/\overline{a}$ and $\varphi_a(1/\overline{a}) = \infty$. Then φ_a is a one-to-one continuous map of $\hat{\mathbb{C}}$ into $\hat{\mathbb{C}}$ whose inverse is φ_{-a}. Also, φ_a is analytic on $\hat{\mathbb{C}} \setminus \{1/\overline{a}\}$ with a simple pole at $1/\overline{a}$ (and a zero of order 1 at a). Thus φ_a is analytic on a neighborhood of the closed disk $\overline{D}$. Finally,

$$\varphi_a(D) = D, \quad \varphi_a(\partial D) = \partial D, \quad \varphi_a'(z) = \frac{1 - |a|^2}{(1 - \overline{a}z)^2}$$

hence

$$\varphi_a'(a) = \frac{1}{1 - |a|^2} \quad \text{and} \quad \varphi_a'(0) = 1 - |a|^2.$$

Proof. Most of the statements follow from the definition of φ_a and the fact that it is a linear fractional transformation. To see that φ_a maps $|z| = 1$ into itself, we compute, for $|z| = 1$,

$$\left|\frac{z-a}{1-\overline{a}z}\right| = \left|\frac{z-a}{\overline{z}(1-\overline{a}z)}\right| = \left|\frac{z-a}{\overline{z}-\overline{a}}\right| = 1.$$

Thus by the maximum principle, φ_a maps D into D. Since $\varphi_a^{-1} = \varphi_{-a}$ (a computation shows that $\varphi_{-a}(\varphi_a(z)) = z$), and $|a| < 1$ iff $|-a| < 1$, it follows that φ_a maps D *onto* D and maps ∂D onto ∂D. The formulas involving the derivative of φ_a are verified by a direct calculation. ♣

4.6.2 Remark

The functions φ_a are useful in factoring out the zeros of a function g on D, because $g(z)$ and $\varphi_a(g(z))$ have the same maximum modulus on D, unlike $g(z)$ and $(z-a)g(z)$. In fact, if g is defined on the closed disk $\overline{D}$, then

$$\left|\frac{z-a}{1-\overline{a}z}g(z)\right| = |g(z)| \text{ for } |z| = 1.$$

This property of the functions φ_a will be applied several times in this section and the problems following it.

We turn now to what is often called Pick's generalization of Schwarz's lemma.

4.6.3 Theorem

Let $f : D \to D$ be analytic. then for any $a \in D$ and any $z \in D$,

$$\left|\frac{f(z)-f(a)}{1-\overline{f(a)}f(z)}\right| \leq \left|\frac{z-a}{1-\overline{a}z}\right| \tag{i}$$

and

$$|f'(a)| \leq \frac{1-|f(a)|^2}{1-|a|^2}. \tag{ii}$$

Furthermore, if equality holds in (i) for some $z \neq a$, or if equality holds in (ii), then f is a linear fractional transformation. In fact, there is a unimodular complex number λ such that with $b = f(a)$, f is the composition $\varphi_{-b} \circ \lambda\varphi_a = \varphi_b^{-1} \circ \lambda\varphi_a$. That is,

$$f(z) = \frac{\lambda\varphi_a(z)+b}{1+\overline{b}\lambda\varphi_a(z)}, \quad |z| < 1.$$

Proof. Let $a \in D$ and set $b = f(a)$. We are going to apply Schwarz's lemma (2.4.16) to the function $g = \varphi_b \circ f \circ \varphi_{-a}$. First, since f maps D into D, so does g. Also,

$$g(0) = \varphi_b(f(\varphi_{-a}(0))) = \varphi_b(f(a)) = \varphi_b(b) = 0.$$

By Schwarz's lemma, $|g(w)| \leq |w|$ for $|w| < 1$, and replacing w by $\varphi_a(z)$ and noting that $g(\varphi_a(z)) = \varphi_b(f(z))$, we obtain (i). Also by (2.4.16), we have $|g'(0) \leq 1$. But by (4.6.1),

$$\begin{aligned} g'(0) &= \varphi_b'(f(\varphi_{-a}(0)))f'(\varphi_{-a}(0))\varphi_{-a}'(0) \\ &= \varphi_b'(f(a))f'(a)(1-|a|^2) \\ &= \frac{1}{1-|f(a)|^2)}f'(a)(1-|a|^2). \end{aligned}$$

Thus the condition $|g'(0)| \leq 1$ implies (ii).

Now if equality holds in (i) for some $z \neq a$, then $|g(\varphi_a(z))| = |\varphi_a(z)|$ for some $z \neq a$, hence $|g(w)| = |w|$ for some $w \neq 0$. If equality holds in (ii), then $|g'(0)| = 1$. In either case, (2.4.16) yields a unimodular complex number λ such that $g(w) = \lambda w$ for $|w| < 1$. Set $w = \varphi_a(z)$ to obtain $\varphi_b(f(z)) = \lambda\varphi_a(z)$, that is, $f(z) = \varphi_{-b}(\lambda\varphi_a(z))$ for $|z| < 1$. ♣

An important application of Theorem 4.6.3 is in characterizing the one-to-one analytic maps of D onto itself as having the form $\lambda\varphi_a$ where $|\lambda| = 1$ and $a \in D$.

4.6.4 Theorem

Suppose f is a one-to-one analytic map of D onto D. then $f = \lambda\varphi_a$ for some unimodular λ and $a \in D$.

Proof. Let $a \in D$ be such that $f(a) = 0$ and let $g = f^{-1}$, so $g(0) = a$. Now since $g(f(z)) = z$, we have $1 = g'(f(z))f'(z)$, in particular, $1 = g'(f(a))f'(a) = g'(0)f'(a)$. Next, (4.6.3ii) implies that $|g'(0)| \leq 1 - |a|^2$ and $|f'(a)| \leq 1/(1-|a|^2)$. Thus

$$1 = |g'(0)||f'(a)| \leq \frac{1-|a|^2}{1-|a|^2} = 1.$$

Necessarily then, $|f'(a)| = 1/(1-|a|^2)$ (and $|g'(0)| = 1 - |a|^2$). Consequently, by the condition for equality in (4.6.3ii), $f = \lambda\varphi_a$, as required. ♣

4.6.5 Remark

One implication of the previous theorem is that any one-to-one analytic map of D onto D actually extends to a homeomorphism of $\overline{D}$ onto $\overline{D}$. We will see when we study the Riemann mapping theorem in the next chapter that more generally, if f maps D onto a special type of region Ω, then f again extends to a homeomorphism of $\overline{D}$ onto $\overline{\Omega}$.

Our final result is a characterization of those continuous functions on $\overline{D}$ which are analytic on D and have constant modulus on the boundary $|z| = 1$. The technique mentioned in (4.6.2) will be used.

4.6.6 Theorem

Suppose f is continuous on $\overline{D}$, analytic on D, and $|f(z)| = 1$ for $|z| = 1$. Then there is a unimodular λ, finitely many points $a_1, \dots, a_n$ in D, and positive integers $k_1, \dots, k_n$, such that

$$f(z) = \lambda \prod_{j=1}^{n} \left(\frac{z - a_j}{1 - \overline{a}_j z} \right)^{k_j}.$$

In other words, f is, to within a multiplicative constant, a finite product of functions of the type φ_a. (If f is constant on $\overline{D}$, the product is empty and we agree that it is identically 1 in this case.)

Proof. First note that $|f(z)| = 1$ for $|z| = 1$ implies that f has at most finitely many zeros in D. If f has no zeros in D, then by the maximum and minimum principles, f is constant on $\overline{D}$. Suppose then that f has its zeros at the points $a_1, \dots, a_n$ with orders $k_1, \dots, k_n$ respectively. Put

$$g(z) = \prod_{j=1}^{n} \left(\frac{z - a_j}{1 - \overline{a}_j z} \right)^{k_j}, \quad z \in D.$$

Then f/g has only removable singularities in D, the analytic extension of f/g has no zeros in D, and $|f/g| = 1$ on ∂D. Again by the maximum and minimum principles, f/g is constant on $D \setminus \{a_1, \dots, a_n\}$. Thus $f = \lambda g$ with $|\lambda| = 1$. ♣

Problems

1. Derive the inequality (4.6.3ii) *directly* from (4.6.3i).
2. Let f be an analytic map of $D(0, 1)$ into the right half plane $\{z : \operatorname{Re} z > 0\}$. Show that

$$\frac{1 - |z|}{1 + |z|} |f(0)| \le |f(z)| \le \frac{1 + |z|}{1 - |z|} |f(0)|, \quad z \in D(0, 1),$$

   and

$$|f'(0)| \le 2 |\operatorname{Re} f(0)|.$$

   Hint: Apply Schwarz's lemma to $T \circ f$, where $T(w) = (w - f(0))/(w + \overline{f(0)})$.
3. Show that if f is an analytic map of $D(0, 1)$ into itself, and f has two or more fixed points, then $f(z) = z$ for all $z \in D(0, 1)$.
4. (a) Characterize the entire functions f such that $|f(z)| = 1$ for $|z| = 1$ [see (4.6.6)].
   (b) Characterize the meromorphic functions f on $\mathbb{C}$ such that $|f(z)| = 1$ for $|z| = 1$. (Hint: If f has a pole of order k at $a \in D(0, 1)$, then $[(z - a)/(1 - \overline{a} z)]^k f(z)$ has a removable singularity at a.)
5. Suppose that in Theorem 4.6.3, the unit disk D is replaced by $D(0, R)$ and $D(0, M)$. That is, suppose $f : D(0, R) \to D(0, M)$. How are the conclusions (i) and (ii) modified in this case? (Hint: Consider $g(z) = f(Rz)/M$.)
6. Suppose $f : \overline{D}(0, 1) \to \overline{D}(0, 1)$ is continuous and f is analytic on $D(0, 1)$. Assume that f has zeros at $z_1, \dots, z_n$ of orders $k_1, \dots, k_n$ respectively. Show that

$$|f(z)| \le \prod_{j=1}^{n} \left| \frac{z - z_j}{1 - \overline{z}_j z} \right|^{k_j}.$$

Suppose equality holds for some $z \in D(0,1)$ with $z \neq z_j, j = 1, \ldots, n$. Find a formula for $f(z)$.

4.7 The Poisson Integral Formula and its Applications

Our aim in this section is to solve the Dirichlet problem for a disk, that is, to construct a solution of Laplace's equation in the disk subject to prescribed boundary values. The basic tool is the Poisson integral formula, which may be regarded as an analog of the Cauchy integral formula for harmonic functions. We will begin by extending Cauchy's theorem and the Cauchy integral formula to functions continuous on a disk and analytic on its interior.

4.7.1 Theorem

Suppose f is continuous on $\overline{D}(0,1)$ and analytic on $D(0,1)$. Then
(i) $\int_{C(0,1)} f(w)\,dw = 0$ and
(ii) $f(z) = \frac{1}{2\pi i}\int_{C(0,1)} \frac{f(w)}{w-z}\,dw$ for all $z \in D(0,1)$.
Proof. For $0 < r < 1$, $\int_{C(0,r)} f(w)\,dw = 0$ by Cauchy's theorem. For $n = 1, 2, \ldots$, put $f_n(z) = f(\frac{n}{n+1}z)$. Then f_n is analytic on $D(0, \frac{n+1}{n})$ and the sequence $\{f_n\}$ converges to f uniformly on $C(0,1)$ [by continuity of f on $\overline{D}(0,1)$]. Hence $\int_{C(0,1)} f_n(w)\,dw \to \int_{C(0,1)} f(w)\,dw$. Since $\int_{C(0,1)} f_n(w)\,dw = \frac{n+1}{n}\int_{C(0,\frac{n}{n+1})} f(w)\,dw = 0$, we have (i). To prove (ii), we apply (i) to the function g, where

$$g(w) = \begin{cases} \frac{f(w)-f(z)}{w-z}, & w \neq z \\ f'(z), & w = z. \end{cases} \clubsuit$$

Note that the same proof works with only minor modifications if $D(0,1)$ is replaced by an arbitrary disk $D(z_0, R)$.

4.7.2 Definition

For $z \in D(0,1)$, define functions P_z and Q_z on the real line $\mathbb{R}$ by

$$P_z(t) = \frac{1-|z|^2}{|e^{it}-z|^2} \quad \text{and} \quad Q_z(t) = \frac{e^{it}+z}{e^{it}-z};$$

$P_z(t)$ is called the *Poisson kernel* and $Q_z(t)$ the *Cauchy kernel*. We have

$$\operatorname{Re}[Q_z(t)] = \operatorname{Re}\left[\frac{(e^{it}+z)(e^{-it}-\overline{z})}{|e^{it}-z|^2}\right] = \operatorname{Re}\left[\frac{1-|z|^2+ze^{-it}-\overline{z}e^{it}}{|e^{it}-z|^2}\right] = P_z(t).$$

Note also that if $z = re^{i\theta}$, then

$$P_z(t) = \frac{1-r^2}{|e^{it}-re^{i\theta}|^2} = \frac{1-r^2}{|e^{i(t-\theta)}-r|^2} = P_r(t-\theta).$$

Since $|e^{i(t-\theta)} - r|^2 = 1 - 2r\cos(t-\theta) + r^2$, we see that

$$P_r(t-\theta) = \frac{1-r^2}{1-2r\cos(t-\theta)+r^2} = \frac{1-r^2}{1-2r\cos(\theta-t)+r^2} = P_r(\theta - t).$$

Thus for $0 \le r < 1$, $P_r(x)$ is an even function of x. Note also that $P_r(x)$ is positive and decreasing on $[0, \pi]$.

After these preliminaries, we can establish the Poisson integral formula for the unit disk, which states that the value of an analytic function at a point inside the disk is a weighted average of its values on the boundary, the weights being given by the Poisson kernel. The precise statement is as follows.

4.7.3 Poisson Integral Formula

Suppose f is continuous on $\overline{D}(0,1)$ and analytic on $D(0,1)$. then for $z \in D(0,1)$ we have

$$f(z) = \frac{1}{2\pi}\int_0^{2\pi} P_z(t) f(e^{it})\,dt$$

and therefore

$$\operatorname{Re} f(z) = \frac{1}{2\pi}\int_0^{2\pi} P_z(t)\operatorname{Re} f(e^{it})\,dt.$$

Proof. By Theorem 4.7.1(ii),

$$f(0) = \frac{1}{2\pi i}\int_{C(0,1)} \frac{f(w)}{w}\,dw = \frac{1}{2\pi}\int_0^{2\pi} f(e^{it})\,dt,$$

hence $f(0) = (1/2\pi)\int_0^{2\pi} P_0(t) f(e^{it})\,dt$ because $P_0(t) \equiv 1$. This takes care of the case $z = 0$. If $z \neq 0$, then again by (4.7.1) we have

$$f(z) = \frac{1}{2\pi i}\int_{C(0,1)} \frac{f(w)}{w-z}\,dw \quad \text{and} \quad 0 = \frac{1}{2\pi i}\int_{C(0,1)} \frac{f(w)}{w - 1/\overline{z}}\,dw,$$

the second equation holding because $1/\overline{z} \notin D(0,1)$. Subtracting the second equation from the first, we get

$$\begin{aligned} f(z) &= \frac{1}{2\pi i}\int_{C(0,1)} [\frac{1}{w-z} - \frac{1}{w-1/\overline{z}}] f(w)\,dw \\ &= \frac{1}{2\pi}\int_0^{2\pi} [\frac{1}{e^{it}-z} - \frac{1}{e^{it}-1/\overline{z}}] e^{it} f(e^{it})\,dt \\ &= \frac{1}{2\pi}\int_0^{2\pi} [\frac{e^{it}}{e^{it}-z} + \frac{\overline{z}e^{it}}{1-\overline{z}e^{it}}] f(e^{it})\,dt \\ &= \frac{1}{2\pi}\int_0^{2\pi} [\frac{e^{it}}{e^{it}-z} + \frac{\overline{z}}{e^{-it}-\overline{z}}] f(e^{it})\,dt \\ &= \frac{1}{2\pi}\int_0^{2\pi} \frac{1-|z|^2}{|e^{it}-z|^2} f(e^{it})\,dt \end{aligned}$$

which proves the first formula. Taking real parts, we obtain the second. ♣

4.7.4 Corollary

For $|z| < 1$, $\frac{1}{2\pi}\int_0^{2\pi} P_z(t)\,dt = 1$.

Proof. Take $f \equiv 1$ in (4.7.3). ♣

Using the formulas just derived for the unit disk $D(0,1)$, we can obtain formulas for functions defined on arbitrary disks.

4.7.5 Poisson Integral Formula for Arbitrary Disks

Let f be continuous on $\overline{D}(z_0,R)$ and analytic on $D(z_0,R)$. Then for $z \in D(z_0,R)$,

$$f(z) = \frac{1}{2\pi}\int_0^{2\pi} P_{(z-z_0)/R}(t) f(z_0 + Re^{it})\,dt.$$

In polar form, if $z = z_0 + re^{i\theta}$, then

$$f(z_0 + re^{i\theta}) = \frac{1}{2\pi}\int_0^{2\pi} P_{r/R}(\theta - t) f(z_0 + Re^{it})\,dt.$$

Proof. Define g on $\overline{D}(0,1)$ by $g(w) = f(z_0 + Rw)$. Then (4.7.3) applies to g, and we obtain

$$g(w) = \frac{1}{2\pi}\int_0^{2\pi} P_w(t) g(e^{it})\,dt, \ |w| < 1.$$

If $z \in D(z_0,R)$, then $w = (z - z_0)/R \in D(0,1)$ and

$$f(z) = g(\frac{z - z_0}{R}) = \frac{1}{2\pi}\int_0^{2\pi} P_{(z-z_0)/R}(t) f(z_0 + Re^{it})\,dt$$

which establishes the first formula. For the second, apply (4.7.2). [See the discussion beginning with "Note also that ... ".] ♣

We now have the necessary machinery available to solve the Dirichlet problem for disks. Again, for notational reasons we will solve the problem for the unit disk $D(0,1)$. If desired, the statement and proof for an arbitrary disk can be obtained by the same technique we used to derive (4.7.5) from (4.7.3).

4.7.6 The Dirichlet Problem

Suppose u_0 is a real-valued continuous function on $C(0,1)$. Define a function u on $\overline{D}(0,1)$ by

$$u(z) = \begin{cases} u_0(z) & \text{for } |z| = 1, \\ \frac{1}{2\pi}\int_0^{2\pi} P_z(t) u_0(e^{it})\,dt & \text{for } |z| < 1. \end{cases}$$

Then u is continuous on $\overline{D}(0,1)$ and harmonic on $D(0,1)$. Furthermore (since P_z is the real part of Q_z), for $z \in D(0,1)$,

$$u(z) = \text{Re}\left[\frac{1}{2\pi}\int_0^{2\pi} Q_z(t)u_0(e^{it})\,dt\right].$$

In particular, the given continuous function u_0 on $C(0,1)$ has a continuous extension to $\overline{D}(0,1)$ which is harmonic on the interior $D(0,1)$.

Proof. The function $z \to \frac{1}{2\pi}\int_0^{2\pi} Q_z(t)u_0(e^{it})\,dt$ is analytic on $D(0,1)$ by (3.3.3), and therefore u is harmonic, hence continuous, on $D(0,1)$. All that remains is to show that u is continuous at points of the boundary $C(0,1)$.

We will actually show that $u(re^{i\theta}) \to u_0(e^{i\theta})$ uniformly in θ as $r \to 1$. Since u_0 is continuous on $C(0,1)$, this will prove that u is continuous at each of point of $C(0,1)$, by the triangle inequality. Thus let θ and r be real numbers with $0 < r < 1$. Then by (4.7.2), (4.7.4) and the definition of $u(z)$,

$$u(re^{i\theta}) - u_0(e^{i\theta}) = \frac{1}{2\pi}\int_0^{2\pi} P_r(\theta - t)[u_0(e^{it}) - u_0(e^{i\theta})]\,dt.$$

Make the change of variable $x = t - \theta$ and recall that P_r is an even function. The above integral becomes

$$\frac{1}{2\pi}\int_{-\theta}^{2\pi-\theta} P_r(x)[u_0(e^{i(\theta+x)}) - u_0(e^{i\theta})]\,dx,$$

and the limits of integration can be changed to $-\pi$ and π, because the integrand has 2π as a period. Now fix δ with $0 < \delta < \pi$ and write the last integral above as the sum,

$$\frac{1}{2\pi}\int_{-\pi}^{-\delta} + \frac{1}{2\pi}\int_{-\delta}^{\delta} + \frac{1}{2\pi}\int_{\delta}^{\pi}.$$

We can estimate each of these integrals. The first and third have absolute value at most $2\sup\{|u_0(e^{it})| : -\pi \le t \le \pi\}P_r(\delta)$, because $P_r(x)$ is a positive and decreasing function on $[0,\pi]$ and $P_r(-x) = P_r(x)$. The middle integral has absolute value at most $\sup\{|u_0(e^{i(\theta+x)}) - u_0(e^{i\theta})| : -\delta \le x \le \delta\}$, by (4.7.4).

But for fixed $\delta > 0$, $P_r(\delta) \to 0$ as $r \to 1$, while $\sup\{|u_0(e^{i(\theta+x)}) - u_0(e^{i\theta})| : -\delta \le x \le \delta\}$ approaches 0 as $\delta \to 0$, uniformly in θ because u_0 is uniformly continuous on $C(0,1)$. Putting this all together, we see that given $\epsilon > 0$ there is an $r_0, 0 < r_0 < 1$, such that for $r_0 < r < 1$ and all θ, we have $|u(re^{i\theta}) - u_0(e^{i\theta})| < \epsilon$. This, along with the continuity of u_0 on $C(0,1)$, shows that u is continuous at each point of $C(0,1)$. ♣

4.7.7 Uniqueness of Solutions to the Dirichlet Problem

We saw in (2.4.15) that harmonic functions satisfy the maximum and minimum principles. Specifically, if u is continuous on $\overline{D}(0,1)$ and harmonic on $D(0,1)$, then

$$\max_{z\in\overline{D}(0,1)} u(z) = \max_{z\in C(0,1)} u(z) \quad\text{and}\quad \min_{z\in\overline{D}(0,1)} u(z) = \min_{z\in C(0,1)} u(z).$$

Thus if $u \equiv 0$ on $C(0,1)$, then $u \equiv 0$ on $\overline{D}(0,1)$.

Now suppose that u_1 and u_2 are solutions to a Dirichlet problem on $\overline{D}(0,1)$ with boundary function u_0. Then $u_1 - u_2$ is continuous on $\overline{D}(0,1)$, harmonic on $D(0,1)$, and identically 0 on $C(0,1)$, hence identically 0 on $\overline{D}(0,1)$. Therefore $u_1 \equiv u_2$, so *the solution to any given Dirichlet problem is unique.*

Here is a consequence of the uniqueness result.

4.7.8 Poisson Integral Formula for Harmonic Functions

Suppose u is continuous on $\overline{D}(0,1)$ and harmonic on $D(0,1)$. Then for $z \in D(0,1)$, we have

$$u(z) = \frac{1}{2\pi}\int_0^{2\pi} P_z(t)u(e^{it})\,dt.$$

More generally, if $D(0,1)$ is replaced by $D(z_0,R)$, then

$$u(z) = \frac{1}{2\pi}\int_0^{2\pi} P_{(z-z_0)/R}(t)u(z_0+Re^{it})\,dt;$$

equivalently,

$$u(z_0+re^{i\theta}) = \frac{1}{2\pi}\int_0^{2\pi} P_{r/R}(\theta-t)u(z_0+Re^{it})\,dt$$

for $0 \le r < R$ and all θ.

Proof. The result for $D(0,1)$ follows from (4.7.6) and (4.7.7). To prove the result for $D(z_0,R)$, we apply (4.7.6) and (4.7.7) to $u^*(w) = u(z_0+Rw), w \in D(0,1)$. If $z = z_0+re^{i\theta}, 0 \le r < R$, then $u(z) = u^*((z-z_0)/R)$, hence

$$u(z) = \frac{1}{2\pi}\int_0^{2\pi} P_{(z-z_0)/R}(t)u^*(e^{it})\,dt = \frac{1}{2\pi}\int_0^{2\pi} P_{r/R}(\theta-t)u(z_0+Re^{it})\,dt$$

as in (4.7.5). ♣

The Poisson integral formula allows us to derive a mean value property for harmonic functions.

4.7.9 Corollary

Suppose u is harmonic on an open set Ω. If $z_0 \in \Omega$ and $\overline{D}(z_0,R) \subseteq \Omega$, then

$$u(z_0) = \frac{1}{2\pi}\int_0^{2\pi} u(z_0+Re^{it})\,dt.$$

That is, $u(z_0)$ is the average of its values on circles with center at z_0.

Proof. Apply (4.7.8) with $r = 0$. ♣

It is interesting that the mean value property characterizes harmonic functions.

4.7.10 Theorem

Suppose φ is a continuous, real-valued function on Ω such that whenever $\overline{D}(z_0, R) \subseteq \Omega$, it is true that $\varphi(z_0) = \frac{1}{2\pi}\int_0^{2\pi} \varphi(z_0 + Re^{it})\,dt$. Then φ is harmonic on Ω.

Proof. Let $D(z_0, R)$ be any disk such that $\overline{D}(z_0, R) \subseteq \Omega$. Let u_0 be the restriction of φ to the circle $C(z_0, R)$ and apply (4.7.6) [for the disk $D(z_0, R)$] to produce a continuous function u on $\overline{D}(z_0, R)$ such that $u = u_0 = \varphi$ on $C(z_0, R)$. We will show that $\varphi = u$ on $D(z_0, R)$, thereby proving that φ is harmonic on $D(z_0, R)$. Since $D(z_0, R)$ is an arbitrary subdisk, this will prove that φ is harmonic on Ω.

The function $\varphi - u$ is continuous on $\overline{D}(z_0, R)$, and hence assumes its maximum and minimum at some points z_1 and z_2 respectively. If both z_1 and z_2 belong to $C(z_0, R)$, then since $u = \varphi$ on $C(z_0, R)$, the maximum and minimum values of $\varphi - u$ are both 0. It follows that $\varphi - u \equiv 0$ on $\overline{D}(z_0, R)$ and we are finished. On the other hand, suppose that (say) z_1 belongs to the open disk $D(z_0, R)$. Define a set A by

$$A = \{z \in D(z_0, R) : (\varphi - u)(z) = (\varphi - u)(z_1)\}.$$

Then A is closed in $D(z_0, R)$ by continuity of $\varphi - u$. We will also show that A is open, and thus conclude by connectedness that $A = D(z_0, R)$. For suppose that $a \in A$ and $r > 0$ is chosen so that $\overline{D}(a, r) \subseteq D(z_0, R)$. Then for $0 < \rho \leq r$ we have

$$\varphi(a) - u(a) = \frac{1}{2\pi}\int_0^{2\pi} [\varphi(a + \rho e^{it}) - u(a + \rho e^{it})]\,dt.$$

Since $\varphi(a + \rho e^{it}) - u(a + \rho e^{it}) \leq \varphi(a) - u(a)$, it follows from Lemma 2.4.11 that $\varphi - u$ is constant on $D(a, r)$. Thus $D(a, r) \subseteq A$, so A is open. A similar argument is used if $z_2 \in D(z_0, R)$. ♣

Remark

The above proof shows that a continuous function with the mean value property that has an absolute maximum or minimum in a region Ω is constant.

Problems

1. Let $Q_z(t)$ be as in (4.7.2). Prove that $\frac{1}{2\pi}\int_0^{2\pi} Q_z(t)\,dt = 1$.
2. Use (4.7.8) to prove *Harnack's inequality*: Suppose u satisfies the hypothesis of (4.7.8), and in addition $u \geq 0$. Then for $0 \leq r < 1$ and all θ,

$$\frac{1-r}{1+r}u(0) \leq u(re^{i\theta}) \leq \frac{1+r}{1-r}u(0).$$

3. Prove the following analog (for harmonic functions) of Theorem 2.2.17. Let $\{u_n\}$ be a sequence of harmonic functions on Ω such that $u_n \to u$ uniformly on compact subsets of Ω. Then u is harmonic on Ω. (Hint: If $\overline{D}(z_0, R) \subseteq \Omega$, the Poisson integral formula holds for u on $D(z_0, R)$.)

4. In Theorem 1.6.2 we showed that every harmonic function is locally the real part of an analytic function. Using results of this section, give a new proof of this fact.

5. Let Ω be a bounded open set and γ a closed path such that the following conditions are satisfied:
(a) $\gamma^* = \partial\Omega$, the boundary of Ω.
(b) There exists z_0 such that for every $\delta, 0 \le \delta < 1$, the path $\gamma_\delta = z_0 + \delta(\gamma - z_0)$ has its range in Ω (see Figure 4.7.1).

If f is continuous on $\overline{\Omega}$ and analytic on Ω, show that

$$\int_\gamma f(w)\,dw = 0 \quad \text{and} \quad n(\gamma, z)f(z) = \frac{1}{2\pi i}\int_\gamma \frac{f(w)}{w-z}\,dw, \ z \in \Omega.$$

Outline:
(i) First show that Ω must be starlike with star center z_0 by showing that if $z \in \Omega$, then the ray $[z_0, z, \infty)$ meets $\partial\Omega$ at some point β. By (a) and (b), $[z_0, \beta) \subseteq \Omega$. Next show that $z \in [z_0, \beta)$, hence $[z_0, z] \subseteq \Omega$.
(ii) The desired conclusions hold with γ replaced by γ_δ; let $\delta \to 1$ to complete the proof.

6. (Poisson integral formula for a half plane). Let f be analytic on $\{z : \operatorname{Im} z > 0\}$ and continuous on $\{z : \operatorname{Im} z \ge 0\}$. If $u = \operatorname{Re} f$, establish the formula

$$u(x, y) = \frac{1}{\pi}\int_{-\infty}^{\infty} \frac{yu(t, 0)}{(t-x)^2 + y^2}\,dt, \ \operatorname{Im} z > 0$$

under an appropriate hypothesis on the growth of f as $z \to \infty$. (Consider the path γ indicated back in Figure 4.2.6. Write, for $\operatorname{Im} z > 0$, $f(z) = (2\pi i)^{-1}\int_\gamma [f(w)/(w-z)]\,dw$ and $0 = (2\pi i)^{-1}\int_\gamma [f(w)/(w-\overline{z})]\,dw$ by using either Problem 5 or a technique similar to that given in the proof of (4.7.1). Then subtract the second equation from the first.)

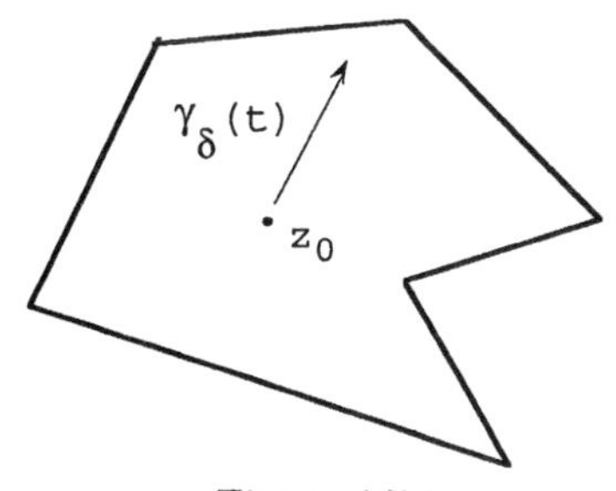

Figure 4.7.1

4.8 The Jensen and Poisson-Jensen Formulas

Suppose f is continuous on $\overline{D}(0, R)$, analytic on $D(0, R)$ and f has no zeros in $\overline{D}(0, R)$. Then we know that f has an analytic logarithm on $D(0, R)$ whose real part $\ln|f|$ is continuous on $\overline{D}(0, R)$ and harmonic on $D(0, R)$. Thus by (4.7.8), the Poisson integral formula for harmonic functions, we have

$$\ln |f(z)| = \frac{1}{2\pi} \int_0^{2\pi} P_{z/R}(t) \ln |f(Re^{it})| \, dt$$

or in polar form,

$$\ln |f(re^{i\theta})| = \frac{1}{2\pi} \int_0^{2\pi} P_{r/R}(\theta - t) \ln |f(Re^{it})| \, dt.$$

If f has zeros in $\overline{D}(0, R)$, then this derivation fails. However, the above formula can be modified to take the zeros of f into account.

4.8.1 Poisson-Jensen Formula

Suppose that f is continuous on $\overline{D}(0, R)$, analytic on $D(0, R)$ and that f has no zeros on $C(0, R)$. Let $a_1, \dots, a_n$ be the distinct zeros of f in $D(0, R)$ with multiplicities $k_1, \dots, k_n$ respectively. Then for $z \in D(0, R)$, z unequal to any of the a_j, we have

$$\ln |f(z)| = \sum_{j=1}^{n} k_j \ln \left| \frac{R(z - a_j)}{R^2 - \overline{a}_j z} \right| + \frac{1}{2\pi} \int_0^{2\pi} P_{z/R}(t) \ln |f(Re^{it}| \, dt.$$

Proof. We first give a proof for the case $R = 1$. By (4.6.2), there is a continuous function g on $\overline{D}(0, 1)$, analytic on $D(0, 1)$, such that g has no zeros in $\overline{D}(0, 1)$ and such that

$$f(z) = \left[\prod_{j=1}^{n} \left(\frac{z - a_j}{1 - \overline{a}_j z} \right)^{k_j} \right] g(z).$$

Since the product has modulus one when $|z| = 1$ we have $|f(z)| = |g(z)|$ for $|z| = 1$. Thus if $f(z) \neq 0$, then

$$\ln |f(z)| = \sum_{j=1}^{n} k_j \ln \left| \frac{z - a_j}{1 - \overline{a}_j z} \right| + \ln |g(z)|.$$

But g has no zeros in $\overline{D}(0, 1)$, so by the discussion in the opening paragraph of this section,

$$\ln |g(z)| = \frac{1}{2\pi} \int_0^{2\pi} P_z(t) \ln |g(e^{it})| \, dt = \frac{1}{2\pi} \int_0^{2\pi} P_z(t) \ln |f(e^{it})| \, dt.$$

This gives the result for $R = 1$. To obtain the formula for arbitrary R, we apply what was just proved to $F(w) = f(Rw)$, $|w| \leq 1$. Thus

$$\ln |F(w)| = \sum_{j=1}^{n} k_j \ln \left| \frac{w - (a_j/R)}{1 - (\overline{a}_j w/R)} \right| + \frac{1}{2\pi} \int_0^{2\pi} P_w(t) \ln |F(e^{it})| \, dt.$$

If we let $z = Rw$ and observe that

$$\frac{w-(a_j/R)}{1-(\overline{a}_j w/R)} = \frac{R(z-a_j)}{R^2-\overline{a}_j z},$$

we have the desired result. ♣

The Poisson-Jensen formula has several direct consequences.

4.8.2 Corollary

Assume that f satisfies the hypothesis of (4.8.1). Then

(a) $\ln|f(z)| \le \frac{1}{2\pi}\int_0^{2\pi} P_{z/R}(t)\ln|f(Re^{it})|\,dt$.

If in addition, $f(0) \neq 0$, then

(b) $\ln|f(0)| = \sum_{j=1}^n k_j \ln|a_j/R| + \frac{1}{2\pi}\int_0^{2\pi}\ln|f(Re^{it})|\,dt$, hence

(c) $\ln|f(0)| \le \frac{1}{2\pi}\int_0^{2\pi}\ln|f(Re^{it})|\,dt$.

Part (b) is known as *Jensen's formula*.

Proof. It follows from (4.6.1) and the proof of (4.8.1) that

$$\left|\frac{R(z-a_j)}{R^2-\overline{a}_j z}\right| < 1, \quad \text{hence} \quad k_j \ln\left|\frac{R(z-a_j)}{R^2-\overline{a}_j z}\right| < 0,$$

proving (a). Part (b) follows from (4.8.1) with $z = 0$, and (c) follows from (b). ♣

Jensen's formula (4.8.2b) does not apply when $f(0) = 0$, and the Poisson-Jensen formula (4.8.1) requires that f have no zeros on $C(0,R)$. It is natural to ask whether any modifications of our formulas are available so that these situations are covered.

First, if f has a zero of order k at 0, with $f(z) \neq 0$ for $|z| = R$, then the left side of Jensen's formula is modified to $k\ln R + \ln|f^{(k)}(0)/k!|$ rather than $\ln|f(0)|$. This can be verified by considering $f(z)/z^k$ and is left as Problem 1 at the end of the section.

However, if $f(z) = 0$ for some $z \in C(0,R)$, then the situation is complicated for several reasons. For example, it is possible that $f(z) = 0$ for infinitely many points on $C(0,R)$ without being identically zero on $\overline{D}(0,R)$ if f is merely assumed continuous on $\overline{D}(0,R)$ and analytic on $D(0,R)$. Thus $\ln|f(z)| = -\infty$ at infinitely many points in $C(0,R)$ and so the Poisson integral of $\ln|f|$ does not à priori exist. It turns out that the integral does exist in the sense of Lebesgue, but Lebesgue integration is beyond the scope of this text. Thus we will be content with a version of the Poisson-Jensen formula requiring analyticity on $\overline{D}(0,R)$, but allowing zeros on the boundary.

4.8.3 Poisson-Jensen Formula, Second Version

Let f be analytic and not identically zero on $\overline{D}(0,R)$. Let $a_1,\dots,a_n$ be the zeros of f in $D(0,R)$, with multiplicities $k_1,\dots,k_n$ respectively. Then for $z \in D(0,R)\setminus Z(f)$,

$$\ln|f(z)| = \sum_{j=1}^n k_j \ln\left|\frac{R(z-a_j)}{R^2-\overline{a}_j z}\right| + \frac{1}{2\pi}\int_0^{2\pi} P_{z/R}(t)\ln|f(Re^{it})|\,dt$$

where the integral exists as an improper Riemann integral.

Proof. Suppose that in addition to $a_1, \dots, a_n$, f has zeros on $C(0, R)$ at $a_{n+1}, \dots, a_m$ with multiplicities $k_{n+1}, \dots, k_m$. There is an analytic function g on $\overline{D}(0, R)$ with no zeros on $C(0, R)$ such that

$$f(z) = (z - a_{n+1})^{k_{n+1}} \cdots (z - a_m)^{k_m} g(z). \tag{1}$$

The function g satisfies the hypothesis of (4.8.1) and has the same zeros as f in $D(0, R)$. Now if $z \in D(0, R) \setminus Z(f)$, then

$$\ln |f(z)| = \sum_{j=1}^{n} k_j \ln |z - a_j| + \ln |g(z)|.$$

But by applying (4.8.1) to g we get

$$\ln |g(z)| = \sum_{j=1}^{n} \ln \left| \frac{R(z - a_j)}{R^2 - \overline{a}_j z} \right| + \frac{1}{2\pi} \int_0^{2\pi} P_{z/R}(t) \ln |g(Re^{it})| \, dt,$$

so the problem reduces to showing that

$$\sum_{j=n+1}^{m} k_j \ln |z - a_j| + \frac{1}{2\pi} \int_0^{2\pi} P_{z/R}(t) \ln |g(Re^{it})| \, dt = \frac{1}{2\pi} \int_0^{2\pi} P_{z/R}(t) \ln |f(Re^{it})| \, dt.$$

Since by (1), $f(Re^{it}) = [\prod_{j=1}^{n} (Re^{it} - a_j)^{k_j}] g(Re^{it}), 0 \le t \le 2\pi$, we see that it is sufficient to show that

$$\ln |z - a_j| = \frac{1}{2\pi} \int_0^{2\pi} P_{z/R}(t) \ln |Re^{it} - a_j| \, dt$$

for $j = n + 1, \dots, m$. In other words, the Poisson integral formula (4.7.8) holds for the functions $u(z) = \ln |z - a|$ when $|a| = R$ (as well as for $|a| < R$. This is essentially the content of the following lemma, where to simplify the notation we have taken $R = 1$ and $a = 1$

4.8.4 Lemma

For $|z| < 1$,

$$\ln |z - 1| = \frac{1}{2\pi} \int_0^{2\pi} P_z(t) \ln |e^{it} - 1| \, dt,$$

where the integral is to be understood as an improper Riemann integral at 0 and 2π. In particular,

$$\frac{1}{2\pi} \int_0^{2\pi} \ln |e^{it} - 1| \, dt = 0.$$

Proof. We note first that the above improper integral exists, because if $0 \le t \le \pi$, then $|e^{it} - 1| = \sqrt{2(1 - \cos t)} = 2\sin(t/2) \ge 2t/\pi$. Therefore

$$P_z(t)\ln|e^{it} - 1| \ge P_z(t)\ln(2t/\pi) = P_z(t)[\ln(2/\pi) + \ln t].$$

Since the improper integral $\int_0^{\pi/2} \ln t\, dt$ exists by elementary calculus and $P_z(t)$ is continuous, the above inequalities imply that

$$\lim_{\delta\to 0^+} \frac{1}{2\pi}\int_\delta^\pi P_z(t)\ln|e^{it} - 1|\, dt > -\infty \text{ and } \lim_{\delta\to 0^+} \frac{1}{2\pi}\int_\pi^{2\pi-\delta} P_z(t)\ln|e^{it} - 1|\, dt > -\infty.$$

Thus it remains to show that the *value* of the improper Riemann integral in the statement of the lemma is $\ln|z - 1|$. We will use a limit argument to evaluate the integral

$$I = \frac{1}{2\pi}\int_0^{2\pi} P_z(t)\ln|e^{it} - 1|\, dt.$$

For $r > 1$, define

$$I_r = \frac{1}{2\pi}\int_0^{2\pi} P_z(t)\ln|e^{it} - r|\, dt.$$

We will show that $I_r \to I$ as $r \to 1^+$. Now for any fixed $r > 1$, the function $z \to \ln|z - r|$ is continuous on $\overline{D}(0,1)$ and harmonic in $D(0,1)$, hence by (4.7.8), $\ln|z - r| = I_r$. Since $\ln|z - r| \to \ln|z - 1|$ as $r \to 1^+$, this will show that $I = \ln|z - 1|$, completing the proof. So consider, for $r > 1$,

$$|I_r - I| = \left|\frac{1}{2\pi}\int_0^{2\pi} P_z(t)\ln\left|\frac{e^{it} - r}{e^{it} - 1}\right| dt\right| = \frac{1}{2\pi}\int_0^{2\pi} P_z(t)\ln\left|\frac{e^{it} - r}{e^{it} - 1}\right| dt.$$

(The outer absolute values may be removed because $|e^{it} - r| > |e^{it} - 1|$ and therefore the integrand is positive.) Using the 2π-periodicity of the integrand, we may write

$$\int_0^{2\pi} = \int_{-\pi}^{\pi} = \int_{-\pi}^{0} + \int_0^{\pi}$$

and since $P_z(-t) = P_z(t)$, this becomes

$$\int_0^{\pi} + \int_0^{\pi} = 2\int_0^{\pi}.$$

Now if $0 \le t \le \pi$, then

$$\frac{e^{it} - r}{e^{it} - 1} = \frac{e^{it} - 1 + 1 - r}{e^{it} - 1} = 1 + \frac{1 - r}{e^{it} - 1}.$$

But as we noted at the beginning of the proof, $|e^{it} - 1| \ge 2t/\pi$, so the above expression is bounded in absolute value by $1 + [\pi(r - 1)/2t]$. Thus

$$0 < \ln\left|\frac{e^{it} - r}{e^{it} - 1}\right| \le \ln\left[1 + \frac{\pi(r - 1)}{2t}\right].$$

Also, the Poisson kernel satisfies

$$P_z(t) = \frac{1-|z|^2}{|e^{it}-z|^2} \le \frac{(1-|z|)(1+|z|)}{(1-|z|)^2} = \frac{1+|z|}{1-|z|},$$

an estimate that was used to establish Harnack's inequality. (See the solution to Section 4.7, Problem 2. Thus we now have

$$0 < |I_r - I| \le \frac{1+|z|}{1-|z|} \cdot \frac{1}{\pi} \int_0^{\pi} \ln\left[1 + \frac{\pi(r-1)}{2t}\right] dt.$$

Now fix $\delta, 0 < \delta < \pi$, and write

$$\int_0^{\pi} \ln\left[1 + \frac{\pi(r-1)}{2t}\right] dt = \int_0^{\delta} \ln\left[1 + \frac{\pi(r-1)}{2t}\right] dt + \int_{\delta}^{\pi} \ln\left[1 + \frac{\pi(r-1)}{2t}\right] dt.$$

Since the integral on the left side is finite (this is essentially the same as saying that $\int_0^{\pi} \ln t \, dt > -\infty$), and the integrand increases as r increases ($r > 1$), the first integral on the right side approaches 0 as $\delta \to 0^+$, uniformly in r. On the other hand, the second integral on the right side is bounded by $(\pi - \delta)\ln(1 + [\pi(r-1)/2\delta])$, which for fixed $\delta > 0$, approaches 0 as $r \to 1^+$. This completes the proof of the lemma, and as we noted earlier, finishes the proof of (4.8.3). ♣

The Poisson-Jensen formula has a number of interesting corollaries, some of which will be stated below. The proof of the next result (4.8.5), as well as other consequences, will be left for the problems.

4.8.5 Jensen's Formula, General Case

Let f be analytic on an open disc $D(0, R)$ and assume that $f \not\equiv 0$. Assume that f has a zero of order $k \ge 0$ at 0 and $a_1, a_2, \ldots$ are the zeros of f in $D(0, R) \setminus \{0\}$, each appearing as often as its multiplicity and arranged so that $0 < |a_1| \le |a_2| \le \cdots$. Then for $0 < r < R$ we have

$$k \ln r + \ln\left|\frac{f^{(k)}(0)}{k!}\right| = \sum_{j=1}^{n(r)} \ln\left|\frac{a_j}{r}\right| + \frac{1}{2\pi}\int_0^{2\pi} \ln|f(re^{it})| \, dt$$

where $n(r)$ is the number of terms of the sequence $a_1, a_2, \ldots$ that are in the disk $D(0, r)$.

Problems

1. Prove (4.8.5).
2. Let f be as in (4.8.2), except that instead of being analytic on all of $D(0, R)$, f has poles at $b_1, \ldots, b_m$ in $D(0, R) \setminus \{0\}$, of orders $l_1, \ldots, l_m$ respectively. State and prove an appropriate version of Jensen's formula in this case.
3. Let $n(r)$ be as in (4.8.5). Show that

$$\int_0^r \frac{n(t)}{t} \, dt = \sum_{j=1}^{n(r)} \ln \frac{r}{|a_j|}.$$

4. With f as in (4.8.5) and $M(r) = \max\{|f(z)| : |z| = r\}$, show that for $0 < r < R$,

$$\int_0^r \frac{n(t)}{t}\,dt \le \ln\left[\frac{M(r)}{|f^{(k)}(0)|r^k/k!}\right].$$

5. Let f be as in (4.8.5). Show that the function

$$r \to \frac{1}{2\pi}\int_0^{2\pi} \ln|f(re^{it})|\,dt$$

is increasing, and discuss the nature of its graph on the interval $(0, R)$.

4.9 Analytic Continuation

In this section we examine the problem of extending an analytic function to a larger domain. An example of this has already been encountered in the Schwarz reflection principle (2.2.15). We first consider a function defined by a power series.

4.9.1 Definition

Let $f(z) = \sum_{n=0}^{\infty} a_n(z - z_0)^n$ have radius of convergence $r, 0 < r < \infty$. Let z^* be a point such that $|z^* - z_0| = r$ and let $r(t)$ be the radius of convergence of the expansion of f about the point $z_1 = (1-t)z_0 + tz^*, 0 < t < 1$. Then $r(t) \ge (1-t)r$ (Figure 4.9.1). If

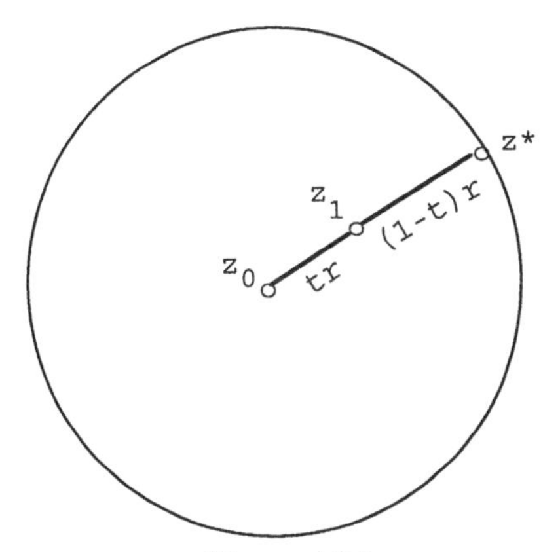

Figure 4.9.1

$r(t) = (1-t)r$ for some (hence for all) $t \in (0,1)$, so that there is *no* function g analytic on an open set containing $D(z_0, r) \cup \{z^*\}$ and such that $g = f$ on $D(z_0, r)$, then z^* is said to be a *singular point* of f. Equivalently, $z^* \in D(0, r)$ is *not* a singular point of f iff f has an analytic extension to an open set containing $D(z_0, r) \cup \{z^*\}$.

We are going to show that there is always at least one singular point on the circle of convergence, although in general, its exact location will not be known. Before doing this, we consider a special case in which it *is* possible to locate a singular point.

4.9.2 Theorem

In (4.9.1), if a_n is real and nonnegative for all n, then $z_0 + r$ is a singular point.

Proof. Fix z_1 between z_0 and $z_0 + r$. Note that since $a_n \geq 0$ for all n and $z_1 - z_0$ is a positive real number, f and its derivatives are nonnegative at z_1. Now assume, to the contrary, that the Taylor series expansion of f about z_1 *does* converge for some z_2 to the right of $z_0 + r$. Then we have

$$\sum_{k=0}^{\infty} \frac{f^{(k)}(z_1)}{k!}(z_2 - z_1)^k < +\infty.$$

But by the remark after (2.2.18),

$$f^{(k)}(z_1) = \sum_{n=k}^{\infty} n(n-1)\cdots(n-k+1)a_n(z_1 - z_0)^{n-k}$$

for $k = 0, 1, 2, \ldots$. Substituting this for $f^{(k)}(z_1)$ in the Taylor expansion of f about z_1 and using the fact that the order of summation in a double series with nonnegative terms can always be reversed, we get

$$\begin{aligned}
+\infty &> \sum_{k=0}^{\infty} \left[\sum_{n=k}^{\infty} n(n-1)\cdots(n-k+1)a_n(z_1 - z_0)^{n-k}\right] \frac{(z_2 - z_1)^k}{k!} \\
&= \sum_{k=0}^{\infty} \left[\sum_{n=k}^{\infty} \binom{n}{k} a_n(z_1 - z_0)^{n-k}\right] (z_2 - z_1)^k \\
&= \sum_{n=0}^{\infty} a_n \left[\sum_{k=0}^{n} \binom{n}{k} (z_1 - z_0)^{n-k}(z_2 - z_1)^k\right] \\
&= \sum_{n=0}^{\infty} a_n(z_2 - z_0)^n
\end{aligned}$$

by the binomial theorem. But this implies that $\sum_{n=0}^{\infty} a_n(z - z_0)^n$ has radius of convergence greater than r, a contradiction. ♣

The preceding theorem is illustrated by the geometric series $1 + z + z^2 + \cdots$, which has radius of convergence equal to 1 and which converges to $1/(1-z)$ for $|z| < 1$. In this case, $z^* = 1$ is the only singular point, but as we will see later, the other extreme is also possible, namely that *every* point on the circle of convergence is a singular point.

4.9.3 Theorem

In (4.9.1), let $\Gamma = \{z : |z - z_0| = r\}$ be the circle of convergence. Then there is at least one singular point on Γ.

Proof. If $z \in \Gamma$ is not a singular point, then there is a function f_z analytic on a disk $D(z, \epsilon_z)$ such that $f_z = f$ on $D(z_0, r) \cap D(z, \epsilon_z)$. Say there are no singular points on Γ.

By compactness, Γ is covered by finitely many such disks, say by $D(z_j, \epsilon_j), j = 1, \dots, n$. Define

$$g(z) = \begin{cases} f(z), & z \in D(z_0, r) \\ f_{z_j}(z), & z \in D(z_j, \epsilon_j), j = 1, \dots, n. \end{cases}$$

We show that g is well defined. If $D(z_j, \epsilon_j) \cap D(z_k, \epsilon_k) \neq \emptyset$, then also $D(z_j, \epsilon_j) \cap D(z_k, \epsilon_k) \cap D(z_0, r) \neq \emptyset$, as is verified by drawing a picture. Now $f_{z_j} - f_{z_k} = f - f = 0$ on $D(z_j, \epsilon_j) \cap D(z_k, \epsilon_k)$ by the identity theorem (2.4.8), proving that g is well defined. Thus g is analytic on $D(z_0, s)$ for some $s > r$, and the Taylor expansion of g about z_0 coincides with that of f since $g = f$ on $D(z_0, r)$. This means that the expansion of f converges in a disk of radius greater than r, a contradiction. ♣

We are now going to construct examples of power series for which the circle of convergence is a *natural boundary,* that is, every point on the circle of convergence is a singular point. The following result will be needed.

4.9.4 Lemma

Let $f_1(w) = (w^p + w^{p+1})/2$, p a positive integer. Then $|w| < 1$ implies $|f_1(w)| < 1$, and if $\Omega = D(0,1) \cup D(1, \epsilon), \epsilon > 0$, then $f_1(D(0,r)) \subseteq \Omega$ for some $r > 1$.

Proof. If $|w| \leq 1$, then $|f_1(w)| = |w|^p |1 + w|/2 \leq |1 + w|/2$, which is less than 1 unless $w = 1$, in which case $f_1(w) = 1$. Thus $|w| < 1$ implies $|f_1(w)| < 1$, and $f_1(\overline{D}(0,1)) \subseteq \Omega$. Hence $f_1^{-1}(\Omega)$ is an open set containing $\overline{D}(0,1)$. Consequently, there exists $r > 1$ such that $D(0,r) \subseteq f_1^{-1}(\Omega)$, from which it follows that $f_1(D(0,r) \subseteq f_1(f_1^{-1}(\Omega)) \subseteq \Omega$. ♣

The construction of natural boundaries is now possible.

4.9.5 Hadamard Gap Theorem

Suppose that $f(z) = \sum_{k=1}^{\infty} a_k z^{n_k}$ and, for some $s > 1$, $n_{k+1}/n_k \geq s$ for all k. (We say that $\sum_k a_k z^{n_k}$ is a *gap series.*) If the radius of convergence of the series is 1, then every point on the circle of convergence is a singular point.

Proof. We will show that 1 is a singular point, from which it will follow (under these hypotheses) that every point on the unit circle is a singular point. Thus assume, to the contrary, that 1 is not a singular point. Then, for some $\epsilon > 0$, f has an analytic extension g to $D(0,1) \cup D(1, \epsilon)$. Let p be a positive integer such that $s > (p+1)/p$, and let f_1 and $r > 1$ be as in Lemma 4.9.4. Then $h(w) = g(f_1(w))$ is analytic on $D(0,r)$, and for

$|w| < 1$,

$$\begin{aligned} g(f_1(w)) = f(f_1(w)) &= \sum_{k=1}^{\infty} a_k (f_1(w))^{n_k} \\ &= \sum_{k=1}^{\infty} a_k 2^{-n_k} (w^p + w^{p+1})^{n_k} \\ &= \sum_{k=1}^{\infty} a_k 2^{-n_k} \sum_{n=0}^{n_k} \binom{n_k}{n} w^{p(n_k-n)} w^{(p+1)n} \\ &= \sum_{k=1}^{\infty} a_k 2^{-n_k} \sum_{n=0}^{n_k} \binom{n_k}{n} w^{pn_k+n}. \end{aligned}$$

Now for each k we have $n_{k+1}/n_k \geq s > (p+1)/p$, so $pn_k + n_k < pn_{k+1}$. Therefore, the highest power of w that appears in $\sum_{n=0}^{n_k} \binom{n_k}{n} w^{pn_k+n}$, namely $w^{pn_k+n_k}$, is less than the lowest power $w^{pn_{k+1}}$ that appears in $\sum_{n=0}^{n_{k+1}} \binom{n_{k+1}}{n} w^{pn_{k+1}+n}$. This means that the series

$$\sum_{k=1}^{\infty} a_k 2^{-n_k} \sum_{n=0}^{n_k} \binom{n_k}{n} w^{pn_k+n}$$

is (with a grouping of terms) precisely the Taylor expansion of h about $w = 0$. But since h is analytic on $D(0,r)$, this expansion converges absolutely on $D(0,r)$, hence (as there are no repetition of powers of w),

$$\sum_{k=1}^{\infty} |a_k| 2^{-n_k} \sum_{n=0}^{n_k} \binom{n_k}{n} |w|^{pn_k+n} < \infty,$$

that is [as in the above computation of $g(f_1(w))$],

$$\sum_{k=1}^{\infty} |a_k| 2^{-n_k} (|w|^p + |w|^{p+1})^{n_k} < \infty$$

for $|w| < r$. But if $1 < |w| < r$, then $2^{-n_k}(|w|^p + |w|^{p+1})^{n_k} = \left[|w|^p(\frac{1+|w|}{2})\right]^{n_k} > 1$. Consequently, $\sum_{k=1}^{\infty} a_k z^{n_k}$ converges for some z with $|z| > 1$, contradicting the assumption that the series defining f has radius of convergence 1.

Finally, if $z^* = e^{i\theta}$ is not a singular point, let $q(z) = f(e^{i\theta}z) = \sum_{k=1}^{\infty} a_k e^{i\theta n_k} z^{n_k}$ (with radius of convergence 1, as before, because $|e^{i\theta n_k}| = 1$). Now f extends to a function analytic on $D(0,1) \cup D(z^*, \epsilon)$ for some $\epsilon > 0$, and thus q extends to a function analytic on $D(0,1) \cup D(1,\epsilon)$, contradicting the above argument. ♣

Some typical examples of gap series are $\sum_{k=1}^{\infty} z^{2^k}$ and $\sum_{k=1}^{\infty} z^{k!}$.

Remarks

The series $\sum_{n=0}^{\infty} z^n$ diverges at every point of the circle of convergence since $|z|^n$ does not approach 0 when $|z| = 1$. However, $z = 1$ is the only singular point since $(1-z)^{-1}$

is analytic except at $z = 1$. On the other hand, $\sum_{n=1}^{\infty} \frac{1}{n!} z^{2^n}$ has radius of convergence 1, for if $a_k = 0, k \neq 2^n; a_{2^n} = 1/n!$, then

$$\limsup_{n\to\infty} |a_n|^{1/n} = \limsup_{n\to\infty} |a_{2^n}|^{1/2^n} = \limsup_{n\to\infty} (1/n!)^{1/2^n} = 1$$

because

$$\ln[(n!)^{1/2^n}] = 2^{-n}\ln(n!) = 2^{-n}\sum_{k=1}^{n} \ln k \leq 2^{-n} n \ln n \to 0.$$

The series converges (as does every series obtained from it by termwise differentiation) at each point of the circle of convergence, and yet by (4.9.5), each such point is singular.

The conclusion of Theorem 4.9.5 holds for any (finite) radius of convergence. For if $\sum a_k z^{n_k}$ has radius of convergence r, then $\sum a_k (rz)^{n_k}$ has radius of convergence 1.

We now consider chains of functions defined by power series.

4.9.6 Definitions

A *function element* in Ω is a pair (f, D), where D is a disk contained in Ω and f is analytic on D. (The convention $D = D(0,1)$ is no longer in effect.) If z is an element of D, then (f, D) is said to be a *function element at* z. Two function elements (f_1, D_1) and (f_2, D_2) in Ω are *direct analytic continuations* of each other (relative to Ω) if $D_1 \cap D_2 \neq \emptyset$ and $f_1 = f_2$ on $D_1 \cap D_2$. Note that in this case, $f_1 \cup f_2$ is an extension of f_1 (respectively f_2) from D_1 (respectively D_2) to $D_1 \cup D_2$. If there is a *chain* $(f_1, D_1), (f_2, D_2), \dots, (f_n, D_n)$ of function elements in Ω, with (f_i, D_i) and (f_{i+1}, D_{i+1}) direct analytic continuations of each other for $i = 1, 2, \dots, n-1$, then (f_1, D_1) and (f_n, D_n) are said to be *analytic continuations* of each other relative to Ω.

Suppose that γ is a curve in Ω, with γ defined on the interval $[a, b]$. If there is a partition $a = t_0 < t_1 < \cdots < t_n = b$, and a chain$(f_1, D_1), \dots, (f_n, D_n)$ of function elements in Ω such that (f_{i+1}, D_{i+1}) is a direct analytic continuation of (f_i, D_i) for $i = 1, 2, \dots, n-1$, and $\gamma(t) \in D_i$ for $t_{i-1} \leq t \leq t_i, i = 1, 2, \dots, n$, then (f_n, D_n) is said to be an *analytic continuation* of (f_1, D_1) *along the curve* γ.

4.9.7 Theorem

Analytic continuation of a *given* function element along a *given* curve is unique, that is, if (f_n, D_n) and (g_m, E_m) are two continuations of (f_1, D_1) along the same curve γ, then $f_n = g_m$ on $D_n \cap E_m$.

Proof. Let the first continuation be $(f_1, D_1), \dots, (f_n, D_n)$, and let the second continuation be $(g_1, E_1), \dots, (g_m, E_m)$, where $g_1 = f_1, E_1 = D_1$. There are partitions $a = t_0 < t_1 < \cdots < t_n = b, a = s_0 < s_1 < \cdots < s_m = b$ such that $\gamma(t) \in D_i$ for $t_{i-1} \leq t \leq t_i, i = 1, 2, \dots, n, \gamma(t) \in E_j$ for $s_{j-1} \leq t \leq s_j, j = 1, 2, \dots, m$.

We claim that if $1 \leq i \leq n, 1 \leq j \leq m$, and $[t_{i-1}, t_i] \cap [s_{j-1}, s_j] \neq \emptyset$, then (f_i, D_i) and (g_j, E_j) are direct analytic continuations of each other. This is true when $i = j = 1$, since $g_1 = f_1$ and $E_1 = D_1$. If it is not true for all i and j, pick from all (i, j) for which the statement is false a pair such that $i+j$ is minimal. Say $t_{i-1} \geq s_{j-1}$ [then $i \geq 2$, for if $i = 1$, then $s_{j-1} = t_0 = a$, hence $j = 1$, and we know that the result holds for the pair (1,1)]. We have $t_{i-1} \leq s_j$ since $[t_{i-1}, t_i] \cap [s_{j-1}, s_j] \neq \emptyset$, hence $s_{j-1} \leq t_{i-1} \leq s_j$. Therefore $\gamma(t_{i-1}) \in D_{i-1} \cap D_i \cap E_j$, in particular, this intersection is not empty. Now (f_i, D_i) is a direct analytic continuation of (f_{i-1}, D_{i-1}), and furthermore (f_{i-1}, D_{i-1}) is a direct analytic continuation of (g_j, E_j) by minimality of $i + j$ (note that $t_{i-1} \in [t_{i-2}, t_{i-1}] \cap [s_{j-1}, s_j]$, so the hypothesis of the claim is satisfied). Since $D_{i-1} \cap D_i \cap E_j \neq \emptyset, (f_i, D_i)$ must be a direct continuation of $g_j, E_j)$, a contradiction. Thus the claim holds for all i and j, in particular for $i = n$ and $j = m$. The result follows. ♣

4.9.8 Definition

Let Ω be an open connected subset of $\mathbb{C}$. The function elements (f_1, D_1) and (f_2, D_2) in Ω are said to be *equivalent* if they are analytic continuations of each other relative to Ω. (It is immediate that this is an equivalence relation.) An equivalence class Φ of function elements in Ω such that for every $z \in \Omega$ there is an element $(f, D) \in \Phi$ with $z \in D$ is called a *generalized analytic function* on Ω.

Note that connectedness of Ω is necessary in this definition if there are to be any generalized analytic functions on Ω at all. For if $z_1, z_2 \in \Omega$, there must exist equivalent function elements (f_1, D_1) and (f_2, D_2) at z_1 and z_2 respectively. This implies that there is a curve in Ω joining z_1 to z_2.

Note also that if g is analytic on Ω, then g determines a generalized analytic function Φ on Ω in the following sense. Take

$$\Phi = \{(f, D) : D \subseteq \Omega \text{ and } f = g|_D\}.$$

However, not every generalized analytic function arises from a single analytic function in this way (see Problem 2). The main result of this section, the monodromy theorem (4.9.11), addresses the question of when a generalized analytic function *is* determined by a single analytic function.

4.9.9 Definition

Let γ_0 and γ_1 be curves in a set $S \subseteq \mathbb{C}$ (for convenience, let γ_0 and γ_1 have common domain $[a, b]$). Assume $\gamma_0(a) = \gamma_1(a) = z_0, \gamma_0(b) = \gamma_1(b) = z_1$, that is, the curves have the same endpoints. Then γ_0 and γ_1 are said to be *homotopic* (in S) if there is a continuous map $H : [a, b] \times [0, 1] \to S$ (called a *homotopy* of γ_0 and γ_1) such that $H(t, 0) = \gamma_0(t), H(t, 1) = \gamma_1(t), a \leq t \leq t; H(a, s) = z_0, H(b, s) = z_1, 0 \leq s \leq 1$. Intuitively, H deforms γ_0 into γ_1 without moving the endpoints or leaving the set S. For $0 \leq s \leq 1$, the curve $t \to H(t, s)$ represents the state of the deformation at "time s".

4.9.10 Theorem

Let Ω be an open connected subset of $\mathbb{C}$, and let γ_0, γ_1 be curves in Ω that are homotopic in Ω. Let (f, D) be a function element at z_0, the initial point of γ_0 and γ_1. Assume that

(f, D) can be continued along all possible curves in Ω, that is, if γ is a curve in Ω joining z_0 to another point z_n, there is an analytic continuation (f_n, D_n) of (f, D) along γ.

If (g_0, D_0) is a continuation of (f, D) along γ_0 and (g_1, D_1) is a continuation of (f, D) along γ_1, then $g_0 = g_1$ on $D_0 \cap D_1$. (Note that $D_0 \cap D_1 \neq \emptyset$ since the terminal point z_1 of γ_0 and γ_1 belongs to $D_0 \cap D_1$.) Thus (g_0, D_0) and (g_1, D_1) are direct analytic continuations of each other.

Proof. Let H be a homotopy of γ_0 and γ_1. By hypothesis, if $0 \leq s \leq 1$, then (f, D) can be continued along the curve $\gamma_s = H(\cdot\ , s)$, say to (g_s, D_s). Fix s and pick one such continuation, say $(h_1, E_1), \dots, (h_n, E_n)$ [with $(h_1, E_1) = (f, D), (h_n, E_n) = (g_s, D_s)$]. There is a partition $a = t_0 < t_1 < \cdots t_n = b$ such that $\gamma_s(t) \in E_i$ for $t_{i-1} \leq t \leq t_i, i = 1, \dots, n$. Let K_i be the compact set $\gamma_s([t_{i-1}, t_i]) \subseteq E_i$, and let

$$\epsilon = \min_{1 \leq i \leq n} \{\text{dist}(K_i, \mathbb{C} \setminus E_i\} > 0.$$

Since H is uniformly continuous, there exists $\delta > 0$ such that if $|s - s_1| < \delta$, then $|\gamma_s(t) - \gamma_{s_1}(t)| < \epsilon$ for all $t \in [a, b]$. In particular, if $t_{i-1} \leq t \leq t_i$, then since $\gamma_s(t) \in K_i$ and $|\gamma_s(t) - \gamma_{s_1}(t)| < \epsilon$, we have $\gamma_{s_1}(t) \in E_i$.

Thus by definition of continuation along a curve, $(h_1, E_1), \dots, (h_n, E_n)$ is also a continuation of (f, D) along γ_{s_1}. But we specified at the beginning of the proof that (f, D) is continued along γ_{s_1} to (g_{s_1}, D_{s_1}). By (4.9.7), $g_s = g_{s_1}$ on $D_s \cap D_{s_1}$. Thus for each $s \in [0, 1]$ there is an open interval I_s such that $g_s = g_{s_1}$ on $D_s \cap D_{s_1}$ whenever $s_1 \in I_s$. Since $[0, 1]$ can be covered by finitely many such intervals, it follows that $g_0 = g_1$ on $D_0 \cap D_1$. ♣

4.9.11 Monodromy Theorem

Let Ω be an open connected subset of $\mathbb{C}$ with the property that every closed curve γ in Ω is homotopic to a point, that is, homotopic (in Ω) to $\gamma_0 \equiv z$, where z is the initial and terminal point of γ. Let Φ be a generalized analytic function on Ω, and assume that each element of Φ can be continued along all possible curves in Ω. Then there is a function g analytic on Ω such that whenever $(f, D) \in \Phi$ we have $g = f$ on D. Thus Φ is determined by a single analytic function.

Proof. If $z \in \Omega$ there is a function element $(f, D) \in \Phi$ such that $z \in D$. Define $g(z) = f(z)$. We must show that g is well defined. If $(f^*, D^*) \in \Phi$ and $z \in D^*$, we have to show that $f(z) = f^*(z)$. But since $(f, D), (f^*, D^*) \in \Phi$, there is a continuation in Ω from (f, D) to (f^*, D^*); since $z \in D \cap D^*$, we can find a curve γ (in fact a polygonal path) in Ω with initial and terminal point z such that the continuation is along γ. But by hypothesis, γ is homotopic to the curve $\gamma_0 \equiv z$. Since (f, D) is a continuation of (f, D) along γ_0, it follows from (4.9.10) that $f = f^*$ on $D \cap D^*$, in particular, $f(z) = f^*(z)$. Since $g = f$ on D, g is analytic on Ω. ♣

Remarks

Some authors refer to (4.9.10), rather than (4.9.11), as the monodromy theorem. Still others attach this title to our next result (4.9.13), which is a corollary of (4.9.11). It is appropriate at this point to assign a name to the topological property of Ω that appears in the hypothesis of (4.9.11).

4.9.12 Definition

Let Ω be a plane region, that is, an open connected subset of $\mathbb{C}$. We say that Ω is (homotopically) *simply connected* if every closed curve in Ω is homotopic (in Ω) to a point. In the next chapter, we will show that the homotopic and homological versions of simple connectedness are equivalent.

Using this terminology, we have the following corollary to the monodromy theorem (4.9.11).

4.9.13 Theorem

Let Ω be simply connected region and let (f, D) be a function element in Ω such that (f, D) can be continued along all curves in Ω whose initial points are in D. Then there is an analytic function g on Ω such that $g = f$ on D.

Proof (outline). Let Φ be the collection of all function elements (h, E) such that (h, E) is a continuation of (f, D). One can then verify that Φ satisfies the hypothesis of (4.9.11). Since $(f, D) \in \Phi$, the result follows. ♣

Alternatively, we need not introduce Φ at all, but instead imitate the proof of (4.9.11).

We conclude this section with an important and interesting application of analytic continuation in simply connected regions.

4.9.14 Theorem

If Ω is a (homotopically) simply connected region, then every harmonic function on Ω has a harmonic conjugate.

Proof. If u is harmonic on Ω, we must produce an analytic function g on Ω such that $u = \operatorname{Re} g$. We make use of previous results for disks; if D is a disk contained in Ω, then by (1.6.2), there is an analytic function f on D such that $\operatorname{Re} f = u$. That is, (f, D) is a function element in Ω with $\operatorname{Re} f = u$ on D.

If $\gamma : [a, b] \to \Omega$ is any curve in Ω such that $\gamma(a) \in D$, we need to show that (f, D) can be continued along γ. As in the proof of (3.1.7), there is a partition $a = t_0 < t_1 < \cdots < t_n = b$ and disks $D_1, \dots, D_n$ with centers at $\gamma(t_1), \dots, \gamma(t_n)$ respectively, such that if $t_{j-1} \le t \le t_j$, then $\gamma(t) \in D_j$. Now $D \cap D_1 \neq \emptyset$, and by repeating the above argument we see that there exists f_1 analytic on D_1 such that $\operatorname{Re} f_1 = u$ on D_1. Since $f - f_1$ is pure imaginary on $D \cap D_1$, it follows (from the open mapping theorem (4.3.1), for example) that $f - f_1$ is a purely imaginary constant on $D \cap D_1$. By adding this constant to f_1 on D_1, we obtain a new f_1 on D_1 such that (f_1, D_1) is a direct continuation of (f, D). Repeating this process with (f_1, D_1) and (f_2, D_2), and so on, we obtain a continuation (f_n, D_n) of (f, D) along γ. Thus by (4.9.13), there is an analytic function g on Ω such that $g = f$ on D. Then $\operatorname{Re} g = u$ on D, and hence by the identity theorem for harmonic functions (2.4.14), $\operatorname{Re} g = u$ on Ω. ♣

In the next chapter we will show that the converse of (4.9.14) holds. However, this will require a closer examination of the connection between homology and homotopy. Also, we can give an alternative (but less constructive) proof of (4.9.14) after proving the Riemann mapping theorem.

Problems

1. Let $f(z) = \sum_{n=0}^{\infty} z^{n!}$, $z \in D(0,1)$. Show directly that f has $C(0,1)$ as its natural boundary without appealing to the Hadamard gap theorem. (Hint: Look at f on radii which terminate at points of the form $e^{i2\pi p/q}$ where p and q are integers.)
2. Let $f(z) = \operatorname{Log} z = \sum_{n=1}^{\infty}(-1)^{n-1}(z-1)^n/n$, $z \in D = D(1,1)$. Let $\Omega = \mathbb{C} \setminus \{0\}$ and let Φ be the equivalence class determined by (f, D).
   (a) Show that Φ is actually a generalized analytic function on Ω, that is, if $z \in \Omega$ then there is an element $(g, E) \in \Phi$ with $z \in E$.
   (b) Show that there is no function h analytic on Ω such that for every $(g, E) \in \Phi$ we have $h = g$ on E.
3. Criticize the following argument: Let $f(z) = \sum_{n=0}^{\infty} a_n(z-z_0)^n$ have radius of convergence r. If $z_1 \in D(z_0, r)$, then by the rearrangement procedure of (4.9.2) we can find the Taylor expansion of f about z_1, namely
$$f(z) = \sum_{k=0}^{\infty}\left[\sum_{n=k}^{\infty}\binom{n}{k}a_n(z_1-z_0)^{n-k}\right](z-z_1)^k.$$
   If the expansion about z_1 converges at some point $z \notin \overline{D}(z_0, r)$, then since power series converge absolutely inside the circle of convergence, we may rearrange the expansion about z_1 to show that the original expansion about z_0 converges at z, a contradiction. Consequently, for any function defined by a power series, the circle of convergence is a natural boundary.
4. (Law of permanence of functional equations). Let $F : \mathbb{C}^{k+1} \to \mathbb{C}$ be such that F and all its first order partial derivatives are continuous. Let $f_1, \dots, f_k$ be analytic on a disk D, and assume that $F(z, f_1(z), \dots, f_k(z)) = 0$ for all $z \in D$. Let $(f_{i1}, D_1), (f_{i2}, D_2), \dots, (f_{in}, D_n)$, with $f_{i1} = f_i, D_1 = D$, form a continuation of $(f_i, D), i = 1, \dots, k$. Show that $F(z, f_{1n}(z), \dots, f_{kn}(z)) = 0$ for all $z \in D_n$. An example: If $e^g = f$ on D and the continuation carries f into f^* and g into g^*, then $e^{g^*} = f^*$ on D_n (take $F(z, w_1, w_2) = w_1 - e^{w_2}, f_1 = f, f_2 = g$).
5. Let (f^*, D^*) be a continuation of (f, D). Show that $(f^{*\prime}, D^*)$ is a continuation of (f', D). ("The derivative of the continuation, that is, $f^{*\prime}$, is the continuation of the derivative.")

Reference

W. Rudin, "Real and Complex Analysis," 3rd ed., McGraw Hill Series in Higher Mathematics, New York, 1987.

Chapter 5

Families of Analytic Functions

In this chapter we consider the linear space $A(\Omega)$ of all analytic functions on an open set Ω and introduce a metric d on $A(\Omega)$ with the property that convergence in the d-metric is uniform convergence on compact subsets of Ω. We will characterize the compact subsets of the metric space $(A(\Omega), d)$ and prove several useful results on convergence of sequences of analytic functions. After these preliminaries we will present a fairly standard proof of the Riemann mapping theorem and then consider the problem of extending the mapping function to the boundary. Also included in this chapter are Runge's theorem on rational approximations and the homotopic version of Cauchy's theorem.

5.1 The Spaces $A(\Omega)$ and $C(\Omega)$

5.1.1 Definitions

Let Ω be an open subset of $\mathbb{C}$. Then $A(\Omega)$ will denote the space of analytic functions on Ω, while $\mathbb{C}(\Omega)$ will denote the space of all continuous functions on Ω. For $n = 1, 2, 3 \ldots$, let

$$K_n = \overline{D}(0, n) \cap \{z : |z - w| \geq 1/n \text{ for all } w \in \mathbb{C} \setminus \Omega\}.$$

By basic topology of the plane, the sequence $\{K_n\}$ has the following three properties:
(1) K_n is compact,
(2) $K_n \subseteq K^o_{n+1}$ (the interior of K_{n+1}),
(3) If $K \subseteq \Omega$ is compact, then $K \subseteq K_n$ for n sufficiently large.

Now fix a nonempty open set Ω, let $\{K_n\}$ be as above, and for $f, g \in \mathbb{C}(\Omega)$, define

$$d(f, g) = \sum_{n=1}^{\infty} \left(\frac{1}{2^n}\right) \frac{\|f - g\|_{K_n}}{1 + \|f - g\|_{K_n}},$$

where

$$\|f - g\|_{K_n} = \begin{cases} \sup\{|f(z) - g(z)| : z \in K_n\}, & K_n \neq \emptyset \\ 0, & K_n = \emptyset \end{cases}$$

5.1.2 Theorem

The assignment $(f,g) \to d(f,g)$ defines a metric on $C(\Omega)$. A sequence $\{f_j\}$ in $C(\Omega)$ is d-convergent (respectively d-Cauchy) iff $\{f_j\}$ is uniformly convergent (respectively uniformly Cauchy) on compact subsets of Ω. Thus $(C(\Omega), d)$ and $(A(\Omega), d)$ are complete metric spaces.

Proof. That d is a metric on $C(\Omega)$ is relatively straightforward. The only troublesome part of the argument is verification of the triangle inequality, whose proof uses the inequality: If a, b and c are nonnegative numbers and $a \le b + c$, then

$$\frac{a}{1+a} \le \frac{b}{1+b} + \frac{c}{1+c}.$$

To see this, note that $h(x) = x/(1+x)$ increases with $x \ge 0$, and consequently $h(a) \le h(b+c) = \frac{b}{1+b+c} + \frac{c}{1+b+c} \le \frac{b}{1+b} + \frac{c}{1+c}$. Now let us show that a sequence $\{f_j\}$ is d-Cauchy iff $\{f_j\}$ is uniformly Cauchy on compact subsets of Ω. Suppose first that $\{f_j\}$ is d-Cauchy, and let K be any compact subset of Ω. By the above property (3) of the sequence $\{K_n\}$, we can choose n so large that $K \subseteq K_n$. Since $d(f_j, f_k) \to 0$ as $j, k \to \infty$, the same is true of $\|f_j - f_k\|_{K_n}$. But $\|f_j - f_k\|_K \le \|f_j - f_k\|_{K_n}$, hence $\{f_j\}$ is uniformly Cauchy on K. Conversely, assume that $\{f_j\}$ is uniformly Cauchy on compact subsets of Ω. Let $\epsilon > 0$ and choose a positive integer m such that $\sum_{n=m+1}^{\infty} 2^{-n} < \epsilon$. Since $\{f_j\}$ is uniformly Cauchy on K_m in particular, there exists $N = N(m)$ such that $j, k \ge N$ implies $\|f_j - f_k\|_{K_m} < \epsilon$, hence

$$\begin{aligned}\sum_{n=1}^{m} \left(\frac{1}{2^n}\right) \frac{\|f_j - f_k\|_{K_n}}{1 + \|f_j - f_k\|_{K_n}} &\le \sum_{n=1}^{m} \left(\frac{1}{2^n}\right) \|f_j - f_k\|_{K_n} \\ &\le \|f_j - f_k\|_{K_m} \sum_{n=1}^{m} \frac{1}{2^n} < \epsilon.\end{aligned}$$

It follows that for $j, k \ge N$,

$$d(f_j, f_k) = \sum_{n=1}^{\infty} \left(\frac{1}{2^n}\right) \frac{\|f_j - f_k\|_{K_n}}{1 + \|f_j - f_k\|_{K_n}} < 2\epsilon.$$

The remaining statements in (5.1.2) follow from the above, Theorem 2.2.17, and completeness of $\mathbb{C}$. ♣

If $\{f_n\}$ is a sequence in $A(\Omega)$ and $f_n \to f$ uniformly on compact subsets of Ω, then we know that $f \in \mathbb{A}(\Omega)$ also. The next few theorems assert that certain other properties of the limit function f may be inferred from the possession of these properties by the f_n. The first results of this type relate the zeros of f to those of the f_n.

5.1.3 Hurwitz's Theorem

Suppose that $\{f_n\}$ is a sequence in $A(\Omega)$ that converges to f uniformly on compact subsets of Ω. Let $\overline{D}(z_0, r) \subseteq \Omega$ and assume that $f(z) \neq 0$ for $|z - z_0| = r$. Then there is a positive integer N such that for $n \ge N$, f_n and f have the same number of zeros in $D(z_0, r)$.

Proof. Let $\epsilon = \min\{|f(z)| : |z - z_0| = r\} > 0$. Then for sufficiently large n, $|f_n(z) - f(z)| < \epsilon \leq |f(z)|$ for $|z - z_0| = r$. By Rouché's theorem (4.2.8), f_n and f have the same number of zeros in $D(z_0, r)$. ♣

5.1.4 Theorem

Let $\{f_n\}$ be a sequence in $A(\Omega)$ such that $f_n \to f$ uniformly on compact subsets of Ω. If Ω is connected and f_n has no zeros in Ω for infinitely many n, then either f has no zeros in Ω or f is identically zero.

Proof. Assume f is not identically zero, but f has a zero at $z_0 \in \Omega$. Then by the identity theorem (2.4.8), there is $r > 0$ such that the hypothesis of (5.1.3) is satisfied. Thus for sufficiently large n, f_n has a zero in $D(z_0, r)$. ♣

5.1.5 Theorem

Let $\{f_n\}$ be a sequence in $A(\Omega)$ such that f_n converges to f uniformly on compact subsets of Ω. If Ω is connected and the f_n are one-to-one on Ω, then either f is constant on Ω or f is one-to-one.

Proof. Assume that f is not constant on Ω, and choose any $z_0 \in \Omega$. The sequence $\{f_n - f_n(z_0)\}$ satisfies the hypothesis of (5.1.4) on the open connected set $\Omega \setminus \{z_0\}$ (because the f_n are one-to-one). Since $f - f(z_0)$ is not identically zero on $\Omega \setminus \{z_0\}$, it follows from (5.1.4) that $f - f(z_0)$ has no zeros in $\Omega \setminus \{z_0\}$. Since z_0 is an arbitrary point of Ω, we conclude that f is one-to-one on Ω. ♣

The next task will be to identify the compact subsets of the space $A(\Omega)$ (equipped with the topology of uniform convergence on compact subsets of Ω). After introducing the appropriate notion of boundedness for subsets $\mathcal{F} \subseteq A(\Omega)$, we show that each sequence of functions in $\mathcal{F}$ has a subsequence that converges uniformly on compact subsets of Ω. This leads to the result that a subset of $A(\Omega)$ is compact iff it is closed and bounded.

5.1.6 Definition

A set $\mathcal{F} \subseteq C(\Omega)$ is *bounded* if for each compact set $K \subseteq \Omega$, $\sup\{\|f\|_K : f \in \mathcal{F}\} < \infty$, that is, the functions in $\mathcal{F}$ are uniformly bounded on each compact subset of Ω.

We will also require the notion of equicontinuity for a family of functions.

5.1.7 Definition

A family $\mathcal{F}$ of functions on Ω is *equicontinuous* at $z_0 \in \Omega$ if given $\epsilon > 0$ there exists $\delta > 0$ such that if $z \in \Omega$ and $|z - z_0| < \delta$, then $|f(z) - f(z_0)| < \epsilon$ for all $f \in \mathcal{F}$.

We have the following relationship between bounded and equicontinuous subsets of $A(\Omega)$.

5.1.8 Theorem

Let $\mathcal{F}$ be a bounded subset of $A(\Omega)$. Then $\mathcal{F}$ is equicontinuous at each point of Ω.

Proof. Let $z_0 \in \Omega$ and choose $r > 0$ such that $\overline{D}(z_0, r) \subseteq \Omega$. Then for $z \in D(z_0, r)$ and $f \in \mathcal{F}$, we have

$$f(z) - f(z_0) = \frac{1}{2\pi i} \int_{C(z_0,r)} \frac{f(w)}{w - z}\, dw - \frac{1}{2\pi i} \int_{C(z_0,r)} \frac{f(w)}{w - z_0}\, dw.$$

Thus

$$|f(z) - f(z_0)| \le \frac{1}{2\pi} \sup \left\{ \left| \frac{f(w)}{w - z} - \frac{f(w)}{w - z_0} \right| : w \in C(z_0, r) \right\} 2\pi r$$

$$= r|z - z_0| \sup \left\{ \left| \frac{f(w)}{(w - z)(w - z_0)} \right| : w \in C(z_0, r) \right\}.$$

But by hypothesis, there exists M_r such that $|f(w)| \le M_r$ for all $w \in C(z_0, r)$ and all $f \in \mathcal{F}$. Consequently, if $z \in D(z_0, r/2)$ and $f \in \mathcal{F}$, then

$$r|z - z_0| \sup \left\{ \left| \frac{f(w)}{(w - z)(w - z_0)} \right| : w \in C(z_0, r) \right\} \le r|z - z_0| \frac{M_r}{(r/2)^2},$$

proving equicontinuity of $\mathcal{F}$. ♣

We will also need the following general fact about equicontinuous families.

5.1.9 Theorem

Suppose $\mathcal{F}$ is an equicontinuous subset of $C(\Omega)$ (that is, each $f \in \mathcal{F}$ is continuous on Ω and $\mathcal{F}$ is equicontinuous at each point of Ω) and $\{f_n\}$ is a sequence from $\mathcal{F}$ such that f_n converges pointwise to f on Ω. Then f is continuous on Ω and $f_n \to f$ uniformly on compact subsets of Ω. More generally, if $f_n \to f$ pointwise on a dense subset of Ω, then $f_n \to f$ on all of Ω and the same conclusion holds.

Proof. Let $\epsilon > 0$. For each $w \in \Omega$, choose a $\delta_w > 0$ such that $|f_n(z) - f_n(w)| < \epsilon$ for each $z \in D(w, \delta_w)$ and all n. It follows that $|f(z) - f(w)| \le \epsilon$ for all $z \in D(w, \delta_w)$, so f is continuous. Let K be any compact subset of Ω. Since $\{D(w, \delta_w) : w \in K\}$ is an open cover of K, there are $w_1, \dots, w_m \in K$ such that $K \subseteq \cup_{j=1}^m D(w_j, \delta_{w_j})$. Now choose N such that $n \ge N$ implies that $|f(w_j) - f_n(w_j)| < \epsilon$ for $j = 1, \dots, m$. Hence if $z \in D(w_j, \delta_{w_j})$ and $n \ge N$, then

$$|f(z) - f_n(z)| \le |f(z) - f(w_j)| + |f(w_j) - f_n(w_j)| + |f_n(w_j) - f_n(z)| < 3\epsilon.$$

In particular, if $z \in K$ and $n \ge N$, then $|f(z) - f_n(z)| < 3\epsilon$, showing that $f_n \to f$ uniformly on K.

Finally, suppose only that $f_n \to f$ pointwise on a dense subset $S \subseteq \Omega$. Then as before, $|f_n(z) - f_n(w)| < \epsilon$ for all n and all $z \in D(w, \delta_w)$. But since S is dense, $D(w, \delta_w)$ contains a point $z \in S$, and for m and n sufficiently large,

$$|f_m(w) - f_n(w)| \le |f_m(w) - f_m(z)| + |f_m(z) - f_n(z)| + |f_n(z) - f_n(w)| < 3\epsilon.$$

Thus $\{f_n(w)\}$ is a Cauchy sequence and therefore converges, hence $\{f_n\}$ converges pointwise on all of Ω and the first part of the theorem applies. ♣

5.1.10 Montel's Theorem

Let $\mathcal{F}$ be a bounded subset of $A(\Omega)$, as in (5.1.6). Then each sequence $\{f_n\}$ from $\mathcal{F}$ has a subsequence $\{f_{n_j}\}$ which converges uniformly om compact subsets of Ω.

Remark

A set $\mathcal{F} \subseteq C(\Omega$ is said to be *relatively compact* if the closure of $\mathcal{F}$ in $C(\Omega)$ is compact. The conclusion of (5.1.10) is equivalent to the statement that $\mathcal{F}$ is a relatively compact subset of $C(\Omega)$, and hence of $A(\Omega)$.

Proof. Let $\{f_n\}$ be any sequence from $\mathcal{F}$ and choose any countable dense subset $S = \{z_1, z_2, \dots\}$ of Ω. The strategy will be to show that $f_n\}$ has a subsequence which converges pointwise on S. Since $\mathcal{F}$ is a bounded subset of $A(\Omega)$, it is equicontinuous on Ω by (5.1.8). Theorem 5.1.9 will then imply that this subsequence converges uniformly on compact subsets of Ω, thus completing the proof. So consider the following bounded sequences of complex numbers:

$$\{f_j(z_1)\}_{j=1}^{\infty}, \{f_j(z_2)\}_{j=1}^{\infty}, \dots .$$

There is a subsequence $\{f_{1j}\}_{j=1}^{\infty}$ of $\{f_j\}_{j=1}^{\infty}$ which converges at z_1. There is a subsequence $\{f_{2j}\}_{j=1}^{\infty}$ of $\{f_{1j}\}_{j=1}^{\infty}$ which converges at z_2 and (necessarily) at z_1 as well. Proceeding inductively, for each $n \geq 1$ and each $k = 1, \dots, n$ we construct sequences $\{f_{kj}\}_{j=1}^{\infty}$ converging at $z_1, \dots, z_k$, each a subsequence of the preceding sequence.

Put $g_j = f_{jj}$. Then $\{g_j\}$ is a subsequence of $\{f_j\}$, and $\{g_j\}$ converges pointwise on $\{z_1, z_2, \dots\}$ since for each n, $\{g_j\}$ is eventually a subsequence of $\{f_{nj}\}_{j=1}^{\infty}$. ♣

5.1.11 Theorem (Compactness Criterion)

Let $\mathcal{F} \subseteq A(\Omega)$. Then $\mathcal{F}$ is compact iff $\mathcal{F}$ is closed and bounded. Also, $\mathcal{F}$ is relatively compact iff $\mathcal{F}$ is bounded.

(See Problem 3 for the second part of this theorem.)

Proof.
If $\mathcal{F}$ is compact, then $\mathcal{F}$ is closed (a general property that holds in any metric space). In order to show that $\mathcal{F}$ is bounded, we will use the following device. Let K be any compact subset of Ω. Then $f \to \|f\|_K$ is a continuous map from $A(\Omega)$ into $\mathbb{R}$. Hence $\{\|f\|_K : f \in \mathcal{F}\}$ is a compact subset of $\mathbb{R}$ and thus is bounded. Conversely, if $\mathcal{F}$ is closed and bounded, then $\mathcal{F}$ is closed and, by Montel's theorem, relatively compact. Therefore $\mathcal{F}$ is compact. ♣

Remark

Problem 6 gives an example which shows that the preceding compactness criterion fails in the larger space $C(\Omega)$. That is, there are closed and bounded subsets of $C(\Omega)$ that are not compact.

5.1.12 Theorem

Suppose $\mathcal{F}$ is a nonempty compact subset of $A(\Omega)$. Then given $z_0 \in \Omega$, there exists $g \in \mathcal{F}$ such that $|g'(z_0)| \geq |f'(z_0)|$ for all $f \in \mathcal{F}$.

Proof. Just note that the map $f \to |f'(z_0)|, f \in A(\Omega)$, is continuous. ♣

Here is a compactness result that will be needed for the proof of the Riemann mapping theorem in the next section.

5.1.13 Theorem

Assume that Ω is connected, $z_0 \in \Omega$, and $\epsilon > 0$. Define

$$\mathcal{F} = \{f \in A(\Omega) : f \text{ is a one-to-one map of } \Omega \text{ into } \overline{D}(0,1) \text{ and } |f'(z_0)| \geq \epsilon\}.$$

Then $\mathcal{F}$ is compact. The same conclusion holds with $\overline{D}(0,1)$ replaced by $D(0,1)$.

Proof. By its definition, $\mathcal{F}$ is bounded, and $\mathcal{F}$ is closed by (5.1.5). Thus by (5.1.11), $\mathcal{F}$ is compact. To prove the last statement of the theorem, note that if $f_n \in \mathcal{F}$ and $f_n \to f$ uniformly on compact subsets of Ω, then (5.1.5) would imply that $f \in \mathcal{F}$, were it not for the annoying possibility that $|f(w)| = 1$ for some $w \in \Omega$. But if this happens, the maximum principle implies that f is constant, contradicting $|f'(z_0)| \geq \epsilon > 0$. ♣

The final result of this section shows that if Ω is connected, then any bounded sequence in $A(\Omega)$ that converges pointwise on a set having a limit point in Ω, must in fact converge uniformly on compact subsets of Ω.

5.1.14 Vitali's Theorem

Let $\{f_n\}$ be a bounded sequence in $A(\Omega)$ where Ω is connected. Suppose that $\{f_n\}$ converges pointwise on $S \subseteq \Omega$ and S has a limit point in Ω. Then $\{f_n\}$ is uniformly Cauchy on compact subsets of Ω, hence uniformly convergent on compact subsets of Ω to some $f \in A(\Omega)$.

Proof. Suppose, to the contrary, that there is a compact set $K \subseteq \Omega$ such that $\{f_n\}$ is not uniformly Cauchy on K. Then for some $\epsilon > 0$, we can find sequences $\{m_j\}$ and $\{n_j\}$ of positive integers such that $m_1 < n_1 < m_2 < n_2 < \cdots$ and for each j, $\|f_{m_j} - f_{n_j}\|_K \geq \epsilon$. Put $\{g_j\} = \{f_{m_j}\}$ and $\{h_j\} = \{f_{n_j}\}$. Now apply Montel's theorem (5.1.10) to $\{g_j\}$ to obtain a subsequence $\{g_{j_r}\}$ converging uniformly on compact subsets of Ω to some $g \in A(\Omega)$, and then apply Montel's theorem to $\{h_{j_r}\}$ to obtain a subsequence converging uniformly on compact subsets of Ω to some $h \in A(\Omega)$. To prevent the notation from getting out of hand, we can say that without loss of generality, we have $g_n \to g$ and $h_n \to h$ uniformly on compact subsets, and $\|g_n - h_n\|_K \geq \epsilon$ for all n, hence $\|g - h\|_K \geq \epsilon$. But by hypothesis, $g = h$ on S and therefore, by (2.4.9), $g = h$ on Ω, a contradiction. ♣

Problems

1. Let $\mathcal{F} = \{f \in A(D(0,1)) : \operatorname{Re} f > 0 \text{ and } |f(0)| \leq 1\}$. Prove that $\mathcal{F}$ is relatively compact. Is $\mathcal{F}$ compact? (See Section 4.6, Problem 2.)

2. Let Ω be (open and) connected and let $\mathcal{F} = \{f \in A(\Omega) : |f(z) - a| \geq r$ for all $z \in \Omega\}$, where $r > 0$ and $a \in \mathbb{C}$ are fixed. Show that $\mathcal{F}$ is a *normal family*, that is, if $f_n \in \mathcal{F}, n = 1, 2, \ldots$, then either there is a subsequence $\{f_{n_j}\}$ converging uniformly on compact subsets to a function $f \in A(\Omega)$ or there is a subsequence $\{f_{n_j}\}$ converging uniformly on compact subsets to ∞. (Hint: Look at the sequence $\{1/(f_n - a)\}$.)
3. (a) If $\mathcal{F} \subseteq C(\Omega)$, show that $\mathcal{F}$ is relatively compact iff each sequence in $\mathcal{F}$ has a convergent subsequence (whose limit need not be in $\mathcal{F}$).
   (b) Prove the last statement in Theorem 5.1.11.
4. Let $\mathcal{F} \subseteq A(D(0,1))$. Show that $\mathcal{F}$ is relatively compact iff there is a sequence of nonnegative real numbers M_n with $\limsup_{n\to\infty}(M_n)^{1/n} \leq 1$ such that for all $f \in \mathcal{F}$ and all $n = 0, 1, 2, \ldots$, we have $|f^{(n)}(0)/n!| \leq M_n$.
5. (a) Suppose that f is analytic on Ω and $\overline{D}(a, R) \subseteq \Omega$. Prove that

$$|f(a)|^2 \leq \frac{1}{\pi R^2} \int_0^{2\pi} \int_0^R |f(a + re^{it})|^2 r \, dr \, dt.$$

   (b) Let $M > 0$ and define $\mathcal{F}$ to be the set

$$\{f \in A(\Omega) : \int_\Omega \int |f(x + iy)|^2 \, dx \, dy \leq M\}.$$

   Show that $\mathcal{F}$ is relatively compact.
6. Let Ω be open and $K = \overline{D}(a, R) \subseteq \Omega$. Define $\mathcal{F}$ to be the set of all $f \in C(\Omega)$ such that $|f(z)| \leq 1$ for all $z \in \Omega$ and $f(z) = 0$ for $z \in \Omega \setminus K\}$. Show that $\mathcal{F}$ is a closed and bounded subset of $C(\Omega)$, but $\mathcal{F}$ is not compact. (Hint: Consider the map from $\mathcal{F}$ to the reals given by

$$f \to \left[\int_K \int (1 - |f(x + iy)|) \, dx \, dy\right]^{-1}.$$

   Show that this map is continuous but not bounded on $\mathcal{F}$.)

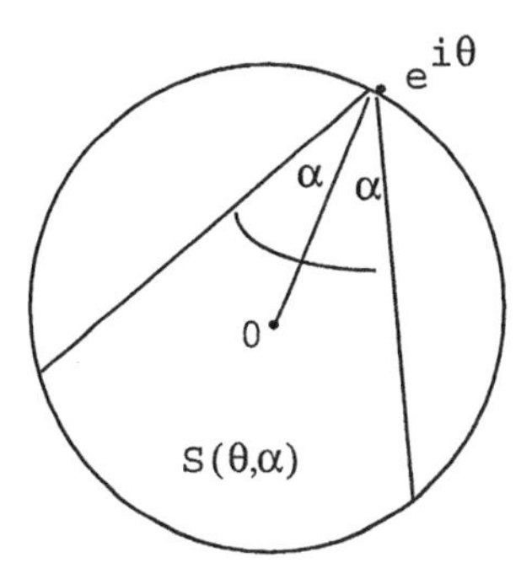

Figure 5.1.1

7. (An application of Vitali's theorem.) Let f be a *bounded* analytic function on $D(0,1)$ with the property that for some θ, $f(re^{i\theta})$ approaches a limit L as $r \to 1^-$. Fix $\alpha \in (0, \pi/2)$ and consider the region $S(\theta, \alpha)$ in Figure 5.1.1. Prove that if $z \in S(\theta, \alpha)$ and $z \to e^{i\theta}$, then $f(z) \to L$. (Suggestion: Look at the sequence of functions defined by $f_n(z) = f(e^{i\theta} + \frac{1}{n}(z - e^{i\theta})), z \in D(0,1)$.)
8. Let L be a multiplicative linear functional on $A(\Omega)$, that is, $L : A(\Omega) \to \mathbb{C}$ such that $L(af + bg) = aL(f) + bL(g)$ and $L(fg) = L(f)L(g)$ for all $a, b \in \mathbb{C}, f, g \in A(\Omega)$. Assume $L \not\equiv 0$. Show that L is a point evaluation, that is, there is some $z_0 \in \Omega$ such that $L(f) = f(z_0)$ for all $f \in A(\Omega)$.
   Outline: First show that for $f \equiv 1$, $L(f) = 1$. Then apply L to the function $I(z) = z$, the identity on Ω, and show that if $L(I) = z_0$, then $z_0 \in \Omega$. Finally, if $f \in A(\Omega)$, apply L to the function
$$g(z) = \begin{cases} \frac{f(z)-f(z_0)}{z-z_0}, & z \neq z_0 \\ f'(z_0), & z = z_0. \end{cases}$$
9. (Osgood's theorem). Let $\{f_n\}$ be a sequence in $A(\Omega)$ such that $f_n \to f$ pointwise on Ω. Show that there is an open set U, dense in Ω, such that $f_n \to f$ uniformly on compact subsets of U. In particular, f is analytic on a dense open subset of Ω.
   (Let $A_n = \{z \in \Omega : |f_k(z)| \leq n \text{ for all } k = 1, 2, \dots\}$. Recall the Baire category theorem: If a complete metric space X is the union of a sequence $\{S_n\}$ of closed subsets, then some S_n contains a nonempty open ball. Use this result to show that some A_n contains a disk D. By Vitali's theorem, $f_n \to f$ uniformly on compact subsets of D. Take U to be the union of all disks D such that $f_n \to f$ uniformly on compact subsets of D.)

5.2 Riemann Mapping Theorem

Throughout this section, Ω will be a nonempty open connected proper subset of $\mathbb{C}$ with the property that every zero-free analytic function has an analytic square root. Later in the section we will prove that *any* open subset Ω such that every zero-free analytic function on Ω has an analytic square root must be (homotopically) simply connected, and conversely. Thus we are considering open, connected and simply connected proper subsets of $\mathbb{C}$. Our objective is to prove the Riemann mapping theorem, which states that there is a one-to-one analytic map of Ω onto the open unit disk D. The proof given is due to Fejer and F.Riesz.

5.2.1 Lemma

There is a one-to-one analytic map of Ω *into* D.

Proof. Fix $a \in \mathbb{C} \setminus \Omega$. Then the function $z - a$ satisfies our hypothesis on Ω and hence there exists $h \in A(\Omega)$ such that $(h(z))^2 = z - a, z \in \Omega$. Note that h is one-to-one and $0 \notin h(\Omega)$. Furthermore, $h(\Omega)$ is open by (4.3.1), the open mapping theorem, hence so is $-h(\Omega) = \{-h(z) : z \in \Omega\}$, and $[h(\Omega] \cap [-h(\Omega] = \emptyset$ (because $0 \notin h(\Omega)$). Now choose $w \in -h(\Omega)$. Since $-h(\Omega)$ is open, there exists $r > 0$ such that $D(w,r) \subseteq -h(\Omega)$, hence $h(\Omega) \cap D(w,r) = \emptyset$. The function $f(z) = 1/(h(z) - w)$, $z \in \Omega$, is one-to-one, and its magnitude is less than $1/r$ on Ω. Thus rf is a one-to-one map of Ω into D. ♣

5.2.2 Riemann Mapping Theorem

Let Ω be as in (5.2.1), that is, a nonempty, proper, open and connected subset of $\mathbb{C}$ such that every zero-free analytic function on Ω has an analytic square root. Then there is a one-to-one analytic map of Ω *onto* D.

Proof. Fix $z_0 \in \Omega$ and a one-to-one analytic map f_0 of Ω into D [f_0 exists by (5.2.1)]. Let $\mathcal{F}$ be the set of all $f \in A(\Omega)$ such that f is a one-to-one analytic map of Ω into D and $|f'(z_0)| \geq |f_0'(z_0)|$. Note that $|f_0'(z_0)| > 0$ by (4.3.1).

Then $\mathcal{F} \neq \emptyset$ (since $f_0 \in \mathcal{F}$) and $\mathcal{F}$ is bounded. Also, $\mathcal{F}$ is closed, for if $\{f_n\}$ is a sequence in $\mathcal{F}$ such that $f_n \to f$ uniformly on compact subsets of Ω, then by (5.1.5), either f is constant on Ω or f is one-to-one. But since $f_n' \to f'$, it follows that $|f'(z_0)| \geq |f_0'(z_0)| > 0$, so f is one-to-one. Also, f maps Ω into D (by the maximum principle), so $f \in \mathcal{F}$. Since $\mathcal{F}$ is closed and bounded, it is compact (Theorem 5.1.1). Hence by (5.1.2), there exists $g \in \mathcal{F}$ such that $|g'(z_0)| \geq |f'(z_0)|$ for all $f \in \mathcal{F}$. We will now show that such a g must map Ω *onto* D. For suppose that there is some $a \in D \setminus g(\Omega)$. Let φ_a be as in (4.6.1), that is,

$$\varphi_a(z) = \frac{z-a}{1-\overline{a}z}, \quad z \in D.$$

Then $\varphi_a \circ g : \Omega \to D$ and $\varphi_a \circ g$ is one-to-one with no zeros in Ω. By hypothesis, there is an analytic square root h for $\varphi_a \circ g$. Note also that $h^2 = \varphi_a \circ g$ is one-to-one, and therefore so is h. Set $b = h(z_0)$ and define $f = \varphi_b \circ h$. Then $f(z_0) = \varphi_b(b) = 0$ and we can write

$$g = \varphi_{-a} \circ h^2 = \varphi_{-a} \circ (\varphi_{-b} \circ f)^2 = \varphi_{-a} \circ (\varphi_{-b}^2 \circ f) = (\varphi_{-a} \circ \varphi_{-b}^2) \circ f.$$

Now

$$\begin{aligned} g'(z_0) &= (\varphi_{-a} \circ \varphi_{-b}^2)'(f(z_0))f'(z_0) \\ &= (\varphi_{-a} \circ \varphi_{-b}^2)'(0)f'(z_0). \end{aligned} \tag{1}$$

The function $\varphi_{-a} \circ \varphi_{-b}^2$ *is* an analytic map of D into D, but it is *not* one-to-one; indeed, it is two-to-one. Hence by the Schwarz-Pick theorem (4.6.3), part (ii), it must be the case that

$$|(\varphi_{-a} \circ \varphi_{-b}^2)'(0)| < 1 - |\varphi_{-a} \circ \varphi_{-b}^2(0)|^2.$$

Since $f'(z_0) \neq 0$, it follows from (1) that

$$|g'(z_0)| < (1 - |\varphi_{-a} \circ \varphi_{-b}^2(0)|^2)|f'(z_0)| \leq |f'(z_0)|.$$

This contradicts our choice of $g \in \mathcal{F}$ as maximizing the numbers $|f'(z_0)|, f \in \mathcal{F}$. Thus $g(\Omega) = D$ as desired. ♣

5.2.3 Remarks

(a) Any function g that maximizes the numbers $\{|f'(z_0)| : f \in \mathcal{F}\}$ must send z_0 to 0.
Proof. Let $a = g(z_0)$. Then $\varphi_a \circ g$ is a one-to-one analytic map of Ω into D. Moreover,

$$\begin{aligned}|(\varphi_a \circ g)'(z_0)| &= |\varphi_a'(g(z_0))g'(z_0)| \\ &= |\varphi_a'(a)g'(z_0)| \\ &= \frac{1}{1-|a|^2}|g'(z_0)| \\ &\geq |g'(z_0)| \geq |f_0'(z_0)|.\end{aligned}$$

Thus $\varphi_a \circ g \in \mathcal{F}$, and since $|g'(z_0)|$ maximizes $|f'(z_0)|$ for $f \in \mathcal{F}$, it follows that equality must hold in the first inequality. Therefore $1/(1-|a|^2) = 1$, so $0 = a = g(z_0)$. ♣
(b) Let f and h be one-to-one analytic maps of Ω onto D such that $f(z_0) = h(z_0) = 0$ and $f'(z_0) = h'(z_0)$ (it is enough that $\operatorname{Arg} f'(z_0) = \operatorname{Arg} h'(z_0)$). Then $f = h$.
Proof. The function $h \circ f^{-1}$ is a one-to-one analytic map of D onto D, and $h \circ f^{-1}(0) = h(z_0) = 0$. Hence by Theorem 4.6.4 (with $a = 0$), there is a unimodular complex number λ such that $h(f^{-1}(z)) = \lambda z, z \in D$. Thus $h(w) = \lambda f(w), w \in D$. But if $h'(z_0) = f'(z_0)$ (which is equivalent to $\operatorname{Arg} h'(z_0) = \operatorname{Arg} f'(z_0)$ since $|h'(z_0)| = |\lambda||f'(z_0)| = |f'(z_0)|$), we have $\lambda = 1$ and $f = h$. ♣
(c) Let f be any analytic map of Ω *into* D (not necessarily one-to-one or onto) with $f(z_0) = 0$. Then with g as in the theorem, $|f'(z_0)| \leq |g'(z_0)|$. Also, equality holds iff $f = \lambda g$ with $|\lambda| = 1$.
Proof. The function $f \circ g^{-1}$ is an analytic map of D into D such that $f \circ g^{-1}(0) = 0$. By Schwarz's lemma (2.4.16), $|f(g^{-1}(z))| \leq |z|$ and $|f'(g^{-1}(0)) \cdot \frac{1}{g'(z_0)}| \leq 1$. Thus $|f'(z_0)| \leq |g'(z_0)|$. Also by (2.4.16), equality holds iff for some unimodular λ we have $f \circ g^{-1}(z) = \lambda z$, that is, $f(z) = \lambda g(z)$, for all $z \in D$. ♣

If we combine (a), (b) and (c), and observe that $\lambda g'(z_0)$ will be real and greater than 0 for appropriately chosen unimodular λ, then we obtain the following existence and uniqueness result.
(d) Given $z_0 \in \Omega$, there is a unique one-to-one analytic map g of Ω onto D such that $g(z_0) = 0$ and $g'(z_0)$ is real and positive.

As a corollary of (d), we obtain the following result, whose proof will be left as an exercise; see Problem 1.
(e) Let Ω_1 and Ω_2 be regions that satisfy the hypothesis of the Riemann mapping theorem. Let $z_1 \in \Omega_1$ and $z_2 \in \Omega_2$. Then there is a unique one-to-one analytic map f of Ω_1 onto Ω_2 such that $f(z_1) = z_2$ and $f'(z_1)$ is real and positive.

Recall from (3.4.6) that if $\Omega \subseteq \mathbb{C}$ and Ω satisfies any one of the six equivalent conditions listed there, then Ω is called (homologically) simply connected. Condition (6) is that every zero-free analytic function on Ω have an analytic n-th root for $n = 1, 2, \ldots$. Thus if Ω is homologically simply connected, then in particular, assuming $\Omega \neq \mathbb{C}$, the Riemann mapping theorem implies that Ω is *conformally equivalent* to D, in other words, there is a one-to-one analytic map of Ω onto D. The converse is also true, but before showing this, we need to take a closer look at the relationship between homological simple connectedness and homotopic simple connectedness [see (4.9.12)].

5.2.4 Theorem

Let γ_0 and γ_1 be closed curves in an open set $\Omega \subseteq \mathbb{C}$. If γ_0 and γ_1 are Ω-homotopic (in other words, homotopic in Ω), then they are Ω-homologous, that is, $n(\gamma_0, z) = n(\gamma_1, z)$ for every $z \in \mathbb{C} \setminus \Omega$.

Proof. We must show that $n(\gamma_0, z) = n(\gamma_1, z)$ for each $z \in \mathbb{C} \setminus \Omega$. Thus let $z \in \mathbb{C} \setminus \Omega$, let H be a homotopy of γ_0 to γ_1, and let θ be a continuous argument of $H - z$. (See Problem 6 of Section 3.2.) That is, θ is a real continuous function on $[a, b] \times [0, 1]$ such that

$$H(t, s) - z = |H(t, s) - z|e^{i\theta(t,s)}$$

for $(t, s) \in [a, b] \times [0, 1]$. Then for each $s \in [0, 1]$, the function $t \to \theta(t, s)$ is a continuous argument of $H(\cdot, s) - z$ and hence

$$n(H(\cdot, s), z) = \frac{\theta(b, s) - \theta(a, s)}{2\pi}.$$

This shows that the function $s \to n(H(\cdot, s), z)$ is continuous, and since it is integer valued, it must be constant. In particular,

$$n(H(\cdot, 0), z) = n(H(\cdot, 1), z).$$

In other words, $n(\gamma_0, z) = n(\gamma_1, z)$. ♣

The above theorem implies that if γ is Ω-homotopic to a point in Ω, then γ must be Ω-homologous to 0. Thus if γ is a closed path in Ω such that γ is Ω-homotopic to a point, then $\int_\gamma f(z)\, dz = 0$ for every analytic function f on Ω. We will state this result formally.

5.2.5 The Homotopic Version of Cauchy's Theorem

Let γ be a closed path in Ω such that γ is Ω-homotopic to a point. Then $\int_\gamma f(z)\, dz = 0$ for every analytic function f on Ω.

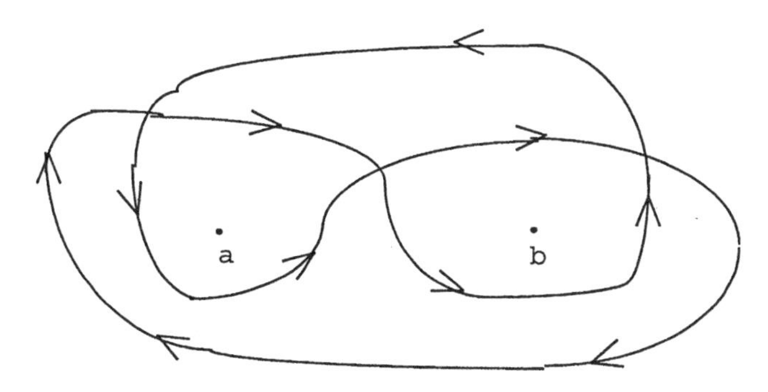

Figure 5.2.1

Remark

The converse of Theorem 5.2.5 is *not* true. In particular, there are closed curves γ and open sets Ω such that γ is Ω-homologous to 0 but γ is not homotopic to a point. Take $\Omega = \mathbb{C} \setminus \{a, b\}$, $a \neq b$, and consider the closed path γ of Figure 5.2.1. Then $n(\gamma, a) = n(\gamma, b) = 0$, hence γ is Ω-homologous to 0. But (intuitively at least) we see that γ cannot be shrunk to a point without passing through a or b. It follows from this example and Theorem 5.2.4 that the homology version of Cauchy's theorem (3.3.1) is actually stronger than the homotopy version (5.2.5). That is, if γ is a closed path to which the homotopy version applies, then so does the homology version, while the homology version applies to the above path, but the homotopy version does not. However, if *every* closed path in Ω is homologous to zero, then every closed path is homotopic to a point, as we now show.

5.2.6 Theorem

Let Ω be an open connected subset of $\mathbb{C}$. The following are equivalent.

(1) Every zero-free $f \in A(\Omega)$ has an analytic square root.

(2) If $\Omega \neq \mathbb{C}$, then Ω is conformally equivalent to D.

(3) Ω is homeomorphic to D.

(4) Ω is homotopically simply connected.

(5) Each closed *path* in Ω is homotopic to a point.

(6) Ω is homologically simply connected.

Proof.

(1) implies (2): This is the Riemann mapping theorem.

(2) implies (3): If $\Omega \neq \mathbb{C}$, this follows because a conformal equivalence is a homeomorphism., while if $\Omega = \mathbb{C}$, then the map $h(z) = z/(1 + |z|)$ is a homeomorphism of $\mathbb{C}$ onto D (see Problem 2).

(3) implies (4): Let $\gamma : [a, b] \to \Omega$ be any closed curve in Ω. By hypothesis there is a homeomorphism f of Ω onto D. Then $f \circ \gamma$ is a closed curve in D, and there is a homotopy H (in D) of $f \circ \gamma$ to the point $f(\gamma(a))$ (see Problem 4). Therefore $f^{-1} \circ H$ is a homotopy in Ω of γ to $\gamma(a)$.

(4) implies (5): Every closed path is a closed curve.

(5) implies (6): Let γ be any closed path in Ω. If γ is Ω-homotopic to a point, then by Theorem 5.2.4, γ is Ω-homologous to zero.

(6) implies (1): This follows from part (6) of (3.4.6).

Remark

If Ω is any open set (not necessarily connected) then the statement of the preceding theorem applies to each component of Ω. Therefore (1), (4), (5) and (6) are equivalent for arbitrary open sets.

Here is yet another condition equivalent to simple connectedness of an open set Ω.

5.2.7 Theorem

Let Ω be a simply connected open set. Then every harmonic function on Ω has a harmonic conjugate. Conversely, if Ω is an open set such that every harmonic function on Ω has a harmonic conjugate, then Ω is simply connected.

Proof. The first assertion was proved as Theorem 4.9.14 using the method of analytic continuation. However, we can also give a short proof using the Riemann mapping theorem, as follows. First note that we can assume that Ω is connected by applying this case to components. If $\Omega = \mathbb{C}$ then every harmonic function on Ω has a harmonic conjugate as in Theorem 1.6.2. Suppose then that $\Omega \neq \mathbb{C}$. By the Riemann mapping theorem, there is a conformal equivalence f of Ω onto D. Let u be harmonic on Ω. Then $u \circ f^{-1}$ is harmonic on D and thus by (1.6.2), there is a harmonic function V on D such that $u \circ f^{-1} + iV$ is analytic on D. Since $(u \circ f^{-1} + iV) \circ f$ is analytic on Ω, there is a harmonic conjugate of u on Ω, namely $v = V \circ f$.

Conversely, suppose that Ω is not simply connected. Then Ω is not homologically simply connected, so there exists $z_0 \in \mathbb{C}\backslash\Omega$ and a closed path γ in Ω such that $n(\gamma, z_0) \neq 0$. Thus by (3.1.9) and (3.2.3), the function $z \to z - z_0$ does not have an analytic logarithm on Ω, hence $z \to \ln|z - z_0|$ does not have a harmonic conjugate. ♣

The final result of this section is Runge's theorem on rational and polynomial approximation of analytic functions. One consequence of the development is another condition that is equivalent to simple connectedness.

5.2.8 Runge's Theorem

Let K be a compact subset of $\mathbb{C}$, and S a subset of $\hat{\mathbb{C}} \setminus K$ that contains at least one point in each component of $\hat{\mathbb{C}} \setminus K$. Define $B(S) = \{f : f$ is a uniform limit on K of rational functions whose poles lie in $S\}$. Then every function f that is analytic on a neighborhood of K is in $B(S)$. That is, there is a sequence $\{R_n\}$ of rational functions whose poles lie in S such that $R_n \to f$ uniformly on K.

Before giving the proof, let us note the conclusion in the special case where $\hat{\mathbb{C}} \setminus K$ is connected. In this case, we can take $S = \{\infty\}$, and our sequence of rational functions will actually be a sequence of polynomials. The proof given is due to Sandy Grabiner (Amer. Math. Monthly, 83 (1976), 807-808) and is based on three lemmas.

5.2.9 Lemma

Suppose K is a compact subset of the open set $\Omega \subseteq \mathbb{C}$. If $f \in A(\Omega)$, then f is a uniform limit on K of rational functions whose poles (in the extended plane!) lie in $\Omega \setminus K$.

5.2.10 Lemma

Let U and V be open subsets of $\mathbb{C}$ with $V \subseteq U$ and $\partial V \cap U = \emptyset$. If H is any component of U and $V \cap H \neq \emptyset$, then $H \subseteq V$.

5.2.11 Lemma

If K is a compact subset of $\mathbb{C}$ and $\lambda \in \mathbb{C} \setminus K$, then $(z - \lambda)^{-1} \in B(S)$.

Let us see how Runge's theorem follows from these three lemmas, and then we will prove the lemmas. First note that if f and g belong to $B(S)$, then so do $f + g$ and fg. Thus by Lemma 5.2.11 (see the partial fraction decomposition of Problem 4.1.7), every rational function with poles in $\hat{\mathbb{C}} \setminus K$ belongs to $B(S)$. Runge's theorem is then a consequence of Lemma 5.2.9. (The second of the three lemmas is used to prove the third.)

Proof of Lemma 5.2.9

Let Ω be an open set containing K. By (3.4.7), there is a cycle γ in $\Omega \setminus K$ such that for every $f \in A(\Omega)$ and $z \in K$,

$$f(z) = \frac{1}{2\pi i} \int_\gamma \frac{f(w)}{w - z} \, dw.$$

Let $\epsilon > 0$ be given. Then $\delta = \operatorname{dist}(\gamma^*, K) > 0$ because γ^* and K are disjoint compact sets. Assume $[0, 1]$ is the domain of γ and let $s, t \in [0, 1], z \in K$. Then

$$\begin{aligned}
&\left| \frac{f(\gamma(t))}{\gamma(t) - z} - \frac{f(\gamma(s))}{\gamma(s) - z} \right| \\
&= \left| \frac{f(\gamma(t))(\gamma(s) - z) - f(\gamma(s))(\gamma(t) - z)}{(\gamma(t) - z)(\gamma(s) - z)} \right| \\
&= \left| \frac{f(\gamma(t))(\gamma(s) - \gamma(t)) + \gamma(t)(f(\gamma(t)) - f(\gamma(s))) - z(f(\gamma(t)) - f(\gamma(s)))}{(\gamma(t) - z)(\gamma(s) - z)} \right| \\
&\leq \frac{1}{\delta^2} (|f(\gamma(t))||\gamma(s) - \gamma(t)| + |\gamma(t)||f(\gamma(t)) - f(\gamma(s))| + |z||f(\gamma(t)) - f(\gamma(s))|).
\end{aligned}$$

Since γ and $f \circ \gamma$ are bounded functions and K is a compact set, there exists $C > 0$ such that for $s, t \in [0, 1]$ and $z \in K$, the preceding expression is bounded by

$$\frac{C}{\delta^2} (|\gamma(s) - \gamma(t)| + |f(\gamma(t)) - f(\gamma(s))|.$$

Thus by uniform continuity of γ and $f \circ \gamma$ on the interval $[0, 1]$, there is a partition $0 = t_0 < t_1 < \cdots < t_n = 1$ such that for $t \in [t_{j-1}, t_j]$ and $z \in K$,

$$\left| \frac{f(\gamma(t))}{\gamma(t) - z} - \frac{f(\gamma(t_j))}{\gamma(t_j) - z} \right| < \epsilon.$$

Define

$$R(z) = \sum_{j=1}^{n} \frac{f(\gamma(t_j))}{\gamma(t_j) - z} (\gamma(t_j) - \gamma(t_{j-1})), \ z \neq \gamma(t_j).$$

Then $R(z)$ is a rational function whose poles are included in the set $\{\gamma(t_1), \dots, \gamma(t_n)\}$, in particular, the poles are in $\Omega \setminus K$. Now for all $z \in K$,

$$\begin{aligned}
|2\pi i f(z) - R(z)| &= \left| \int_\gamma \frac{f(w)}{w-z}\, dw - \sum_{j=1}^n \frac{f(\gamma(t_j))}{\gamma(t_j) - z}(\gamma(t_j) - \gamma(t_{j-1})) \right| \\
&= \left| \sum_{j=1}^n \int_{t_{j-1}}^{t_j} \left(\frac{f(\gamma(t))}{\gamma(t) - z} - \frac{f(\gamma(t_j))}{\gamma(t_j) - z} \right) \gamma'(t)\, dt \right| \\
&\le \epsilon \int_0^1 |\gamma'(t)|\, dt = \epsilon \cdot \text{length of } \gamma.
\end{aligned}$$

Since the length of γ is independent of ϵ, the lemma is proved. ♣

Proof of Lemma 5.2.10

Let H be any component of U such that $V \cap H \neq \emptyset$. We must show that $H \subseteq V$. let $s \in V \cap H$ and let G be that component of V that contains s. It suffices to show that $G = H$. Now $G \subseteq H$ since G is a connected subset of U containing s and H is the union of all subsets with this property. Write

$$H = G \cup (H \setminus G) = G \cup [(\partial G \cap H) \cup (H \setminus \overline{G})].$$

But $\partial G \cap H = \emptyset$, because otherwise the hypothesis $\partial V \cap U = \emptyset$ would be violated. Thus $H = G \cup (H \setminus \overline{G})$, the union of two disjoint open sets. Since H is connected and $G \neq \emptyset$, we have $G = H$ as required. ♣

Proof of Lemma 5.2.11

Suppose first that $\infty \in S$. Then for sufficiently large $|\lambda_0|$, with λ_0 in the unbounded component of $\mathbb{C} \setminus K$, the Taylor series for $(z - \lambda_0)^{-1}$ converges uniformly on K. Thus $(z - \lambda_0)^{-1} \in B(S)$, and it follows that

$$B((S \setminus \{\infty\}) \cup \{\lambda_0\}) \subseteq B(S).$$

(If $f \in B((S \setminus \{\infty\}) \cup \{\lambda_0\})$ and R is a rational function with poles in $(S \setminus \{\infty\}) \cup \{\lambda_0\}$ that approximates f, write $R = R_1 + R_2$ where all the poles (if any) of R_1 lie in $S \setminus \{\infty\}$ and the pole (if any) of R_0 is at λ_0. But R_0 can be approximated by a polynomial P_0, hence $R_1 + P_0$ approximates f and has its poles in S, so $f \in B(S)$.) Thus it is sufficient to establish the lemma for sets $S \subseteq \mathbb{C}$. We are going to apply Lemma 5.2.10. Put $U = \mathbb{C} \setminus K$ and define

$$V = \{\lambda \in U : (z - \lambda)^{-1} \in B(S)\}.$$

Recall that by hypothesis, $S \subseteq U$ and hence $S \subseteq V \subseteq U$. To apply (5.2.10) we must first show that V is open. Suppose $\lambda \in V$ and μ is such that $0 < |\lambda - \mu| < \text{dist}(\lambda, K)$. Then $\mu \in \mathbb{C} \setminus K$ and for all $z \in K$,

$$\frac{1}{z - \mu} = \frac{1}{(z - \lambda)[1 - \frac{\mu - \lambda}{z - \lambda}]}.$$

Since $(z-\lambda)^{-1} \in B(S)$, it follows from the remarks preceding the proof of Lemma 5.2.9 that $(z-\mu)^{-1} \in B(S)$. Thus $\mu \in V$, proving that V is open. Next we'll show that $\partial V \cap U = \emptyset$. Let $w \in \partial V$ and let $\{\lambda_n\}$ be a sequence in V such that $\lambda_n \to w$. Then as we noted earlier in this proof, $|\lambda_n - w| < \text{dist}(\lambda_n, K)$ implies $w \in V$, so it must be the case that $|\lambda_n - w| \geq \text{dist}(\lambda_n, K)$ for all n. Since $|\lambda_n - w| \to 0$, the distance from w to K must be 0, so $w \in K$. Thus $w \notin U$, proving that $\partial V \cap U = \emptyset$, as desired. Consequently, V and U satisfy the hypotheses of (5.2.10).

Let H be any component of U. By definition of S, there exists $s \in S$ such that $s \in H$. Now $s \in V$ because $S \subseteq V$. Thus $H \cap V \neq \emptyset$, and Lemma 5.2.10 implies that $H \subseteq V$. We have shown that every component of U is a subset of V, and consequently $U \subseteq V$. Since $V \subseteq U$, we conclude that $U = V$. ♣

5.2.12 Remarks

Theorem 5.2.8 is often referred to as Runge's theorem for compact sets. Other versions of Runge's theorem appear as Problems 6(a) and 6(b).

We conclude this section by collecting a long list of conditions, all equivalent to simple connectedness.

5.2.13 Theorem

If Ω is an open subset of $\mathbb{C}$, the following are equivalent.

(a) $\hat{\mathbb{C}} \setminus \Omega$ is connected.

(b) $n(\gamma, z) = 0$ for each closed path (or cycle) γ in Ω and each point $z \in \mathbb{C} \setminus \Omega$.

(c) $\int_\gamma f(z)\,dz = 0$ for each $f \in A(\Omega)$ and each closed path γ in Ω.

(d) $n(\gamma, z) = 0$ for each closed curve γ in Ω and each $z \in \mathbb{C} \setminus \Omega$.

(e) Every analytic function on Ω has a primitive.

(f) Every zero-free analytic function on Ω has an analytic logarithm.

(g) Every zero-free analytic function on Ω has an analytic n-th root for $n = 1, 2, 3, \ldots$.

(h) Every zero-free analytic function on Ω has an analytic square root.

(i) Ω is homotopically simply connected.

(j) Each closed path in Ω is homotopic to a point.

(k) If Ω is connected and $\Omega \neq \mathbb{C}$, then Ω is conformally equivalent to D.

(l) If Ω is connected, then Ω is homeomorphic to D.

(m) Every harmonic function on Ω has a harmonic conjugate.

(n) Every analytic function on Ω can be uniformly approximated on compact sets by polynomials.

Proof. See (3.4.6), (5.2.4), (5.2.6), (5.2.7), and Problem 6(b) in this section. ♣

Problems

1. Prove (5.2.3e).
2. Show that $h(z) = z/(1+|z|)$ is a homeomorphism of $\mathbb{C}$ onto D.
3. Let $\gamma : [a, b] \to \Omega$ be a closed curve in a convex set Ω. Prove that

$$H(t, s) = s\gamma(a) + (1 - s)\gamma(t), \ t \in [a, b], \ s \in [0, 1]$$

   is an Ω-homotopy of γ to the point $\gamma(a)$.
4. Show directly, using the techniques of Problem 3, that a starlike open set is homotopically simply connected.
5. This problem is in preparation for other versions of Runge's theorem that appear in Problem 6. Let Ω be an open subset of $\mathbb{C}$, and let $\{K_n\}$ be as in (5.1.1). Show that in addition to the properties (1), (2) and (3) listed in (5.1.1), the sequence $\{K_n\}$ has an additional property:
   (4) Each component of $\hat{\mathbb{C}} \setminus K_n$ contains a component of $\hat{\mathbb{C}} \setminus \Omega$.
6. Prove the following versions of Runge's theorem:
   (a) Let Ω be an open set and let S be a set containing at least one point in each component of $\hat{\mathbb{C}} \setminus \Omega$. Show that if $f \in A(\Omega)$, then there is a sequence $\{R_n\}$ of rational functions with poles in S such that $R_n \to f$ uniformly on compact subsets of Ω.
   (b) Let Ω be an open subset of $\mathbb{C}$. Show that Ω is simply connected if and only if for each $f \in A(\Omega)$, there is a sequence $\{P_n\}$ of polynomials converging to f uniformly on compact subsets of Ω.

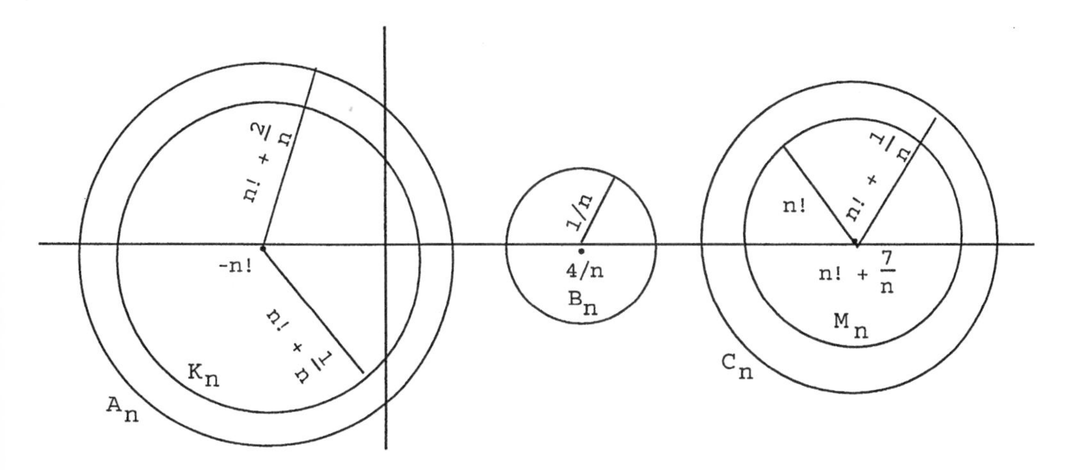

Figure 5.2.2

7. Define sequences of sets as follows:

$$A_n = \{z : |z + n!| < n! + \frac{2}{n}\}, \quad B_n = \{z : |z - \frac{4}{n}| < \frac{1}{n}\},$$

$$C_n = \{z : |z - (n! + \frac{7}{n})| < n! + \frac{1}{n}\}, \quad K_n = \{z : |z + n!| \leq n! + \frac{1}{n}\},$$

$$L_n = \{\frac{4}{n}\}, \quad M_n = \{z : |z - (n! + \frac{7}{n})| \leq n!\}$$

(see Figure 5.2.2). Define

$$f_n(z) = \begin{cases} 0, & z \in A_n \\ 1, & z \in B_n \\ 0, & z \in C_n \end{cases} \quad \text{and} \quad g_n(z) = \begin{cases} 0, & z \in A_n \\ 1, & z \in C_n \end{cases}$$

(a) By approximating f_n by polynomials (see Problem 6), exhibit a sequence of polynomials converging pointwise to 0 on all of $\mathbb{C}$, but not uniformly on compact subsets.
(b) By approximating g_n by polynomials, exhibit a sequence of polynomials converging pointwise on all of $\mathbb{C}$ to a discontinuous limit.

5.3 Extending Conformal Maps to the Boundary

Let Ω be a proper simply connected region in $\mathbb{C}$. By the Riemann mapping theorem, there is a one-to-one analytic map of Ω onto the open unit disk D. In this section we will consider the problem of extending f to a homeomorphism of the closure $\overline{\Omega}$ of Ω onto $\overline{D}$. Note that if f is extended, then $\overline{\Omega}$ must be compact. Thus we assume in addition that Ω is bounded. We will see that $\partial\Omega$ plays an essential role in determining whether such an extension is possible. We begin with some results of a purely topological nature.

5.3.1 Theorem

Suppose Ω is an open subset of $\mathbb{C}$ and f is a homeomorphism of Ω onto $f(\Omega) = V$. Then a sequence $\{z_n\}$ in Ω has no limit point in Ω iff the sequence $\{f(z_n)\}$ has no limit point in V.

Proof. Assume $\{z_n\}$ has a limit point $z \in \Omega$. There is a subsequence $\{z_{n_j}\}$ in Ω such that $z_{n_j} \to z$. By continuity, $f(z_{n_j}) \to f(z)$, and therefore the sequence $\{f(z_n)\}$ has a limit point in V. The converse is proved by applying the preceding argument to f^{-1}. ♣

5.3.2 Corollary

Suppose f is a conformal equivalence of Ω onto D. If $\{z_n\}$ is a sequence in Ω such that $z_n \to \beta \in \partial\Omega$, then $|f(z_n)| \to 1$.

Proof. Since $\{z_n\}$ has no limit point in Ω, $\{f(z_n)\}$ has no limit point in D, hence $|f(z_n)| \to 1$. ♣

Let us consider the problem of extending a conformal map f to a single boundary point $\beta \in \partial\Omega$. As the following examples indicate, the relationship of Ω and β plays a crucial role.

5.3.3 Examples

(1) Let $\Omega = \mathbb{C}\setminus(-\infty, 0]$ and let $\sqrt{z}$ denote the analytic square root of z such that $\sqrt{1} = 1$. Then $\sqrt{z}$ is a one-to-one analytic map of Ω onto the right half plane. The linear fractional transformation $T(z) = (z-1)/(z+1)$ maps the right half plane onto the unit disk D, hence $f(z) = (\sqrt{z}-1)/(\sqrt{z}+1)$ is a conformal equivalence of Ω and D. Now T maps $\operatorname{Re} z = 0$ onto $\partial D\setminus\{1\}$, so if $\{z_n\}$ is a sequence in $\operatorname{Im} z > 0$ that converges to $\beta \in (-\infty, 0)$, then $\{f(z_n)\}$ converges to a point $w \in \partial D$ with $\operatorname{Im} w > 0$. On the other hand, if $\{z_n\}$ lies in $\operatorname{Im} z < 0$ and $z_n \to \beta$, then $\{f(z_n)\}$ converges to a point $w \in \partial D$ with $\operatorname{Im} w < 0$. Thus f does not have a continuous extension to $\Omega \cup \{\beta\}$ for any β on the negative real axis.

(2) To get an example of a *bounded* simply connected region Ω with boundary points to which the mapping functions are not extendible, let

$$\Omega = [(0,1)\times(0,1)]\setminus\{\{1/n\}\times(0,1/2] : n = 2,3,\dots\}.$$

Thus Ω is the open unit square with vertical segments of height 1/2 removed at each of the points $1/2, 1/3, \dots$ on the real axis; see Figure 5.3.1. Then $\hat{\mathbb{C}}\setminus\Omega$ is seen to be

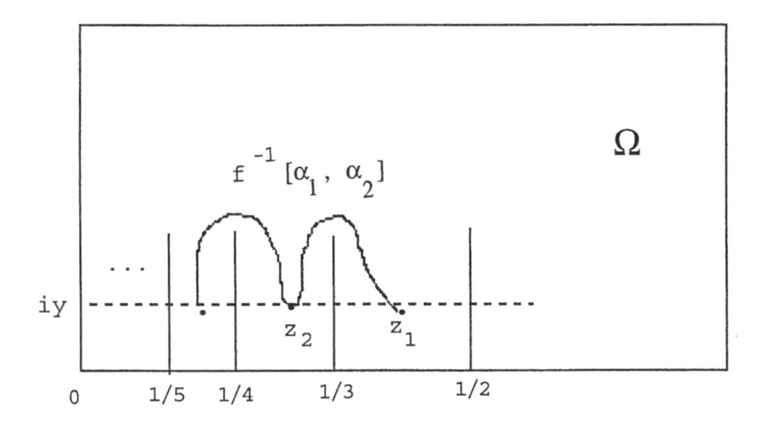

Figure 5.3.1

connected, so that Ω is simply connected. Let $\beta = iy$ where $0 < y < 1/2$, and choose a sequence $\{z_n\}$ in Ω such that $z_n \to iy$ and $\operatorname{Im} z_n = y, n = 1,2,3,\dots$. Let f be any conformal map of Ω onto D. Since by (5.3.2), $|f(z_n)| \to 1$, there is a subsequence $\{z_{n_k}\}$ such that $\{f(z_{n_k})\}$ converges to a point $w \in \partial D$. For simplicity assume that $\{f(z_n)\}$ converges to w. Set $\alpha_n = f(z_n)$ and in D, join α_n to α_{n+1} with the straight line segment $[\alpha_n, \alpha_{n+1}], n = 1,2,3,\dots$. Then $f^{-1}([\alpha_n, \alpha_{n+1}])$ is a curve in Ω joining z_n to $z_{n+1}, n = 1,2,3,\dots$. It follows that every point of $[iy, i/2]$ is a limit point of $\cup_n f^{-1}([\alpha_n, \alpha_{n+1}])$. Hence f^{-1}, in this case, cannot be extended to be continuous at $w \in \partial D$.

As we now show, if $\beta \in \partial\Omega$ is such that sequences of the type $\{z_n\}$ in the previous example are ruled out, then any mapping function can be extended to $\Omega \cup \{\beta\}$.

5.3.4 Definition

A point $\beta \in \partial\Omega$ is called *simple* if to each sequence $\{z_n\}$ in Ω such that $z_n \to \beta$, there corresponds a curve $\gamma : [0,1] \to \Omega \cup \{\beta\}$ and a strictly increasing sequence $\{t_n\}$ in [0,1) such that $t_n \to 1, \gamma(t_n) = z_n$, and $\gamma(t) \in \Omega$ for $0 \le t < 1$.

Thus a boundary point is simple iff for any sequence $\{z_n\}$ that converges to β, there is a curve γ in Ω that contains the points z_n and terminates at β. In Examples 1 and 2 of (5.3.3), none of the boundary points β with $\beta \in (-\infty, 0)$ or $\beta \in (0, i/2)$ is simple.

5.3.5 Theorem

Let Ω be a bounded simply connected region in $\mathbb{C}$, and let $\beta \in \partial\Omega$ be simple. If f is a conformal equivalence of Ω onto D, then f has a continuous extension to $\Omega \cup \{\beta\}$.

To prove this theorem, we will need a lemma due to Lindelöf.

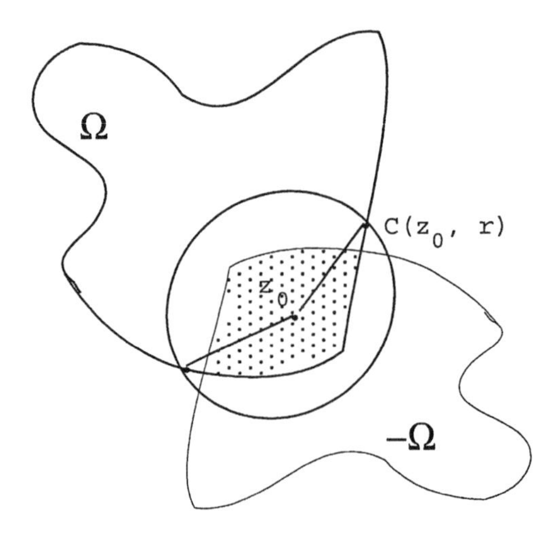

Figure 5.3.2

5.3.6 Lemma

Suppose Ω is an open set in $\mathbb{C}$, $z_0 \in \Omega$, and the circle $C(z_0, r)$ has an arc lying in the complement of $\overline{\Omega}$ which subtends an angle greater than π at z_0 (see Figure 5.3.2). Let g be any continuous function on $\overline{\Omega}$ which is analytic on Ω. If $|g(z)| \le M$ for all $z \in \overline{\Omega}$ while $|g(z)| \le \epsilon$ for all $z \in D(z_0, r) \cap \partial\Omega$, then $|g(z_0)| \le \sqrt{\epsilon M}$.

Proof. Assume without loss of generality that $z_0 = 0$. Put $U = \Omega \cap (-\Omega) \cap D(0, r)$. (This is the shaded region in Figure 5.3.2.) Define h on $\overline{U}$ by $h(z) = g(z)g(-z)$. We claim first that $\overline{U} \subseteq \overline{\Omega} \cap (-\overline{\Omega}) \cap D(0, r)$. For by general properties of the closure operation, $\overline{U} \subseteq \overline{\Omega} \cap (-\overline{\Omega}) \cap \overline{D}(0, r)$. Thus it is enough to show that if $z \in \partial\overline{D}(0, r)$, that is, $|z| = r$, then $z \notin \overline{\Omega}$ or $z \notin (-\overline{\Omega})$. But this is a consequence of our assumption that $C(0, r)$ has an arc lying in the complement of $\overline{\Omega}$ that subtends an angle greater than π at $z_0 = 0$, from which it follows that the entire circle $C(0,r)$ lies in the complement of $\overline{\Omega} \cap (-\overline{\Omega})$. Consequently, we conclude that if $z \in \partial U$, then $z \in \partial\Omega \cap D(0,r)$ or $z \in \partial(-\Omega) \cap D(0,r)$. Therefore, for all $z \in \partial U$, hence for all $z \in \overline{U}$ by the maximum principle, we have

$$|h(z)| = |g(z)||g(-z)| \leq \epsilon M.$$

In particular, $|h(0)| = |g(0)|^2 \leq \epsilon M$, and the lemma is proved. ♣

We now proceed to prove Theorem 5.3.5. Assume the statement of the theorem is false. This implies that there is a sequence $\{z_n\}$ in Ω converging to β, and distinct complex numbers w_1 and w_2 of modulus 1, such that $f(z_{2j-1}) \to w_1$ while $f(z_{2j}) \to w_2$. (Proof: There is a sequence $\{z_n\}$ in Ω such that $z_n \to \beta$ while $\{f(z_n)\}$ does not converge. But $\{f(z_n)\}$ is bounded, hence it has at least two convergent subsequences with different limits w_1, w_2 and with $|w_1| = |w_2| = 1$.) Let p be the midpoint of the positively oriented arc of ∂D from w_1 to w_2. Choose points a and b, interior to this arc, equidistant from p and close enough to p for Figure 5.3.3 to obtain. Let γ and $\{t_n\}$ be as in the definition

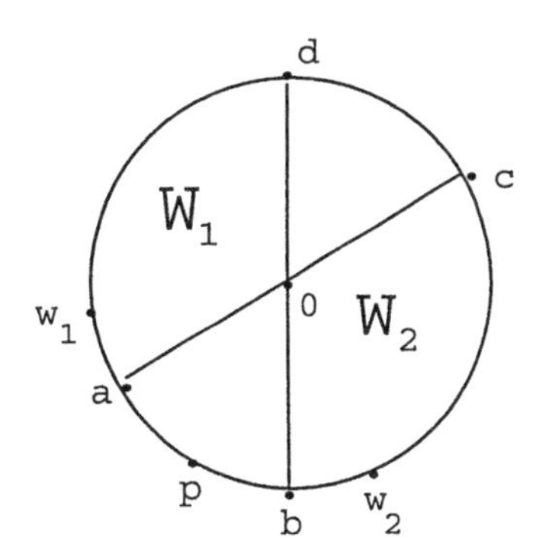

Figure 5.3.3

of simple boundary point. No loss of generality results if we assume that $f(z_{2j-1}) \in W_1$ and $f(z_{2j}) \in W_2$ for all j, and that $|f(\gamma(t))| > 1/2$ for all t. Since $f(\gamma(t_{2j-1})) \in W_1$ and $f(\gamma(t_{2j})) \in W_2$ for each j, there exist x_j and y_j with $t_{2j-1} < x_j < y_j < t_{2j}$ such that one of the following holds:

(1) $f(\gamma(x_j)) \in (0,a)$, $f(\gamma(y_j)) \in (0,b)$, and $f(\gamma(t))$ is in the open sector $a0ba$ for all t such that $x_j < t < y_j$, or

(2) $f(\gamma(x_j)) \in (0,d)$, $f(\gamma(y_j)) \in (0,c)$, and $f(\gamma(t))$ is in the open sector $d0cd$ for all t such that $x_j < t < y_j$.

See Figure 5.3.4 for this and details following. Thus (1) holds for infinitely many j or (2) holds for infinitely many j. Assume that the former is the case, and let J be the set of all j such that (1) is true. For $j \in J$ define γ_j on $[0, 1]$ by

$$\gamma_j(t) = \begin{cases} \frac{t}{x_j} f(\gamma(x_j)), & 0 \le t \le x_j, \\ f(\gamma(t)), & x_j \le t \le y_j, \\ \frac{1-t}{1-y_j} f(\gamma(y_j)), & y_j \le t \le 1. \end{cases}$$

Thus γ_j is the closed path whose trajectory γ_j^* consists of

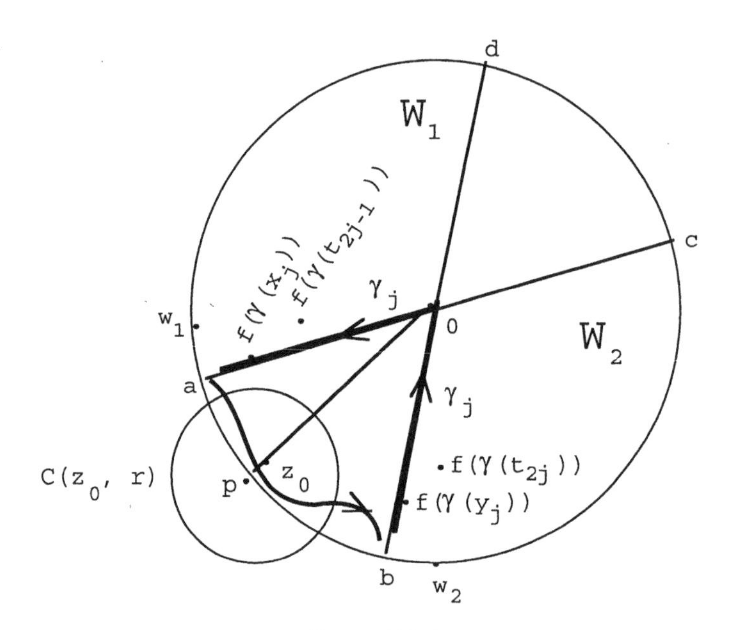

Figure 5.3.4

$$[0, f(\gamma(x_j))] \cup \{f(\gamma(t) : x_j \le t \le y_j\} \cup [f(\gamma(y_j)), 0].$$

Let Ω_j be that component of $\mathbb{C} \setminus \gamma_j^*$ such that $\frac{1}{2}p \in \Omega_j$. Then $\partial\Omega_j \subseteq \gamma_j^*$. Furthermore, $\Omega_j \subseteq D$, for if we compute the index $n(\gamma_j, \frac{1}{2}p)$, we get 1 because $|\gamma_j(t)| > \frac{1}{2}$ for $x_j \le t \le y_j$, while the index of any point in $\mathbb{C} \setminus D$ is 0. Let r be a positive number with $r < \frac{1}{2}|a - b|$ and choose a point z_0 on the open radius $(0, p)$ so close to p that the circle $C(0, r)$ meets the complement of $\overline{D}$ in an arc of length greater than πr. For sufficiently large $j \in J$, $|f(\gamma(t))| > |z_0|$ for all $t \in [t_{2j-1}, t_{2j}]$; so for these j we have $z_0 \in \Omega_j$. Further, if $z \in \partial\Omega_j \cap D(z_0, r)$, then $z \in \{f(\gamma(t)) : t_{2j-1} \le t \le t_{2j}\}$ and hence $f^{-1}(z) \in \gamma([t_{2j-1}, t_{2j}]\}$. Define

$$\epsilon_j = \sup\{|f^{-1}(z) - \beta| : z \in \partial\Omega_j \cap D(z_0, r)\} \le \sup\{|\gamma(t) - \beta| : t \in [t_{2j-1}, t_{2j}]\}$$

and

$$M = \sup\{|f^{-1}(z) - \beta| : z \in D\}.$$

Since $M \geq \sup\{|f^{-1}(z) - \beta| : z \in \overline{\Omega}_j\}$, Lemma 5.3.6 implies that $|f^{-1}(z_0) - \beta| \leq \sqrt{\epsilon_j M}$. Since ϵ_j can be made as small as we please by taking $j \in J$ sufficiently large, we have $f^{-1}(z_0) = \beta$. This is a contradiction since $f^{-1}(z_0) \in \Omega$, and the proof is complete. ♣

We next show that if β_1 and β_2 are simple boundary points and $\beta_1 \neq \beta_2$, then any continuous extension f to $\Omega \cup \{\beta_1, \beta_2\}$ that results from the previous theorem is one-to-one, that is, $f(\beta_1) \neq f(\beta_2)$. The proof requires a lemma that expresses the area of the image of a region under a conformal map as an integral. (Recall that a one-to-one analytic function is conformal.)

5.3.7 Lemma

Let g be a conformal map of an open set Ω. Then the area (Jordan content) of $g(\Omega)$ is $\int\int_\Omega |g'|^2\, dx\, dy$.

Proof. Let $g = u + iv$ and view g as a transformation from $\Omega \subseteq \mathbb{R}^2$ into $\mathbb{R}^2$. Since g is analytic, u and v have continuous partial derivatives (of all orders). Also, the Jacobian determinant of the transformation g is

$$\frac{\partial(u,v)}{\partial(x,y)} = \begin{vmatrix} \frac{\partial u}{\partial x} & \frac{\partial u}{\partial y} \\ \frac{\partial v}{\partial x} & \frac{\partial v}{\partial y} \end{vmatrix} = \frac{\partial u}{\partial x}\frac{\partial v}{\partial y} - \frac{\partial v}{\partial x}\frac{\partial u}{\partial y} = (\frac{\partial u}{\partial x})^2 + (\frac{\partial v}{\partial x})^2 = |g'|^2$$

by the Cauchy-Riemann equations. Since the area of $g(\Omega)$ is $\int\int_\Omega \frac{\partial(u,v)}{\partial(x,y)}\, dx\, dy$, the statement of the lemma follows. ♣

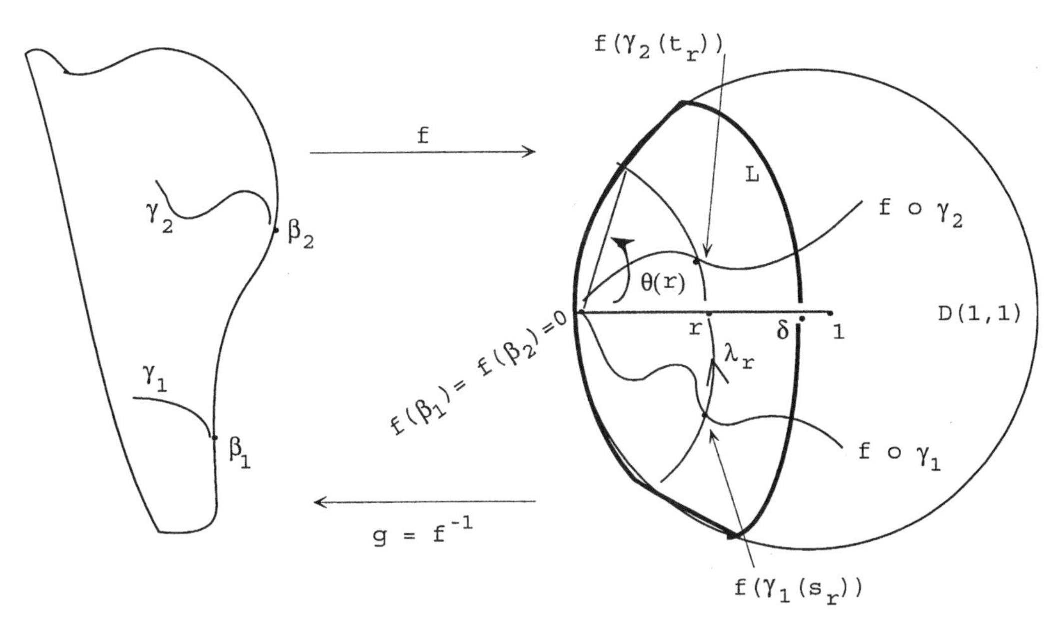

Figure 5.3.5

5.3.8 Theorem

Let Ω be a bounded, simply connected region and f a conformal map of Ω onto D. If β_1 and β_2 are distinct simple boundary points of Ω and f is extended continuously to $\Omega \cup \{\beta_1, \beta_2\}$, then $f(\beta_1) \neq f(\beta_2)$.

Proof. Assume that β_1 and β_2 are simple boundary points of Ω and $f(\beta_1) = f(\beta_2)$. We will show that $\beta_1 = \beta_2$. It will simplify the notation but result in no loss of generality if we replace D by $D(1,1)$ and assume that $f(\beta_1) = f(\beta_2) = 0$.

Since β_1 and β_2 are simple boundary points, for $j = 1, 2$ there are curves γ_j in $\Omega \cup \{\beta_j\}$ such that $\gamma_j([0,1)) \subseteq \Omega$ and $\gamma_j(1) = \beta_j$. Put $g = f^{-1}$. By continuity, there exists $\tau < 1$ such that $\tau < s, t < 1$ implies

$$|\gamma_2(t) - \gamma_1(s)| \geq \frac{1}{2}|\beta_2 - \beta_1| \tag{1}$$

and there exists δ, $0 < \delta < 1$, such that for $t \leq \tau$ we have $f(\gamma_j(t)) \notin \overline{D}(0, \delta), j = 1, 2$. Also, for each r such that $0 < r \leq \delta$, we can choose s_r and $t_r > \tau$ such that $f(\gamma_1(s_r))$ and $f(\gamma_2(t_r))$ meet the circle $C(0, r)$; see Figure 5.3.5. Let $\theta(r)$ be the principal value of the argument of the point of intersection in the upper half plane of $C(0, r)$ and $C(1, 1)$. Now $g(f(\gamma_2(t_r))) - g(f(\gamma_1(s_r)))$ is the integral of g' along the arc λ_r of $C(0, r)$ from $f(\gamma_1(s_r))$ to $f(\gamma_2(t_r))$. It follows from this and (1) that

$$\begin{aligned}
\frac{1}{2}|\beta_2 - \beta_1| &\leq |\gamma_2(t_r) - \gamma_1(s_r)| \\
&= |g(f(\gamma_2(t_r))) - g(f(\gamma_1(s_r)))| \\
&= |\int_{\lambda_r} g'(z)\, dz| \\
&\leq \int_{-\theta(r)}^{\theta(r)} |g'(re^{i\theta})| r\, d\theta.
\end{aligned} \tag{2}$$

(Note: The function $\theta \to |g'(re^{i\theta})|$ is positive and continuous on the open interval $(-\theta(r), \theta(r))$, but is not necessarily bounded. Thus the integral in (2) may need to be treated as an improper Riemann integral. In any case (2) remains correct and the calculations that follow are also seen to be valid.)

Squaring in (2) and applying the Cauchy-Schwarz inequality for integrals we get

$$\frac{1}{4}|\beta_2 - \beta_1|^2 \leq 2\theta(r) r^2 \int_{-\theta(r)}^{\theta(r)} |g'(re^{i\theta})|^2\, d\theta.$$

(The factor $2\theta(r)$ comes from integrating $1^2 d\theta$ from $-\theta(r)$ to $\theta(r)$.) Since $\theta(r) \leq \pi/2$, we have

$$\frac{|\beta_2 - \beta_1|^2}{4\pi r} \leq r \int_{-\theta(r)}^{\theta(r)} |g'(re^{i\theta})|^2\, d\theta. \tag{3}$$

Now integrate the right hand side of (3) with respect to r from $r = 0$ to $r = \delta$. We obtain

$$\int_0^{\delta} \int_{-\theta(r)}^{\theta(r)} |g'(re^{i\theta})|^2 r\, d\theta\, dr \leq \int\int_L |g'(x + iy)|^2\, dx\, dy$$

where L is the lens-shaped open set whose boundary is formed by arcs of $C(0, \delta)$ and $C(1,1)$; see Figure 5.3.5. By (5.3.7), $\int\int_L |g'|^2\,dx\,dy$ is the area (or Jordan content) of $g(L)$. Since $g(L) \subseteq \Omega$ and Ω is bounded, $g(L)$ has finite area. But the integral from 0 to δ of the left hand side of (3) is $+\infty$ unless $\beta_1 = \beta_2$. Thus $f(\beta_1) = f(\beta_2)$ implies that $\beta_1 = \beta_2$. ♣

We can now prove that $f : \Omega \to D$ extends to a homeomorphism of $\overline{\Omega}$ and $\overline{D}$ if *every* boundary point of Ω is simple.

5.3.9 Theorem

Suppose Ω is a bounded, simply connected region with the property that every boundary point of Ω is simple. If $f : \Omega \to D$ is a conformal equivalence, then f extends to a homeomorphism of $\overline{\Omega}$ onto $\overline{D}$.

Proof. By Theorem 5.3.5, for each $\beta \in \partial\Omega$ we can extend f to $\Omega \cup \{\beta\}$ so that f is continuous on $\Omega \cup \{\beta\}$. . Assume this has been done. Thus (the extension of) f is a map of $\overline{\Omega}$ into $\overline{D}$, and Theorem 5.3.8 implies that f is one-to-one. Furthermore, f is continuous at each point $\beta \in \partial\Omega$, for if $\{z_n\}$ is any sequence in $\overline{\Omega}$ such that $z_n \to \beta$ then for each n there exists $w_n \in \Omega$ with $|z_n - w_n| < 1/n$ and also $|f(z_n) - f(w_n)| < 1/n$, by Theorem 5.3.5. But again by (5.3.5), $f(w_n) \to f(\beta)$ because $w_n \to \beta$ and $w_n \in \Omega$. Hence $f(z_n) \to \beta$, proving that f is continuous on $\overline{\Omega}$. Now $D \subseteq f(\overline{\Omega}) \subseteq \overline{D}$, and since $f(\overline{\Omega})$ is compact, hence closed, $f(\overline{\Omega}) = \overline{D}$. Consequently, f is a one-to-one continuous map of $\overline{\Omega}$ onto $\overline{D}$, from which it follows that f^{-1} is also continuous. ♣

Theorem 5.3.9 has various applications, and we will look at a few of these in the sequel.

In the proof of (5.3.8), we used the fact that for open subsets $L \subseteq D$,

$$\int\int_L |g'|^2\,dx\,dy \tag{1}$$

is precisely the area of $g(L)$, where g is a one-to-one analytic function on D. Suppose that $g(z) = \sum_{n=0}^{\infty} a_n z^n$, $z \in D$. Then $g'(z) = \sum_{n=1}^{\infty} n a_n z^{n-1}$. Now in polar coordinates the integral in (1), with L replaced by D, is given by

$$\int\int_D |g'(re^{i\theta})|^2 r\,dr\,d\theta = \int_0^1 r\,dr \int_{-\pi}^{\pi} |g'(re^{i\theta})|^2\,d\theta.$$

But for $0 \le r < 1$,

$$\begin{aligned} |g'(re^{i\theta})|^2 &= g'(re^{i\theta})\overline{g'(re^{i\theta})} \\ &= \sum_{n=1}^{\infty} n a_n r^{n-1} e^{i(n-1)\theta} \sum_{m=1}^{\infty} m \overline{a_m} r^{m-1} e^{-i(m-1)\theta} \\ &= \sum_{j=1}^{\infty} \sum_{m+n=j} nm a_n \overline{a_m} r^{m+n-2} e^{i(n-m)\theta}. \end{aligned} \tag{2}$$

Since

$$\int_{-\pi}^{\pi} e^{i(n-m)\theta}\,d\theta = \begin{cases} 2\pi, & n = m \\ 0, & n \neq m \end{cases}$$

and the series in (2) converges uniformly in θ, we can integrate term by term to get

$$\int_{-\pi}^{\pi} |g'(re^{i\theta})|^2\, d\theta = 2\pi \sum_{k=1}^{\infty} k^2 |a_k|^2 r^{2k-2}.$$

Multiplying by r and integrating with respect to r, we have

$$\int_0^1 r\, dr \int_{-\pi}^{\pi} |g'(re^{i\theta})|^2\, d\theta = \lim_{\rho\to 1^-} 2\pi \sum_{k=1}^{\infty} \frac{k^2|a_k|^2 \rho^{2k}}{2k}$$

If this limit, which is the area of $g(D)$, is finite, then

$$\pi \sum_{k=1}^{\infty} k|a_k|^2 < \infty.$$

We have the following result.

5.3.10 Theorem

Suppose $g(z) = \sum_{n=0}^{\infty} a_n z^n$ is one-to-one and analytic on D. If $g(D)$ has finite area, then $\sum_{n=1}^{\infty} n|a_n|^2 < \infty$.

Now we will use the preceding result to study the convergence of the power series for $g(z)$ when $|z| = 1$. Here is a result on uniform convergence.

5.3.11 Theorem

Let $g(z) = \sum_{n=0}^{\infty} a_n z^n$ be a one-to-one analytic map of D onto a bounded region Ω such that every boundary point of Ω is simple. Then the series $\sum_{n=0}^{\infty} a_n z^n$ converges uniformly on $\overline{D}$ to (the extension of) g on $\overline{D}$.

Proof. By the maximum principle, it is sufficient to show that $\sum_{n=0}^{\infty} a_n z^n$ converges uniformly to $g(z)$ for $|z| = 1$; in other words, $\sum_{n=0}^{\infty} a_n e^{in\theta}$ converges uniformly in θ to $g(e^{i\theta})$. So let $\epsilon > 0$ be given. Since g is uniformly continuous on $\overline{D}$, $|g(e^{i\theta}) - g(re^{i\theta})| \to 0$ uniformly in θ as $r \to 1^-$. If m is any positive integer and $0 < r < 1$, we have

$$|g(e^{i\theta}) - \sum_{n=0}^{m} a_n e^{in\theta}| \le |g(e^{i\theta}) - g(re^{i\theta})| + |g(re^{i\theta}) - \sum_{n=0}^{m} a_n e^{in\theta}|.$$

The first term on the right hand side tends to 0 as $r \to 1^-$, uniformly in θ, so let us consider the second term. If k is any positive integer less than m, then since $g(re^{i\theta}) =$

$\sum_{n=0}^{\infty} a_n r^n e^{in\theta}$, we can write the second term as

$$
\begin{aligned}
&|\sum_{n=0}^{k} a_n(r^n-1)e^{in\theta} + \sum_{n=k+1}^{m} a_n(r^n-1)e^{in\theta} + \sum_{n=m+1}^{\infty} a_n r^n e^{in\theta}| \\
&\le \sum_{n=0}^{k}(1-r^n)|a_n| + \sum_{n=k+1}^{m}(1-r^n)|a_n| + \sum_{n=m+1}^{\infty}|a_n|r^n \\
&\le \sum_{n=0}^{k} n(1-r)|a_n| + \sum_{n=k+1}^{m} n(1-r)|a_n| + \sum_{n=m+1}^{\infty}|a_n|r^n
\end{aligned}
$$

(since $\frac{1-r^n}{1-r} = 1+r+\cdots+r^{n-1} < n$). We continue the bounding process by observing that in the first of the three above terms, we have $n \le k$. In the second term, we write $n(1-r)|a_n| = [\sqrt{n}(1-r)][\sqrt{n}|a_n|]$ and apply Schwarz's inequality. In the third term, we write $|a_n|r^n = [\sqrt{n}|a_n|][r^n/\sqrt{n}]$ and again apply Schwarz's inequality. Our bound becomes

$$
\begin{aligned}
&k(1-r)\sum_{n=0}^{k}|a_n| + \left\{\sum_{n=k+1}^{m} n(1-r)^2\right\}^{1/2}\left\{\sum_{n=k+1}^{m} n|a_n|^2\right\}^{1/2} \\
&+\left\{\sum_{n=m+1}^{\infty} n|a_n|^2\right\}^{1/2}\left\{\sum_{n=m+1}^{\infty}\frac{r^{2n}}{n}\right\}^{1/2}. \qquad (1)
\end{aligned}
$$

Since $\sum_{n=0}^{\infty} n|a_n|^2$ is convergent, there exists $k > 0$ such that $\{\sum_{n=k+1}^{\infty} n|a_n|^2\}^{1/2} < \epsilon/3$. Fix such a k. For $m > k$ put $r_m = (m-1)/m$. Now the first term in (1) is less than $\epsilon/3$ for m sufficiently large and $r = r_m$. Also, since $\{\sum_{n=k+1}^{m} n(1-r_m)^2\}^{1/2} = \{\sum_{n=k+1}^{m} n(1/m)^2\}^{1/2} = (1/m)\{\sum_{n=k+1}^{m} n\}^{1/2} < (1/m)\{m(m+1)/2\}^{1/2} < 1$, the middle term in (1) is also less than $\epsilon/3$. Finally, consider

$$
\left\{\sum_{n=m+1}^{\infty}\frac{r_m^{2n}}{n}\right\}^{1/2} \le \left\{\frac{r_m^{2(m+1)}}{m+1}\sum_{n=0}^{\infty} r_m^{2n}\right\}^{1/2} \le \left\{\frac{1}{m+1}\sum_{n=0}^{\infty} r_m^{n}\right\}^{1/2}
$$

which evaluates to

$$
\left\{\frac{1}{m+1}\frac{1}{1-r_m}\right\}^{1/2} = \left\{\frac{m}{m+1}\right\}^{1/2} < 1.
$$

Thus the last term in (1) is also less than $\epsilon/3$ for all sufficiently large $m > k$ and $r = r_m = (m-1)/m$. Thus $|g(e^{i\theta}) - \sum_{n=0}^{m} a_n e^{in\theta}| \to 0$ uniformly in θ as $m \to \infty$. ♣

The preceding theorem will be used to produce examples of uniformly convergent power series that are not absolutely convergent. That is, power series that converge uniformly, but to which the Weierstrass M-test does not apply. One additional result will be needed.

5.3.12 Theorem

Suppose $f(z) = \sum_{n=0}^{\infty} a_n z^n$ for $z \in D$, and $\sum_{n=0}^{\infty} |a_n| < +\infty$. Then for each θ,

$$\int_0^{1^-} |f'(re^{i\theta})|\, dr = \lim_{\rho \to 1^-} \int_0^{\rho} |f'(re^{i\theta})|\, dr < +\infty.$$

Proof. If $0 \le r \le \rho < 1$ then for any θ we have $f'(re^{i\theta}) = \sum_{n=1}^{\infty} n a_n r^{n-1} e^{i(n-1)\theta}$. Thus

$$\int_0^{\rho} |f'(re^{i\theta})|\, dr \le \sum_{n=1}^{\infty} |a_n| \rho^n \le \sum_{n=1}^{\infty} |a_n| < +\infty. \clubsuit$$

5.3.13 Remark

For each θ, $\int_0^1 |f'(re^{i\theta})|\, dr$ is the length of the image under f of the radius $[0, e^{i\theta}]$ of $\overline{D}$. For if $\gamma(r) = re^{i\theta}, 0 \le r \le 1$, then the length of $f \circ \gamma$ is given by

$$\int_0^1 |(f \circ \gamma)'(r)|\, dr = \int_0^1 |f'(re^{i\theta})e^{i\theta}|\, dr = \int_0^1 |f'(re^{i\theta})|\, dr.$$

Thus in geometric terms, the conclusion of (5.3.12) is that f maps every radius of $\overline{D}$ onto an arc of finite length.

We can now give a method for constructing uniformly convergent power series that are not absolutely convergent. Let Ω be the bounded, connected, simply connected region that appears in Figure 5.3.6. Then each boundary point of Ω is simple with the possible exception of 0, and the following argument shows that 0 is also a simple boundary point. Let $\{z_n\}$ be any sequence in Ω such that $z_n \to 0$. For $n = 1, 2, \ldots$ put $t_n = (n-1)/n$. Then for each n there is a polygonal path $\gamma_n : [t_n, t_{n+1}] \to \Omega$ such that $\gamma_n(t_n) = z_n, \gamma_n(t_{n+1}) = z_{n+1}$, and such that for $t_n \le t \le t_{n+1}, \operatorname{Re} \gamma_n(t)$ is between $\operatorname{Re} z_n$ and $\operatorname{Re} z_{n+1}$. If we define $\gamma = \cup \gamma_n$, then γ is a continuous map of $[0, 1)$ into Ω, and $\gamma(t) = \gamma_n(t)$ for $t_n \le t \le t_{n+1}$. Furthermore, $\gamma(t) \to 0$ as $t \to 1^-$. Thus by definition, 0 is simple boundary point of Ω.

Hence by (5.3.9) and the Riemann mapping theorem (5.2.2), there is a homeomorphism f of $\overline{D}$ onto $\overline{\Omega}$ such that f is analytic on D. Write $f(z) = \sum a_n z^n,\ z \in D$. By (5.3.11), this series converges uniformly on $\overline{D}$. Now let $e^{i\theta}$ be that point in ∂D such that $f(e^{i\theta}) = 0$. Since f is a homeomorphism, f maps the radius of $\overline{D}$ that terminates at $e^{i\theta}$ onto an arc in $\Omega \cup \{0\}$ that terminates at 0. Further, the image arc in $\Omega \cup \{0\}$ cannot have finite length. Therefore by (5.3.12) we have $\sum |a_n| = +\infty$.

Additional applications of the results in this section appear in the exercises.

Problems

1. Let Ω be a bounded simply connected region such that every boundary point of Ω is simple. Prove that the Dirichlet problem is solvable for Ω. That is, if u_0 is a real-valued continuous function on $\partial\Omega$, then u_0 has a continuous extension u to $\overline{\Omega}$ such that u is harmonic on Ω.

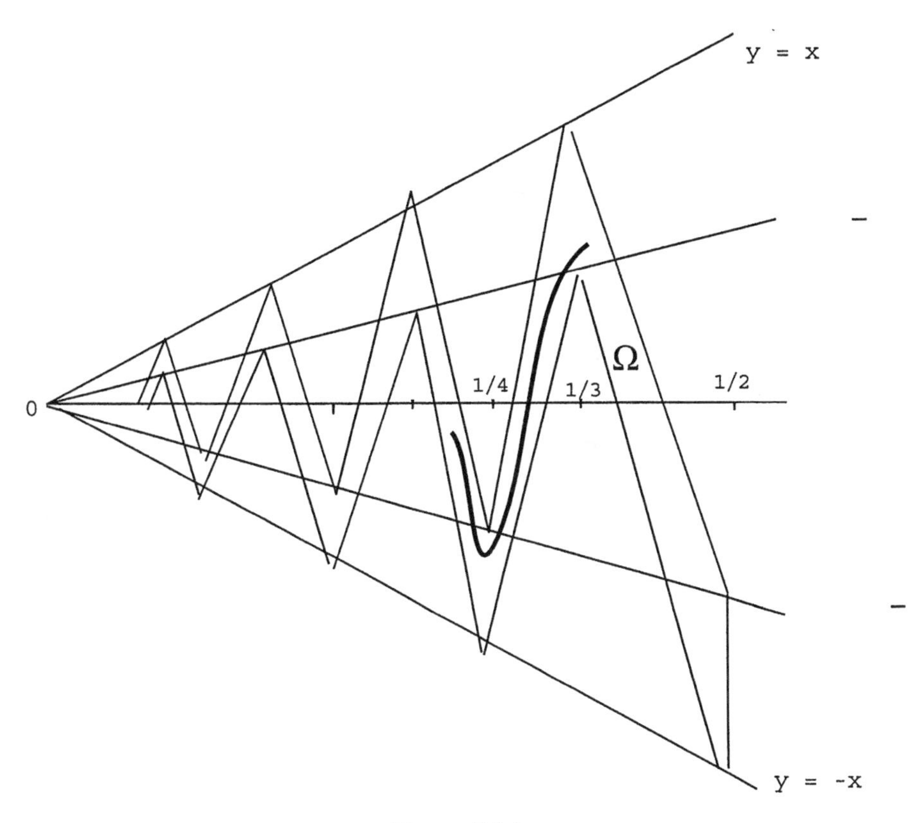

Figure 5.3.6

2. Let $\Omega = \{x + iy : 0 < x < 1 \text{ and } -x^2 < y < x^2\}$. Show that
   (a) The identity mapping $z \to z$ has a continuous argument u on $\overline{\Omega}$ (necessarily harmonic on Ω).
   (b) There is a homeomorphism f of $\overline{D}$ onto $\overline{\Omega}$ which is analytic on D.
   (c) $u \circ f$ is continuous on $\overline{D}$ and harmonic on D.
   (d) No harmonic conjugate V for $u \circ f$ can be bounded on D.

3. Let Ω be a bounded, simply connected region such that every boundary point of Ω is simple. Show that every zero-free continuous function f on $\overline{\Omega}$ has a continuous logarithm g. In addition, show that if f is analytic on Ω, then so is g.

References

1. C. Carathéodory, "Conformal Representation," 2nd ed., Cambridge Tracts in Mathematics and Mathematical Physics no. 28, Cambridge University Press, London, 1952.

2. W.P. Novinger, "An Elementary Approach to the Problem of Extending Conformal Maps to the Boundary," American Mathematical Monthly, 82(1975), 279-282.

3. W.P. Novinger, "Some Theorems from Geometric Function Theory: Applications," American Mathematical Monthly, 82(1975), 507-510.

4. W.A. Veech, "A Second Course in Complex Analysis," Benjamin, New York, 1967.

Chapter 6

Factorization of Analytic Functions

In this chapter we will consider the problems of factoring out the zeros of an analytic function f on a region Ω (à la polynomials), and of decomposing a meromorphic function (à la partial fractions for rational functions). Suppose f is analytic on a region Ω and $f \not\equiv 0$. What can be said about $Z(f)$? Theorem 2.4.8, the identity theorem, asserts that $Z(f)$ has no limit point in Ω. It turns out that no more can be said in general. That is, if A is any subset of Ω with no limit point in Ω, then there exists $f \in A(\Omega)$ whose set of zeros is precisely A. Furthermore, we can prescribe the order of the zero which f shall have at each point of A. Now if A is a finite subset of Ω, say $\{z_1, \dots, z_n\}$, and $m_1, \dots, m_n$ are the corresponding desired multiplicities, then the finite product

$$f(z) = (z - z_1)^{m_1} \cdots (z - z_n)^{m_n}$$

would be such a function. However, in general the construction of such an f is accomplished using *infinite products*, which we now study in detail.

6.1 Infinite Products

Let $\{z_n\}$ be a sequence of complex numbers and put $P_n = \prod_{k=1}^{n} z_k$, the n-th *partial product.* We say that the infinite product $\prod_{n=1}^{\infty} z_n$ *converges* if the sequence $\{P_n\}$ is convergent to a complex number P, and in this case we write $P = \prod_{n=1}^{\infty} z_n$.

This particular definition of convergence of infinite products is a natural one if the usual definition of convergence of infinite series is extended directly to products. Many textbook authors, however, find this approach objectionable, primarily for the following two reasons.

(a) If one of the factors is zero, then the product converges to zero, no matter what the other factors are, and a "correct" notion of convergence should presumably depend on *all* (but possibly finitely many) of the factors.

(b) It is possible for a product to converge to zero without *any* of the factors being zero, unlike the situation for a finite product.

Nevertheless, we have chosen to take the *naive* approach, and will deal with the above if and when they are relevant.

Note that if $P_n \to P \neq 0$, then $z_n = P_n/P_{n-1} \to P/P = 1$ as $n \to \infty$. Thus a necessary (but not sufficient) condition for convergence of the infinite product to a nonzero limit is that $z_n \to 1$.

A natural approach to the study of an infinite product is to formally convert the product into a sum by taking logarithms. In fact this approach is quite fruitful, as the next result shows.

6.1.1 Lemma

Suppose that $z_n \neq 0, n = 1, 2, \ldots$. Then $\prod_{n=1}^{\infty} z_n$ converges to a nonzero limit iff the series $\sum_{n=1}^{\infty} \operatorname{Log} z_n$ converges. (Recall that Log denotes the particular branch of the logarithm such that $-\pi \leq \operatorname{Im}(\operatorname{Log} z) < \pi$.)

Proof. Let $P_n = \prod_{k=1}^{n} z_k$ and $S_n = \sum_{k=1}^{n} \operatorname{Log} z_k$. If $S_n \to S$, then $P_n = e^{S_n} \to e^S \neq 0$. Conversely, suppose that $P_n \to P \neq 0$. Choose any θ such that $\arg_\theta$ is continuous at P (see Theorem 3.1.2). Then $\log_\theta P_n = \ln |P_n| + i \arg_\theta(P_n) \to \ln |P| + i \arg_\theta(P) = \log_\theta P$. Since $e^{S_n} = P_n$, we have $S_n = \log_\theta P_n + 2\pi i l_n$ for some integer l_n. But $S_n - S_{n-1} = \operatorname{Log} z_n \to \operatorname{Log} 1 = 0$. Consequently, $\log_\theta P_n - \log_\theta P_{n-1} + 2\pi i(l_n - l_{n-1}) \to 0$. Since $\log_\theta P_n - \log_\theta P_{n-1} \to \log_\theta P - \log_\theta P = 0$ and $l_n - l_{n-1}$ is an integer, it follows that $l_n - l_{n-1}$ is eventually zero. Therefore l_n is eventually a constant l. Thus $S_n \to \log_\theta P + 2\pi i l$. ♣

6.1.2 Lemma

If $a_n \geq 0$ for all n, then $\prod_{n=1}^{\infty}(1 + a_n)$ converges iff $\sum_{n=1}^{\infty} a_n$ converges.

Proof. Since $1 + x \leq e^x$, we have, for every $n = 1, 2, \ldots$,

$$a_1 + \cdots + a_n \leq (1 + a_1) \cdots (1 + a_n) \leq e^{a_1 + \cdots + a_n}. \quad ♣$$

Lemma 6.1.2 suggests the following useful notion of absolute convergence for infinite products.

6.1.3 Definition

The infinite product $\prod_{n=1}^{\infty}(1 + z_n)$ is said to converge *absolutely* if $\prod_{n=1}^{\infty}(1 + |z_n|)$ converges. Thus by (6.1.2), absolute convergence of $\prod_{n=1}^{\infty}(1 + z_n)$ is equivalent to absolute convergence of the series $\sum_{n=1}^{\infty} z_n$.

With this definition of absolute convergence, we can state and prove a result analogous to a well known property of infinite series.

6.1.4 Lemma

If the infinite product $\prod_{n=1}^{\infty}(1+z_n)$ converges absolutely, then it converges.

Proof. By Lemma 6.1.2, convergence of $\prod_{n=1}^{\infty}(1+|z_n|)$ implies that of $\sum_{n=1}^{\infty}|z_n|$, hence $|z_n| \to 0$ in particular. So we can assume that $|z_n| < 1$ for all n. Now for $|z| < 1$, we have

$$\operatorname{Log}(1+z) = z - \frac{z^2}{2} + \frac{z^3}{3} - \frac{z^4}{4} + \cdots = zh(z)$$

where $h(z) = 1 - \frac{z}{2} + \frac{z^2}{3} - \frac{z^3}{4} + \cdots \to 1$ as $z \to 0$. Consequently, for $m \le p$,

$$|\sum_{n=m}^{p} \operatorname{Log}(1+z_n)| \le \sum_{n=m}^{p} |z_n||h(z_n)|.$$

Since $\{h(z_n) : n = 1, 2, \dots\}$ is a bounded set and $\sum_{n=1}^{\infty}|z_n|$ converges, it follows from the preceding inequality that $|\sum_{n=m}^{p}\operatorname{Log}(1+z_n)| \to 0$ as $m, p \to \infty$. Thus $\sum_{n=1}^{\infty}\operatorname{Log}(1+z_n)$ is convergent, which by (6.1.1) implies that $\prod_{n=1}^{\infty}(1+z_n)$ converges.

The preceding result may be combined with (6.1.2) to obtain a rearrangement theorem for absolutely convergent products.

6.1.5 Theorem

If $\prod_{n=1}^{\infty}(1+z_n)$ converges absolutely, then so does every rearrangement, and to the same limit. That is, if $\prod_{n=1}^{\infty}(1+|z_n|)$ converges and $P = \prod_{n=1}^{\infty}(1+z_n)$, then for every permutation $k \to n_k$ of the positive integers, $\prod_{k=1}^{\infty}(1+z_{n_k})$ also converges to P.

Proof. Since $\prod_{n=1}^{\infty}(1+|z_n|)$ converges, so does $\sum_{n=1}^{\infty}|z_n|$ by (6.1.2). But then every rearrangement of this series converges, so by (6.1.2) again, $\prod_{k=1}^{\infty}(1+|z_{n_k}|)$ converges. Thus it remains to show that $\prod_{k=1}^{\infty}(1+z_{n_k})$ converges to the same limit as does $\prod_{n=1}^{\infty}(1+z_n)$. To this end let $\epsilon > 0$ and for $j = 1, 2, \dots$, let Q_j be the j-th partial product of $\prod_{k=1}^{\infty}(1+z_{n_k})$. Choose N so large that $\sum_{n=N+1}^{\infty}|z_n| < \epsilon$ and J so large that $j \ge J$ implies that $\{1, 2, \dots, N\} \subseteq \{n_1, n_2, \dots, n_j\}$. (The latter is possible because $j \to n_j$ is a permutation of the positive integers.) Then for $j \ge J$ we have

$$\begin{aligned} |Q_j - P| &\le |Q_j - P_N| + |P_N - P| \\ &= |P_N||\prod_k (1+z_{n_k}) - 1| + |P_N - P| \end{aligned} \tag{1}$$

where the product is taken over those $k \le j$ such that $n_k > N$. Now for any complex numbers $w_1, \dots, w_n$ we have (by induction) $|\prod_{k=1}^{n}(1+w_k) - 1| \le \prod_{k=1}^{n}(1+|w_k|) - 1$. Using this, we get from (1) that

$$\begin{aligned} |Q_j - P| &\le |P_N|(\prod_k (1+|z_{n_k}|) - 1) + |P_N - P| \\ &\le |P_N|(e^{\epsilon} - 1) + |P_N - P|. \end{aligned}$$

But the right side of the above inequality can be made as small as we wish by choosing ϵ sufficiently small and N sufficiently large. Therefore $Q_j \to P$ also, and the proof is complete. ♣

6.1.6 Proposition

Let $g_1, g_2, \ldots$ be a sequence of bounded complex-valued functions, each defined on a set S. If the series $\sum_{n=1}^{\infty} |g_n|$ converges uniformly on S, then the product $\prod_{n=1}^{\infty}(1+g_n)$ converges absolutely and uniformly on S. Furthermore, if $f(z) = \prod_{n=1}^{\infty}(1+g_n(z)), z \in S$, then $f(z) = 0$ for some $z \in S$ iff $1 + g_n(z) = 0$ for some n.

Proof. Absolute convergence of the product follows from (6.1.2). If $\sum |g_n|$ converges uniformly on S, there exists N such that $n \geq N$ implies $|g_n(z)| < 1$ for all $z \in S$. Now for any $r \geq N$,

$$\prod_{n=1}^{r}(1+g_n(z)) = \prod_{n=1}^{N-1}(1+g_n(z)) \prod_{n=N}^{r}(1+g_n(z)).$$

As in the proof of (6.1.4), with the same h and with $m, p \geq N$,

$$|\sum_{n=m}^{p} \operatorname{Log}(1+g_n(z))| \leq \sum_{n=m}^{p} |g_n(z)||h(g_n(z))| \to 0$$

uniformly on S as $m, p \to \infty$. Therefore $\sum_{n=N}^{\infty} \operatorname{Log}(1+g_n(z))$ converges uniformly on S. Since the functions $g_N, g_{N+1}, \ldots$ are bounded on S, it follows that the series $\sum_{n=N}^{\infty} |g_n(z)||h(g_n(z))|$ is bounded on S and thus by the above inequality, the same is true of $\sum_{n=N}^{\infty} \operatorname{Log}(1+g_n(z))$. However, the exponential function is uniformly continuous on bounded subsets of $\mathbb{C}$, so we may infer that

$$\exp\left\{\sum_{n=N}^{r} \operatorname{Log}(1+g_n(z))\right\} \to \exp\left\{\sum_{n=N}^{\infty} \operatorname{Log}(1+g_n(z))\right\} \neq 0$$

uniformly on S as $r \to \infty$. This proves uniform convergence on S of $\prod_{n=N}^{\infty}(1+g_n(z))$. Now $1 + g_n(z)$ is never 0 on S for $n \geq N$, so if $f(z) = \prod_{n=1}^{\infty}(1+g_n(z))$, then $f(z) = 0$ for some $z \in S$ iff $1 + g_n(z) = 0$ for some $n < N$. ♣

Remark

The product $\prod_{n=1}^{\infty}(1+|g_n|)$ also converges uniformly on S, as follows from the inequality

$$\prod_{n=m}^{p}(1+|g_n|) \leq \exp\left\{\sum_{n=m}^{p} |g_n|\right\}$$

or by applying (6.1.6) to $|g_1|, |g_2|, \ldots$.

Proposition (6.1.6) supplies the essential ingredients for an important theorem on products of analytic functions.

6.1.7 Theorem

Let $f_1, f_2, \ldots$ be analytic on Ω. If $\sum_{n=1}^{\infty} |f_n - 1|$ converges uniformly on compact subsets of Ω, then $f(z) = \prod_{n=1}^{\infty} f_n(z)$ defines a function f that is analytic on Ω. Furthermore, for any $z \in \Omega$ we have $f(z) = 0$ iff $f_n = 0$ for some n.

Proof. By (6.1.6) with $g_n = f_n - 1$, the product $\prod_{n=1}^{\infty} f_n(z)$ converges uniformly on compact subsets of Ω, hence f is analytic on Ω. The last statement of the theorem is also a direct consequence of (6.1.6). ♣

Problems

1. Let $f_1, f_2, \dots$ and f be as in Theorem 6.1.7. Assume in addition that no f_n is identically zero on any component of Ω. Prove that for each $z \in \Omega, m(f,z) = \sum_{n=1}^{\infty} m(f_n, z)$. (Recall that $m(f,z)$ is the order of the zero of f at z; $m(f,z) = 0$ if $f(z) \neq 0$.)

2. Show that $-\ln(1-x) = x + g(x)x^2, |x| < 1$, where $g(x) \to 1/2$ as $x \to 0$. Conclude that if $a_1, a_2, \dots$ are real numbers and $\sum_{n=1}^{\infty} a_n$ converges, then the infinite product $\prod_n (1 - a_n)$ converges to a nonzero limit iff $\sum_{n=1}^{\infty} a_n^2 < \infty$. Also, if $\sum_{n=1}^{\infty} a_n^2 < \infty$, then $\prod_n (1-a_n)$ converges to a nonzero limit iff $\sum_{n=1}^{\infty} a_n$ converges.

3. Determine whether or not the following infinite products are convergent.
(a) $\prod_n (1 - 2^{-n})$, (b) $\prod_n (1 - \frac{1}{n+1})$, (c) $\prod_n (1 + \frac{(-1)^n}{\sqrt{n}})$, $\prod_n (1 - \frac{1}{n^2})$.

4. (a) Give an example of an infinite product $\prod_n (1 + a_n)$ such that $\sum a_n$ converges but $\prod_n (1+a_n)$ diverges.
(b) Give an example of an infinite product $\prod_n (1 + a_n)$ such that $\sum a_n$ diverges but $\prod_n (1 + a_n)$ converges to a nonzero limit.

5. Show that the following infinite products define entire functions.
(a) $\prod_{n=1}^{\infty} (1 + a^n z)$, $|a| < 1$, (b) $\prod_{n \in \mathbb{Z}, n \neq 0} (1 - z/n) e^{z/n}$,

(c) $\prod_{n=2}^{\infty} [1 + \frac{z}{n(\ln n)^2}]$.

6. Criticize the following argument. We know that $\prod_n (1+z_n)$ converges to a nonzero limit iff $\sum_n \text{Log}(1 + z_n)$ converges. The Taylor expansion of $\text{Log}(1 + z)$ yields $\text{Log}(1 + z) = zg(z)$, where $g(z) \to 1$ as $z \to 0$. If $z_n \to 0$, then $g(z_n)$ will be arbitrarily close to 1 for large n, and thus $\sum_n z_n g(z_n)$ will converge iff $\sum_n z_n$ converges. Consequently, $\prod_n (1 + z_n)$ converges to a nonzero limit iff $\sum_n z_n$ converges.

6.2 Weierstrass Products

In this section we will consider the problem of constructing an analytic function f with a prescribed sequence of complex numbers as its set of zeros, as was discussed at the beginning of the chapter. A naive approach is simply to write $\prod_n (z - a_n)^{m_n}$ where $a_1, a_2, \dots$ is the sequence of (distinct) desired zeros and m_n is the specified multiplicity of the zero, that is, $m(f, a_n) = m_n$. But if $a_1, a_2, \dots$ is an infinite sequence, then the infinite product $\prod_n (z - a_n)^{m_n}$ need not converge. A more subtle approach is required, one that achieves convergence by using factors more elaborate than $(z - a_n)$. These "primary factors" were introduced by Weierstrass.

6.2.1 Definition

Define $E_0(z) = 1 - z$ and for $m = 1, 2, \ldots,$

$$E_m(z) = (1 - z)\exp\left[z + \frac{z^2}{2} + \cdots + \frac{z^m}{m}\right].$$

Note that if $|z| < 1$, then as $m \to \infty$, $E_m(z) \to (1 - z)\exp[-\operatorname{Log}(1 - z)] = 1$. Indeed, $E_m(z) \to 1$ uniformly on compact subsets of the unit disk D. Also, the E_m are entire functions, and E_m has a zero of order 1 at $z = 1$, and no other zeros.

6.2.2 Lemma

$|1 - E_m(z)| \leq |z|^{m+1}$ for $|z| \leq 1$.

Proof. If $m = 0$, equality holds, so assume $m \geq 1$. Then a calculation shows that

$$E'_m(z) = -z^m \exp\left[z + \frac{z^2}{2} + \cdots + \frac{z^m}{m}\right]$$

so that

$$(1 - E_m(z))' = z^m \exp\left[z + \frac{z^2}{2} + \cdots + \frac{z^m}{m}\right]. \tag{1}$$

This shows that the derivative of $1 - E_m$ has a zero of order m at 0. Since $1 - E_m(0) = 0$, it follows that $1 - E_m$ has a zero of order $m + 1$ at $z = 0$. Thus $(1 - E_m(z))/z^{m+1}$ has a removable singularity at 0 and so has a Taylor expansion $\sum_{n=0}^{\infty} a_n z^n$ valid everywhere on $\mathbb{C}$. Equation (1) shows also that the derivative of $1 - E_m$ has nonnegative Taylor coefficients and hence the same must be true of $(1 - E_m(z))/z^{m+1}$. Thus $a_n \geq 0$ for all n. Consequently,

$$\left|\frac{1 - E_m(z)}{z^{m+1}}\right| \leq \sum_{n=0}^{\infty} |a_n||z|^n \leq \sum_{n=0}^{\infty} a_n \text{ if } |z| \leq 1.$$

But $\sum_{n=0}^{\infty} a_n = [(1 - E_m(1)]/1^{m+1} = 1$, and the result follows. ♣

Weierstrass' primary factors E_m will now be used to construct functions with prescribed zeros. We begin by constructing *entire* functions with given zeros.

6.2.3 Theorem

Let $\{z_n\}$ be a sequence of nonzero complex numbers such that $|z_n| \to \infty$. Then there is a sequence $\{m_n\}$ of nonnegative integers such that the infinite product $\prod_{n=1}^{\infty} E_{m_n}(z/z_n)$ defines an entire function f. Furthermore, $f(z) = 0$ iff $z = z_n$ for some n. Thus it is possible to construct an entire function having zeros precisely at the z_n, with prescribed multiplicities. (If a appears k times in the sequence $\{z_n\}$, then f has a zero of order k at a. Also, a zero at the origin is handled by multiplying the product by z^m.)

Proof. Let $\{m_n\}$ be a sequence of nonnegative integers with the property that

$$\sum_{n=1}^{\infty}\left(\frac{r}{|z_n|}\right)^{m_n+1} < \infty$$

for every $r > 0$. (One such sequence is $m_n = n-1$ since for any $r > 0$, $r/|z_n|)$ is eventually less than $1/2$.) For fixed $r > 0$, (6.2.2) implies that

$$|1 - E_{m_n}(z/z_n)| \leq |z/z_n|^{m_n+1} \leq (r/z_n)^{m_n+1}$$

for all $z \in D(0,r)$. Thus the series $\sum |1 - E_{m_n}(z/z_n)|$ converges uniformly on $D(0,r)$. Since r is arbitrary, the series converges uniformly on compact subsets of $\mathbb{C}$. The result follows from (6.1.7). ♣

6.2.4 Remark

Let $\{z_n\}$ be as in (6.2.3). If $|z_n|$ grows sufficiently rapidly, it may be possible to take $\{m_n\}$ to be a constant sequence. For example, if $|z_n| = n$, then we may choose $m_n \equiv 1$. The corresponding product is $\prod_{n=1}^{\infty} E_1(z/z_n) = \prod_{n=1}^{\infty}(1 - z/z_n)e^{z/z_n}$. In this case, $m = 1$ is the *smallest* nonnegative integer for which $\sum_{n=1}^{\infty}(r/|z_n|)^{m+1} < \infty$ for all $r > 0$, and $\prod_{n=1}^{\infty} E_m(z/z_n)$ can be viewed as the *canonical product* associated with the sequence $\{z_n\}$. On the other hand, if $|z_n| = \ln n$, then $\sum_{n=1}^{\infty}(1/|z_n|)^m = +\infty$ for every nonnegative integer m, so no constant sequence suffices. These concepts arise in the study of the order of growth of entire functions, but we will not pursue this area further.

Theorem 6.2.3 allows us to factor out the zeros of an entire function.Specifically, we have a representation of an entire function as a product involving the primary factors E_m.

6.2.5 Weierstrass Factorization Theorem

Let f be an entire function, $f \not\equiv 0$, and let $k \geq 0$ be the order of the zero of f at 0. Let the remaining zeros of f be at $z_1, z_2, \ldots$, where each z_n is repeated as often as its multiplicity. Then

$$f(z) = e^{g(z)} z^k \prod_n E_{m_n}(z/z_n)$$

for some entire function g and nonnegative integers m_n.

Proof. If f has finitely many zeros, the result is immediate, so assume that there are infinitely many z_n. Since $f \not\equiv 0$, $|z_n| \to \infty$. By (6.2.3) there is a sequence $\{m_n\}$ such that

$$h(z) = f(z)/[z^k \prod_{n=1}^{\infty} E_{m_n}(z/z_n)]$$

has a zero-free extension to an entire function, which we will persist in calling h. But now h has an analytic logarithm g on $\mathbb{C}$, hence $h(z) = e^{g(z)}$ and we have the desired representation. ♣

More generally, versions of (6.2.3) and its consequence (6.2.5) are available for any *proper* open subset of $\hat{\mathbb{C}}$. We begin with the generalization of (6.2.3).

6.2.6 Theorem

Let Ω be a proper open subset of $\hat{\mathbb{C}}$, $A = \{a_n : n = 1, 2, \ldots\}$ a set of distinct points in Ω with no limit point in Ω, and $\{m_n\}$ a sequence of positive integers. Then there exists $f \in A(\Omega)$ such that $Z(f) = A$ and such that for each n we have $m(f, a_n) = m_n$.

Proof. We first show that it is sufficient to prove the theorem in the special case where Ω is a deleted neighborhood of ∞ in $\hat{\mathbb{C}}$ and $\infty \notin A$. For suppose that the theorem has been established in this special case. Then let Ω_1 and A_1 be arbitrary but as in the hypothesis of the the theorem. Choose a point $a \neq \infty$ in $\Omega_1 \setminus A_1$ and define $T(z) = 1/(z - a)$, $z \in \hat{\mathbb{C}}$. Then T is a linear fractional transformation of $\hat{\mathbb{C}}$ onto $\hat{\mathbb{C}}$ and thus is a one-to-one continuous map of the open set Ω_1 in $\hat{\mathbb{C}}$ onto an open set Ω. Further, if $A = \{T(a_n) : n = 1, 2 \ldots\}$ then Ω and A satisfy the hypotheses of the special case. Having assumed the special case, there exists f analytic on Ω such that $Z(f) = A$ and $m(f, T(a_n)) = m_n$. Now consider the function $f_1 = f \circ T$. Since T is analytic on $\Omega_1 \setminus \{a\}$, so is f_1. But as $z \to a$, $T(z) \to \infty$, and since f is analytic at ∞, $f(T(z))$ approaches a nonzero limit as $z \to a$. Thus f_1 has a removable singularity at a with $f_1(a) \neq 0$. The statement regarding the zeros of f_1 and their multiplicities follows from the fact that T is one-to-one.

Now we must establish the special case. First, if A is a finite set $\{a_1, \ldots, a_n\}$, then we can simply take

$$f(z) = \frac{(z - a_1)^{m_1} \cdots (z - a_n)^{m_n}}{(z - b)^{m_1 + \cdots + m_n}}$$

where $b \in \mathbb{C} \setminus \Omega$. The purpose of the denominator is to assure that f is analytic and nonzero at ∞.

Now suppose that $A = \{a_1, a_2, \ldots\}$ is an infinite set. Let $\{z_n\}$ be a sequence whose range is A but such that for each j, we have $z_n = a_j$ for exactly m_j values of n. Since $\mathbb{C} \setminus \Omega$ is a *nonempty* compact subset of $\mathbb{C}$, for each $n \geq 1$ there exists a point w_n in $\mathbb{C} \setminus \Omega$ such that $|w_n - z_n| = \operatorname{dist}(z_n, \mathbb{C} \setminus \Omega)$. Note that $|w_n - z_n| \to 0$ as $n \to \infty$ because the sequence $\{z_n\}$ has no limit point in Ω. Let $\{f_n\}$ be the sequence of functions on Ω defined by

$$f_n(z) = E_n\left(\frac{z_n - w_n}{z - w_n}\right),$$

where $f_n(\infty) = E_n(0) = 1$. Then f_n has a simple zero at z_n and no other zeros. Furthermore, $\sum |f_n - 1|$ converges uniformly on compact subsets of Ω. For if $K \subseteq \Omega$, K compact, then eventually $|z_n - w_n|/|z - w_n|$ is uniformly bounded by $1/2$ on K. Thus by Lemma 6.2.2,

$$|f_n(z) - 1| = \left|1 - E_n\left(\frac{z_n - w_n}{z - w_n}\right)\right| \leq \left|\frac{z_n - w_n}{z - w_n}\right|^{n+1} \leq (1/2)^{n+1}$$

for each $z \in K$. The statement of the theorem then follows from (6.1.7) by setting $f(z) = \prod_{n=1}^{\infty} f_n(z)$. ♣

It is interesting to see what the preceding argument yields in the special case $\Omega = \mathbb{C}$, a case which was established directly in (6.2.3). Specifically, suppose that $A = \{a_1, a_2, \dots\}$ is an infinite set of distinct points in $\mathbb{C}$ (with no limit point in $\mathbb{C}$), and assume that $0 \notin A$. Let $\{m_j\}$ and $\{z_n\}$ be as in the preceding proof. We are going to reconstruct the proof in the case where $\infty \in \Omega \setminus A$. In order to do this, consider the transformation $T(z) = 1/z$. This maps $\mathbb{C}$ onto $\hat{\mathbb{C}} \setminus \{0\}$ and the sequence $\{z_n\}$ in $\mathbb{C} \setminus \{0\}$ onto the sequence $\{1/z_n\}$ in $T(\mathbb{C})$. The points w_n obtained in the proof of (6.2.6) are all 0, and the corresponding functions f_n would be given by

$$f_n(z) = E_n(1/z_n z), \quad z \in \mathbb{C} \setminus \{0\}.$$

Thus $f(z) = \prod_{n=1}^{\infty} f_n(z)$ is analytic on $\mathbb{C} \setminus \{0\}$ and f has a zero of order m_j at $1/a_j$. Transforming $\hat{\mathbb{C}} \setminus \{0\}$ back to $\mathbb{C}$, it follows that

$$F(z) = f(1/z) = \prod_{n=1}^{\infty} E_n(z/z_n)$$

is an entire function with zeros of order m_j at a_j and no other zeros. That is, we obtain (6.2.3) with $m_n = n$. (Note that this m_n from (6.2.3) is unrelated to the sequence $\{m_j\}$ above.)

The fact that we can construct analytic functions with prescribed zeros has an interesting consequence, which was referred to earlier in (4.2.5).

6.2.7 Theorem

Let h be meromorphic on the open set $\Omega \subseteq \mathbb{C}$. Then $h = f/g$ where f and g are analytic on Ω.

Proof. Let A be the set of poles of h in Ω. Then A satisfies the hypothesis in (6.2.6). Let g be an analytic function on Ω with zeros precisely at the points in A and such that for each $a \in A$, the order of the zero of g at a equals the order of the pole of h at a. Then gh has only removable singularities in Ω and thus can be extended to an analytic function $f \in A(\Omega)$. ♣

Problems

1. Determine the canonical products associated with each of the following sequences. [See the discussion in (6.2.4).]
   (a) $z_n = 2^n$, (b) $z_n = n^b, b > 0$, (c) $z_n = n(\ln n)^2$.
2. Apply Theorem 6.2.6 to construct an analytic function f on the unit disk D such that f has no proper analytic extension to a region $\Omega \supset D$. (Hint: Construct a countable set $A = \{a_n : n = 1, 2, \dots\}$ in D such that every point in ∂D is an accumulation point of A.) Compare this approach to that in Theorem 4.9.5, where essentially the same result is obtained by quite different means.

6.3 Mittag-Leffler's Theorem and Applications

Let Ω be an open subset of $\mathbb{C}$ and let $A = \{a_n : n = 1, 2, \dots\}$ be a set of distinct points in Ω with no limit point in Ω. If $\{m_n\}$ is a sequence of positive integers, then Theorem 6.2.6 implies (by using $1/f$) that there is a meromorphic function f on Ω such that f has poles of order precisely m_n at precisely the points a_n. The theorem of Mittag-Leffler, which we will prove next, states that we can actually specify the coefficients of the principal part at each pole a_n. The exact statement follows; the proof requires Runge's theorem.

6.3.1 Mittag-Leffler's Theorem

Let Ω be an open subset of $\mathbb{C}$ and B a subset of Ω with no limit point in Ω. Thus $B = \{b_j : j \in J\}$ where J is some finite or countably infinite index set. Suppose that to each $j \in J$ there corresponds a rational function of the form

$$S_j(z) = \frac{a_{j1}}{z - b_j} + \frac{a_{j2}}{(z - b_j)^2} + \cdots + \frac{a_{jn_j}}{(z - b_j)^{n_j}}.$$

Then there is a meromorphic function f on Ω such that f has poles at precisely the points b_j and such that the principal part of the Laurent expansion of f at b_j is exactly S_j.

Proof. Let $\{K_n\}$ be the sequence of compact sets defined in (5.1.1). Recall that $\{K_n\}$ has the properties that $K_n \subseteq K_{n+1}^o$ and $\cup K_n = \Omega$. Furthermore, by Problem 5.2.5, each component of $\mathbb{C} \setminus K_n$ contains a component of $\mathbb{C} \setminus \Omega$, in particular, $\mathbb{C} \setminus \Omega$ *meets* each component of $\mathbb{C} \setminus K_n$. Put $K_0 = \emptyset$ and for $n = 1, 2, \dots$, define

$$J_n = \{j \in J : b_j \in K_n \setminus K_{n-1}\}.$$

The sets J_n are pairwise disjoint (possibly empty), each J_n is finite (since B has no limit point in Ω), and $\cup J_n = J$. For each n, define Q_n by

$$Q_n(z) = \sum_{j \in J_n} S_j(z)$$

where $Q_n \equiv 0$ if J_n is empty. Then Q_n is a rational function whose poles lie in $K_n \setminus K_{n-1}$. In particular, Q_n is analytic on a neighborhood of K_{n-1}. Hence by Runge's theorem (5.2.8) with $S = \mathbb{C} \setminus \Omega$, there is a rational function R_n whose poles lie in $\mathbb{C} \setminus \Omega$ such that

$$|Q_n(z) - R_n(z)| \leq (1/2)^n, \ z \in K_{n-1}.$$

It follows that for any fixed $m \geq 1$, the series $\sum_{n=m+1}^{\infty}(Q_n - R_n)$ converges uniformly on K_m to a function which is analytic on $K_m^o \supseteq K_{m-1}$. Thus it is meaningful to define a function $f : \Omega \to \mathbb{C}$ by

$$f(z) = Q_1(z) + \sum_{n=2}^{\infty}(Q_n(z) - R_n(z)), \ z \in \Omega.$$

Indeed, note that for any fixed m, f is the sum of the rational function $Q_1 + \sum_{n=2}^{m}(Q_n - R_n)$ and the series $\sum_{n=m+1}^{\infty}(Q_n - R_n)$, which is analytic on K_m^o. Therefore f is meromorphic on Ω, as well as analytic on $\Omega \setminus B$. It remains to show that f has the required principal part at each point $b \in B$. But for any $b_j \in B$, we have $f(z) = S_j(z)$ plus a function that is analytic on a neighborhood of b_j. Thus f has a pole at b_j with the required principal part S_j. ♣

6.3.2 Remark

Suppose g is analytic at the complex number b and g has a zero of order $m \geq 1$ at b. Let $c_1, c_2, \dots, c_m$ be given complex numbers, and let R be the rational function given by

$$R(z) = \frac{c_1}{z-b} + \cdots + \frac{c_m}{(z-b)^m}.$$

Then gR has a removable singularity at b, so there exist complex numbers $a_0, a_1, a_2, \dots$ such that for z in a neighborhood of b,

$$g(z)R(z) = a_0 + a_1(z-b) + \cdots + a_{m-1}(z-b)^{m-1} + \cdots .$$

Furthermore, if we write the Taylor series expansion

$$g(z) = b_0(z-b)^m + b_1(z-b)^{m+1} + \cdots + b_{m-1}(z-b)^{2m-1} + \cdots ,$$

then the coefficients $a_0, a_1, \dots$ for gR must satisfy

$$\begin{aligned}
a_0 &= b_0 c_m \\
a_1 &= b_0 c_{m-1} + b_1 c_m \\
&\vdots \\
a_{m-1} &= b_0 c_1 + b_1 c_2 + \cdots + b_{m-1} c_m
\end{aligned}$$

That is, if $c_1, c_2, \dots, c_m$ are *given*, then $a_0, a_1, \dots, a_{m-1}$ are *determined* by the above equations. Conversely, if g is given as above, and $a_0, a_1, \dots, a_{m-1}$ are *given* complex numbers, then since $b_0 \neq 0$, one can sequentially solve the equations to obtain, in order, $c_m, c_{m-1}, \dots, c_1$. This observation plays a key role in the next result, where it is shown that not only is it possible to construct analytic functions with prescribed zeros and with prescribed orders at these zeros, as in (6.2.3) and (6.2.6), but we can specify the values of f and finitely many of its derivatives in an arbitrary way. To be precise, we have the following extension of (6.2.6).

6.3.3 Theorem

Let Ω be an open subset of $\mathbb{C}$ and B a subset of Ω with no limit point in Ω. Index B by J, as in Mittag-Leffler's theorem, so $B = \{b_j : j \in J\}$. Suppose that corresponding to each $j \in J$, there is a nonnegative integer n_j and complex numbers $a_{0j}, a_{1j}, \dots, a_{n_j,j}$. Then there exists $f \in A(\Omega)$ such that for each $j \in J$,

$$\frac{f^{(k)}(b_j)}{k!} = a_{kj}, \; 0 \leq k \leq n_j.$$

Proof. First apply (6.2.6) to produce a function $g \in A(\Omega)$ such that $Z(g) = B$ and for each j, $m(g, b_j) = n_j + 1 = m_j$, say. Next apply the observations made above in (6.3.2) to obtain, for each $b_j \in B$, complex numbers $c_{1j}, c_{2j}, \dots, c_{m_j,j}$ such that

$$g(z)\sum_{k=1}^{m_j} \frac{c_{kj}}{(z-b_j)^k} = a_{0j} + a_{1j}(z-b_j) + \cdots + a_{n_j,j}(z-b_j)^{n_j} + \cdots$$

for z near b_j. Finally, apply Mittag-Leffler's theorem to obtain h, meromorphic on Ω, such that for each j,

$$h - \sum_{k=1}^{m_j} \frac{c_{kj}}{(z-b_j)^k}$$

has a removable singularity at b_j. It follows that the analytic extension of gh to Ω is the required function f. (To see this, note that

$$gh = g\left(h - \sum_{k=1}^{m_j} \frac{c_{kj}}{(z-b_j)^k}\right) + g\sum_{k=1}^{m_j} \frac{c_{kj}}{(z-b_j)^k}$$

and $m(g, b_j) > n_j$.) ♣

6.3.4 Remark

Theorem 6.3.3 will be used to obtain a number of *algebraic* properties of the ring $A(\Omega)$. This theorem, together with most of results to follow, were obtained (in the case $\Omega = \mathbb{C}$) by Olaf Helmer, Duke Mathematical Journal, volume 6, 1940, pp.345-356.

Assume in what follows that Ω is connected. Thus by Problem 2.4.11, $A(\Omega)$ is an integral domain. Recall that in a ring, such as $A(\Omega)$, g *divides* f if $f = gq$ for some $q \in A(\Omega)$. Also, g is a *greatest common divisor* of a *set* $\mathcal{F}$ if g is a divisor of each $f \in \mathcal{F}$ and if h divides each $f \in \mathcal{F}$, then h divides g.

6.3.5 Proposition

Each nonempty subfamily $\mathcal{F} \subseteq A(\Omega)$ has a greatest common divisor, provided $\mathcal{F} \neq \{0\}$.

Proof. Put $B = \cap\{Z(f) : f \in \mathcal{F}\}$. Apply Theorem 6.2.6 to obtain $g \in A(\Omega)$ such that $Z(g) = B$ and for each $b \in B, m(g,b) = \min\{m(f,b) : f \in \mathcal{F}\}$. Then $f \in \mathcal{F}$ implies that $g|f$ (g divides f). Furthermore, if $h \in A(\Omega)$ and $h|f$ for each $f \in \mathcal{F}$, then $Z(h) \subseteq B$ and for each $b \in B, m(h,b) \leq \min\{m(f,b) : f \in \mathcal{F}\} = m(g,b)$. Thus $h|g$, and consequently g is a greatest common divisor of $\mathcal{F}$. ♣

6.3.6 Definitions

A *unit* in $A(\Omega)$ is a function $f \in A(\Omega)$ such that $1/f \in A(\Omega)$. Thus f is a unit iff f has no zeros in Ω. If $f, g \in A(\Omega)$, we say that f and g are *relatively prime* if each greatest common divisor of f and g is a unit. It follows that f and g are relatively prime iff $Z(f) \cap Z(g) = \emptyset$. (Note that f and g have a common zero iff they have a nonunit common factor.)

6.3.7 Proposition

If the functions $f_1, f_2 \in A(\Omega)$ are relatively prime, then there exist $g_1, g_2 \in A(\Omega)$ such that $f_1g_1 + f_2g_2 \equiv 1$.

Proof. By the remarks above, $Z(f_1) \cap Z(f_2) = \emptyset$. By working backwards, i.e., solving $f_1g_1 + f_2g_2 = 1$ for g_1, we see that it suffices to obtain g_2 such that $(1 - f_2g_2)/f_1$ has only removable singularities. But this entails obtaining g_2 such that $Z(f_1) \subseteq Z(1 - f_2g_2)$ and such that for each $a \in Z(f_1), m(f_1, a) \leq m(1 - f_2g_2, a)$. However, the latter condition may be satisfied by invoking (6.3.3) to obtain $g_2 \in A(\Omega)$ such that for each $a \in Z(f_1)$ (recalling that $f_2(a) \neq 0$),

$$\begin{aligned}
0 &= 1 - f_2(a)g_2(a) = (1 - f_2g_2)(a) \\
0 &= f_2(a)g_2'(a) + f_2'(a)g_2(a) = (1 - f_2g_2)'(a) \\
0 &= f_2(a)g_2''(a) + 2f_2'(a)g_2'(a) + f_2''(a)g_2(a) = (1 - f_2g_2)''(a) \\
&\vdots \\
0 &= f_2(a)g_2^{(m-1)}(a) + \cdots + f_2^{(m-1)}(a)g_2(a) = (1 - f_2g_2)^{(m-1)}(a)
\end{aligned}$$

where $m = m(f_1, a)$. [Note that these equations successively determine $g_2(a), g_2'(a), \dots, g_2^{(m-1)}(a)$.] This completes the proof of the proposition. ♣

The preceding result can be generalized to an arbitrary finite collection of functions.

6.3.8 Proposition

If $\{f_1, f_2, \dots, f_n\} \subseteq A(\Omega)$ and d is a greatest common divisor for this set, then there exist $g_1, g_2, \dots, g_n \in A(\Omega)$ such that $f_1g_1 + f_2g_2 + \cdots + f_ng_n = d$.

Proof. Use (6.3.7) and induction. The details are left as an exercise (Problem 1). ♣

Recall that an *ideal* $I \subseteq A(\Omega)$ is a subset that is closed under addition and subtraction and has the property that if $f \in A(\Omega)$ and $g \in I$, then $fg \in I$.

We are now going to show that $A(\Omega)$ is what is referred to in the literature as a *Bezout domain.* This means that each finitely generated ideal in the integral domain $A(\Omega)$ is a principal ideal. A *finitely generated* ideal is an ideal of the form $\{f_1g_1 + \cdots + f_ng_n : g_1, \dots, g_n \in A(\Omega)\}$ where $\{f_1, \dots, f_n\}$ is some fixed finite set of elements in $A(\Omega)$. A *principal ideal* is an ideal that is generated by a *single* element f_1. Most of the work has already been done in preceding two propositions.

6.3.9 Theorem

Let $f_1, \dots, f_n \in A(\Omega)$ and let $I = \{f_1g_1 + \cdots + f_ng_n : g_1, \dots, g_n \in A(\Omega)\}$ be the ideal generated by $f_1, \dots, f_n$. Then there exists $f \in A(\Omega)$ such that $I = \{fg : g \in A(\Omega)\}$. In other words, I is a principal ideal.

Proof. If $f \in I$ then $f = f_1h_1 + \cdots + f_nh_n$ for some $h_1, \dots, h_n \in A(\Omega)$. If d is a greatest common divisor for $\{f_1, \dots, f_n\}$, then d divides each f_j, hence d divides f. Thus f is a multiple of d. On the other hand, by (6.3.8), there exist $g_1, \dots, g_n \in A(\Omega)$ such that

$d = f_1g_1 + \cdots + f_ng_n$. Therefore d and hence every multiple of d belongs to I. Thus I is the ideal generated by the single element d. ♣

A *principal ideal domain* is an integral domain in which every ideal is principal. Problem 2 asks you to show that $A(\Omega)$ is *never* a principal ideal domain, regardless of the region Ω. There is another class of (commutative) rings called *Noetherian*; these are rings in which every ideal is finitely generated. Problem 2, when combined with (6.3.9), also shows that $A(\Omega)$ is never Noetherian.

Problems

1. Supply the details to the proof of (6.3.8). (Hint: Use induction, (6.3.7), and the fact that if d is a greatest common divisor (gcd) for $\{f_1, \ldots, f_n\}$ and d_1 is a gcd for $\{f_1, \ldots, f_{n-1}\}$, then d is a gcd for the set $\{d_1, f_n\}$. Also note that 1 is a gcd for $\{f_1/d, \ldots, f_n/d\}$.)
2. Show that $A(\Omega)$ is never a principal ideal domain. that is, there always exists ideals I that are not principal ideals, and thus by (6.3.9) are not finitely generated. (Hint: Let $\{a_n\}$ be a sequence of distinct points in Ω with no limit point in Ω. For each n, apply (6.2.6) to the set $\{a_n, a_{n+1}, \ldots\}$.)

Chapter 7

The Prime Number Theorem

In this final chapter we will take advantage of an opportunity to apply many of the ideas and results from earlier chapters in order to give an analytic proof of the famous prime number theorem: If $\pi(x)$ is the number of primes less than or equal to x, then $x^{-1}\pi(x)\ln x \to 1$ as $x \to \infty$. That is, $\pi(x)$ is asymptotically equal to $x/\ln x$ as $x \to \infty$. (In the sequel, prime will be taken to mean *positive* prime.)

Perhaps the first recorded property of $\pi(x)$ is that $\pi(x) \to \infty$ as $x \to \infty$, in other words, the number of primes is infinite. This appears in Euclid's "Elements". A more precise result that was established much later by Euler (1737) is that the series of reciprocals of the prime numbers,

$$\frac{1}{2} + \frac{1}{3} + \frac{1}{5} + \frac{1}{7} + \frac{1}{11} + \cdots,$$

is a divergent series. This can be interpreted in a certain sense as a statement about how fast $\pi(x) \to \infty$ as $x \to \infty$. Later, near the end of the 18-th century, mathematicians, including Gauss and Legendre, through mainly empirical considerations, put forth conjectures that are equivalent to the above statement of the prime number theorem (PNT). However, it was not until nearly 100 years later, after much effort by numerous 19-th century mathematicians, that the theorem was finally established (independently) by Hadamard and de la Vallée Poussin in 1896. The quest for a proof led Riemann, for example, to develop complex variable methods to attack the PNT and related questions. In the process, he made a remarkable and as yet unresolved conjecture known as the Riemann hypothesis, whose precise statement will be given later. Now it is not clear on the surface that there is a connection between complex analysis and the distribution of prime numbers. But in fact, every proof of the PNT dating from Hadamard and de la Vallée Poussin, up to 1949 when P. Erdös and A.Selberg succeeded in finding "elementary" proofs, has used the methods of complex variables in an essential way. In 1980, D.J. Newman published a new proof of the PNT which, although still using complex analysis, nevertheless represents a significant simplification of previous proofs. It is Newman's proof, as modified by J. Korevaar, that we present in this chapter.

There are a number of preliminaries that must be dealt with before Newman's method can be applied to produce the theorem. The proof remains far from trivial but the steps along the way are of great interest and importance in themselves. We begin by introducing the Riemann zeta function, which arises via Euler's product formula and forms a key link between the sequence of prime numbers and the methods of complex variables.

7.1 The Riemann Zeta function

The *Riemann zeta function* is defined by

$$\zeta(z) = \sum_{n=1}^{\infty} \frac{1}{n^z}$$

where $n^z = e^{z \ln n}$. Since $|n^z| = n^{\operatorname{Re} z}$, the given series converges absolutely on $\operatorname{Re} z > 1$ and uniformly on $\{z : \operatorname{Re} z \geq 1 + \delta\}$ for every $\delta > 0$. Let $p_1, p_2, p_3, \dots$ be the sequence 2,3,5, ... of prime numbers and note that for $j = 1, 2, \dots$ and $\operatorname{Re} z > 1$, we have

$$\frac{1}{1 - 1/p_j^z} = 1 + \frac{1}{p_j^z} + \frac{1}{p_j^{2z}} + \cdots.$$

Now consider the partial product

$$\prod_{j=1}^{m} \frac{1}{1 - p_j^{-z}} = \prod_{j=1}^{m} (1 + \frac{1}{p_j^z} + \frac{1}{p_j^{2z}} + \cdots).$$

By multiplying the finitely many absolutely convergent series on the right together, rearranging, and applying the fundamental theorem of arithmetic, we find that the product is the same as the sum $\sum_{n \in P_m} \frac{1}{n^z}$, where P_m consists of 1 along with those positive integers whose prime factorization uses only primes from the set $\{p_1, \dots, p_m\}$. Therefore

$$\prod_{j=1}^{\infty} \frac{1}{1 - p_j^{-z}} = \sum_{n=1}^{\infty} \frac{1}{n^z}, \quad \operatorname{Re} z > 1.$$

We now state this formally.

7.1.1 Euler's Product formula

For $\operatorname{Re} z > 1$, the Riemann zeta function $\zeta(z) = \sum_{n=1}^{\infty} 1/n^z$ is given by the product

$$\prod_{j=1}^{\infty} \left(\frac{1}{1 - p_j^{-z}} \right)$$

where $\{p_j\}$ is the (increasing) sequence of prime numbers.

The above series and product converge uniformly on compact subsets of $\operatorname{Re} z > 1$, hence ζ is analytic on $\operatorname{Re} z > 1$. Furthermore, the product representation of ζ shows that ζ has no zeros in $\operatorname{Re} z > 1$ (Theorem 6.1.7). Our proof of the PNT requires a number of additional properties of ζ. The first result is concerned with extending ζ to a region larger than $\operatorname{Re} z > 1$.

7.1.2 Extension Theorem for Zeta

The function $\zeta(z) - 1/(z-1)$ has an analytic extension to the right half plane $\operatorname{Re} z > 0$. Thus ζ has an analytic extension to $\{z : \operatorname{Re} z > 0, z \neq 1\}$ and has a simple pole with residue 1 at $z = 1$.

Proof. For $\operatorname{Re} z > 1$, apply the summation by parts formula (Problem 2.2.7) with $a_n = n$ and $b_n = 1/n^z$ to obtain

$$\sum_{n=1}^{k-1} n\left[\frac{1}{(n+1)^z} - \frac{1}{n^z}\right] = \frac{1}{k^{z-1}} - 1 - \sum_{n=1}^{k-1} \frac{1}{(n+1)^z}.$$

Thus

$$1 + \sum_{n=1}^{k-1} \frac{1}{(n+1)^z} = \frac{1}{k^{z-1}} - \sum_{n=1}^{k-1} n\left[\frac{1}{(n+1)^z} - \frac{1}{n^z}\right].$$

But

$$n\left[\frac{1}{(n+1)^z} - \frac{1}{n^z}\right] = -nz\int_n^{n+1} t^{-z-1}\,dt = -z\int_n^{n+1} [t]t^{-z-1}\,dt$$

where $[t]$ is the largest integer less than or equal to t. Hence we have

$$\begin{aligned}\sum_{n=1}^{k} \frac{1}{n^z} = 1 + \sum_{n=1}^{k-1} \frac{1}{(n+1)^z} &= \frac{1}{k^{z-1}} + z\sum_{n=1}^{k-1} \int_n^{n+1} [t]t^{-z-1}\,dt \\ &= \frac{1}{k^{z-1}} + z\int_1^k [t]t^{-z-1}\,dt.\end{aligned}$$

Letting $k \to \infty$, we obtain the integral formula

$$\zeta(z) = z\int_1^\infty [t]t^{-z-1}\,dt \tag{1}$$

for $\operatorname{Re} z > 1$. Consider, however, the closely related integral

$$z\int_1^\infty tt^{-z-1}\,dt = z\int_1^\infty t^{-z}\,dt = \frac{z}{z-1} = 1 + \frac{1}{z-1}.$$

Combining this with (1) we can write

$$\zeta(z) - \frac{1}{z-1} = 1 + z\int_1^\infty ([t] - t)t^{-z-1}\,dt.$$

Now fix $k > 1$ and consider the integral $\int_1^k ([t] - t)t^{-z-1}\,dt$. By (3.3.3), this integral is an entire function of z. furthermore, if $\operatorname{Re} z > 0$, then

$$|\int_1^k ([t] - t)t^{-z-1}\,dt| \le \int_1^k t^{-\operatorname{Re}(z+1)}\,dt \le \int_1^\infty t^{-1-\operatorname{Re} z}\,dt = \frac{1}{\operatorname{Re} z}.$$

This implies that the sequence $f_k(z) = \int_1^k ([t]-t)t^{-z-1}\,dt$ of analytic functions on $\operatorname{Re} z > 0$ is uniformly bounded on compact subsets. Hence by Vitali's theorem (5.1.14), the limit function

$$f(z) = \int_1^\infty ([t]-t)t^{-z-1}\,dt$$

(as the uniform limit on compact subsets of $\operatorname{Re} z > 0$) is analytic, and thus the function

$$1 + z\int_1^\infty ([t]-t)t^{-z-1}\,dt$$

is also analytic on $\operatorname{Re} z > 0$. But this function agrees with $\zeta(z) - \frac{1}{z-1}$ for $\operatorname{Re} z > 1$, and consequently provides the required analytic extension of ζ to $\operatorname{Re} z > 0$. This completes the proof of the theorem. ♣

We have seen that Euler's formula (7.1.1) implies that ζ has no zeros in the half plane $\operatorname{Re} z > 1$, but how about zeros of (the extension of) ζ in $0 < \operatorname{Re} z \leq 1$? The next theorem asserts that ζ has no zeros on the line $\operatorname{Re} z = 1$. This fact is crucial to our proof of the PNT.

7.1.3 Theorem

The Riemann zeta function has no zeros on $\operatorname{Re} z = 1$, so $(z-1)\zeta(z)$ is analytic and zero-free on a neighborhood of $\operatorname{Re} z \geq 1$.

Proof. Fix a real number $y \neq 0$ and consider the auxiliary function

$$h(x) = \zeta^3(x)\zeta^4(x+iy)\zeta(x+i2y)$$

for x real and $x > 1$. By Euler's product formula, if $\operatorname{Re} z > 1$ then

$$\ln|\zeta(z)| = -\sum_{j=1}^\infty \ln|1-p_j^{-z}| = -\operatorname{Re}\sum_{j=1}^\infty \operatorname{Log}(1-p_j^{-z}) = \operatorname{Re}\sum_{j=1}^\infty\sum_{n=1}^\infty \frac{1}{n}p_j^{-nz}$$

where we have used the expansion $-\operatorname{Log}(1-w) = \sum_{n=1}^\infty w^n/n$, valid for $|w| < 1$. Hence

$$\begin{aligned}\ln|h(x)| &= 3\ln|\zeta(x)| + 4\ln|\zeta(x+iy)| + \ln|\zeta(x+i2y)| \\ &= 3\operatorname{Re}\sum_{j=1}^\infty\sum_{n=1}^\infty \frac{1}{n}p_j^{-nx} + 4\operatorname{Re}\sum_{j=1}^\infty\sum_{n=1}^\infty \frac{1}{n}p_j^{-nx-iny} \\ &\quad + \operatorname{Re}\sum_{j=1}^\infty\sum_{n=1}^\infty \frac{1}{n}p_j^{-nx-i2ny} \\ &= \sum_{j=1}^\infty\sum_{n=1}^\infty \frac{1}{n}p_j^{-nx}\operatorname{Re}(3+4p_j^{-iny}+p_j^{-i2ny}).\end{aligned}$$

But $p_j^{-iny} = e^{-iny\ln p_j}$ and $p_j^{-i2ny} = e^{-i2ny\ln p_j}$. Thus $\operatorname{Re}(3+4p_j^{-iny}+p_j^{-i2ny})$ has the form

$$3 + 4\cos\theta + \cos 2\theta = 3 + 4\cos\theta + 2\cos^2\theta - 1 = 2(1+\cos\theta)^2 \geq 0.$$

Therefore $\ln|h(x)| \geq 0$ and consequently

$$|h(x)| = |\zeta^3(x)||\zeta^4(x+iy)||\zeta(x+i2y)| \geq 1.$$

Thus

$$\frac{|h(x)|}{x-1} = |(x-1)\zeta(x)|^3 \left|\frac{\zeta(x+iy)}{x-1}\right|^4 |\zeta(x+i2y)| \geq \frac{1}{x-1}.$$

But if $\zeta(1+iy) = 0$, then the left hand side of this inequality would approach a finite limit $|\zeta'(1+iy)|^4|\zeta(1+i2y)|$ as $x \to 1^+$ since ζ has a simple pole at 1 with residue 1. However, the right hand side of the inequality contradicts this. We conclude that $\zeta(1+iy) \neq 0$. Since y is an arbitrary nonzero real number, ζ has no zeros on $\operatorname{Re} z = 1$. ♣

Remark

The ingenious introduction of the auxiliary function h is due to Mertens (1898). We now have shown that any zeros of ζ in $\operatorname{Re} z > 0$ must lie in the strip $0 < \operatorname{Re} z < 1$. The study of the zeros of ζ has long been the subject of intensive investigation by many mathematicians. Riemann had stated in his seminal 1859 paper that he considered it "very likely" that all the zeros of ζ in the above strip, called the *critical strip*, lie on the line $\operatorname{Re} z = 1/2$. This assertion is now known as the *Riemann hypothesis*, and remains as yet unresolved. However, a great deal *is* known about the distribution of the zeros of ζ in the critical strip, and the subject continues to capture the attention of eminent mathematicians. To state just one such result, G.H. Hardy proved in 1915 that ζ has infinitely many zeros on the line $\operatorname{Re} z = 1/2$. Those interested in learning more about this fascinating subject may consult, for example, the book *Riemann's Zeta Function* by H.M. Edwards. Another source is http://mathworld.wolfram.com/RiemannHypothesis.html.

We turn next to zeta's logarithmic derivative ζ'/ζ, which we know is analytic on $\operatorname{Re} z > 1$. In fact, more is true, for by (7.1.3), ζ'/ζ is analytic on a neighborhood of $\{z : \operatorname{Re} z \geq 1 \text{ and } z \neq 1\}$. Since ζ has a simple pole at $z = 1$, so does ζ'/ζ, with residue $\operatorname{Res}(\zeta'/\zeta, 1) = -1$. [See the proof of (4.2.7).] We next obtain an integral representation for ζ'/ζ that is similar to the representation (1) above for ζ. [See the proof of (7.1.2).] But first, we must introduce the *von Mangoldt function* Λ, which is defined by

$$\Lambda(n) = \begin{cases} \ln p & \text{if } n = p^m \text{ for some } m, \\ 0 & \text{otherwise.} \end{cases}$$

Thus $\Lambda(n)$ is $\ln p$ if n is a power of the prime p, and is 0 if not. Next define ψ on $x \geq 0$ by

$$\psi(x) = \sum_{n \leq x} \Lambda(n). \tag{2}$$

An equivalent expression for ψ is

$$\psi(x) = \sum_{p \leq x} m_p(x) \ln p,$$

where the sum is over primes $p \le x$ and $m_p(x)$ is the largest integer such that $p^{m_p(x)} \le x$. (For example, $\psi(10.4) = 3\ln 2 + 2\ln 3 + \ln 5 + \ln 7$.) Note that $p^{m_p(x)} \le x$ iff $m_p(x)\ln p \le \ln x$ iff $m_p(x) \le \frac{\ln x}{\ln p}$. Thus $m_p(x) = \left[\frac{\ln x}{\ln p}\right]$ where as before, [ ] denotes the greatest integer function. The function ψ will be used to obtain the desired integral representation for ζ'/ζ.

7.1.4 Theorem

For $\operatorname{Re} z > 1$,

$$-\frac{\zeta'(z)}{\zeta(z)} = z\int_1^\infty \psi(t)t^{-z-1}\,dt \tag{3}$$

where ψ is defined as above.

Proof. In the formulas below, p and q range over primes. If $\operatorname{Re} z > 1$, we have $\zeta(z) = \prod_p (1-p^{-z})^{-1}$ by (7.1.1), hence

$$\begin{aligned}
\zeta'(z) &= \sum_p \frac{-p^{-z}\ln p}{(1-p^{-z})^2}\prod_{q\ne p}\frac{1}{1-q^{-z}} \\
&= \zeta(z)\sum_p \frac{-p^{-z}\ln p}{(1-p^{-z})^2}(1-p^{-z}) \\
&= \zeta(z)\sum_p \frac{-p^{-z}\ln p}{1-p^{-z}}.
\end{aligned}$$

Thus

$$-\frac{\zeta'(z)}{\zeta(z)} = \sum_p \frac{p^{-z}\ln p}{1-p^{-z}} = \sum_p\sum_{n=1}^\infty p^{-nz}\ln p.$$

The iterated sum is absolutely convergent for $\operatorname{Re} z > 1$, so it can be rearranged as a double sum

$$\sum_{(p,n),n\ge 1} (p^n)^{-z}\ln p = \sum_k k^{-z}\ln p$$

where $k = p^n$ for some n. Consequently,

$$-\frac{\zeta'(z)}{\zeta(z)} = \sum_{k=1}^\infty k^{-z}\Lambda(k) = \sum_{k=1}^\infty k^{-z}(\psi(k) - \psi(k-1))$$

by the definitions of Λ and ψ. But using partial summation once again we obtain, with $a_k = k^{-z}$, $b_{k+1} = \psi(k)$, and $b_1 = \psi(0) = 0$ in Problem 2.2.7,

$$\sum_{k=1}^M k^{-z}(\psi(k) - \psi(k-1)) = \psi(M)(M+1)^{-z} + \sum_{k=1}^M \psi(k)(k^{-z} - (k+1)^{-z}).$$

Now from the definition (2) of $\psi(x)$ we have $\psi(x) \leq x \ln x$, so if $\operatorname{Re} z > 1$ we have $\psi(M)(M+1)^{-z} \to 0$ as $M \to \infty$. Moreover, we can write

$$\begin{aligned}\sum_{k=1}^{M} \psi(k)(k^{-z} - (k+1)^{-z}) &= \sum_{k=1}^{M} \psi(k) z \int_k^{k+1} t^{-z-1}\, dt \\ &= \sum_{k=1}^{M} z \int_k^{k+1} \psi(t) t^{-z-1}\, dt \\ &= z \int_1^M \psi(t) t^{-z-1}\, dt\end{aligned}$$

because ψ is constant on each interval $[k, k+1)$. Taking limits as $M \to \infty$, we finally get

$$-\frac{\zeta'(z)}{\zeta(z)} = z \int_1^{\infty} \psi(t) t^{-z-1}\, dt, \ \operatorname{Re} z > 1. \ \clubsuit$$

7.2 An Equivalent Version of the Prime Number Theorem

The function ψ defined in (2) above provides yet another connection, through (3), between the Riemann zeta function and properties of the prime numbers. The integral that appears in (3) is called the *Mellin transform* of ψ and is studied in its own right. We next establish a reduction, due to Chebyshev, of the prime number theorem to a statement involving the function ψ.

7.2.1 Theorem

The prime number theorem holds, that is, $x^{-1}\pi(x) \ln x \to 1$, iff $x^{-1}\psi(x) \to 1$ as $x \to \infty$.

Proof. Recall that

$$\begin{aligned}\psi(x) &= \sum_{p \leq x} \left[\frac{\ln x}{\ln p}\right] \ln p \\ &\leq \sum_{p \leq x} \frac{\ln x}{\ln p} \ln p \\ &= \ln x \sum_{p \leq x} 1 \\ &= (\ln x)\pi(x).\end{aligned} \tag{1}$$

However, if $1 < y < x$, then

$$\begin{aligned}\pi(x) &= \pi(y) + \sum_{y<p\le x} 1 \\ &\le \pi(y) + \sum_{y<p\le x} \frac{\ln p}{\ln y} \qquad (2)\\ &< y + \frac{1}{\ln y}\sum_{y<p\le x} \ln p \\ &\le y + \frac{1}{\ln y}\psi(x).\end{aligned}$$

Now take $y = x/(\ln x)^2$ in (2), and we get

$$\pi(x) \le \frac{x}{(\ln x)^2} + \frac{1}{\ln x - 2\ln\ln x}\psi(x).$$

Thus

$$\pi(x)\frac{\ln x}{x} \le \frac{1}{\ln x} + \frac{\ln x}{\ln x - 2\ln\ln x}\frac{\psi(x)}{x}. \qquad (3)$$

It now follows from (1) and (3) that

$$\frac{\psi(x)}{x} \le \frac{\ln x}{x}\pi(x) \le \frac{1}{\ln x} + \frac{\ln x}{\ln x - 2\ln\ln x}\frac{\psi(x)}{x}$$

and from this we can see that $x^{-1}\psi(x) \to 1$ iff $x^{-1}\pi(x)\ln x \to 1$ as $x \to \infty$. ♣

The goal will now be to show that $\psi(x)/x \to 1$ as $x \to \infty$. A necessary intermediate step for our proof is to establish the following weaker estimate on the asymptotic behavior of $\psi(x)$.

7.2.2 Lemma

There exists $C > 0$ such that $\psi(x) \le Cx, x > 0$. For short, $\psi(x) = O(x)$.

Proof. Again recall that $\psi(x) = \sum_{p\le x}[\frac{\ln x}{\ln p}]\ln p,\ x > 0$. Fix $x > 0$ and let m be an integer such that $2^m < x \le 2^{m+1}$. Then

$$\begin{aligned}\psi(x) &= \psi(2^m) + \psi(x) - \psi(2^m) \\ &\le \psi(2^m) + \psi(2^{m+1}) - \psi(2^m) \\ &= \sum_{p\le 2^m}\left[\frac{\ln 2^m}{\ln p}\right]\ln p + \sum_{2^m<p\le 2^{m+1}}\left[\frac{\ln 2^{m+1}}{\ln p}\right]\ln p.\end{aligned}$$

Consider, for any positive integer n,

$$\sum_{n<p\le 2n}\ln p = \ln\prod_{n<p\le 2n} p.$$

Now for any prime p such that $n < p \leq 2n$, p divides $(2n)!/n! = n!\binom{2n}{n}$. Since such a p does not divide $n!$, it follows that p divides $\binom{2n}{n}$. Hence

$$\prod_{n<p\leq 2n} p \leq \binom{2n}{n} < (1+1)^{2n} = 2^{2n},$$

and we arrive at

$$\sum_{n<p\leq 2n} \ln p < 2n \ln 2.$$

Therefore

$$\sum_{p\leq 2^m} \ln p = \sum_{k=1}^{m} \left(\sum_{2^{k-1}<p\leq 2^k} \ln p \right) < \sum_{k=1}^{m} 2^k \ln 2 < 2^{m+1} \ln 2$$

and

$$\sum_{2^m<p\leq 2^{m+1}} \ln p < 2^{m+1} \ln 2.$$

But if $p \leq x$ is such that $\left[\frac{\ln x}{\ln p}\right] > 1$, then $\frac{\ln x}{\ln p} \geq 2$ and hence $x \geq p^2$ so that $\sqrt{x} \geq p$. Thus those terms in the sum $\sum_{p\leq x} \left[\frac{\ln x}{\ln p}\right] \ln p$ where $\left[\frac{\ln x}{\ln p}\right] > 1$ occur only when $p \leq \sqrt{x}$, and the sum of terms of this form contribute no more than

$$\sum_{p\leq\sqrt{x}} \frac{\ln x}{\ln p} \ln p = \pi(\sqrt{x}) \ln x.$$

It follows from the above discussion that if $2^m < x \leq 2^{m+1}$, then

$$\begin{aligned}
\psi(x) &\leq 2^{m+1} \ln 2 + 2^{m+1} \ln 2 + \pi(\sqrt{x}) \ln x \\
&= 2^{m+2} \ln 2 + \pi(\sqrt{x}) \ln x \\
&< 4x \ln 2 + \pi(\sqrt{x}) \ln x \\
&\leq 4x \ln 2 + \sqrt{x} \ln x \\
&= (4 \ln 2 + \frac{1}{\sqrt{x}} \ln x) x.
\end{aligned}$$

Since $\frac{1}{\sqrt{x}} \ln x \to 0$ as $x \to \infty$, we conclude that $\psi(x) = O(x)$, which proves the lemma. ♣

7.3 Proof of the Prime Number Theorem

Our approach to the prime number theorem has been along traditional lines, but at this stage we will apply D.J. Newman's method (*Simple Analytic Proof of the Prime Number Theorem*, American Math. Monthly 87 (1980), 693-696) as modified by J. Korevaar (*On*

Newman's Quick Way to the Prime Number Theorem, Math. Intelligencer 4 (1982), 108-115). Korevaar's approach is to apply Newman's ideas to obtain properties of certain Laplace integrals that lead to the prime number theorem.

Our plan is to deduce the prime number theorem from a "Tauberian" theorem (7.3.1) and its corollary (7.3.2). Then we will prove (7.3.1) and (7.3.2).

7.3.1 Auxiliary Tauberian Theorem

Let F be bounded and piecewise continuous on $[0, +\infty)$, so that its Laplace transform

$$G(z) = \int_0^\infty F(t)e^{-zt}\, dt$$

exists and is analytic on $\operatorname{Re} z > 0$. Assume that G has an analytic extension to a neighborhood of the imaginary axis, $\operatorname{Re} z = 0$. Then $\int_0^\infty F(t)\, dt$ exists as an improper integral and is equal to $G(0)$. [In fact, $\int_0^\infty F(t)e^{-iyt}\, dt$ converges for every $y \in \mathbb{R}$ to $G(iy)$.]

Results like (7.3.1) are named for A. Tauber, who is credited with proving the first theorem of this type near the end of the 19th century. The phrase "Tauberian theorem" was coined by G.H. Hardy, who along with J.E. Littlewood made a number of contributions in this area. Generally, Tauberian theorems are those in which some type of "ordinary" convergence (e.g., convergence of $\int_0^\infty F(t)e^{-iyt}\, dt$ for each $y \in \mathbb{R}$), is deduced from some "weaker" type of convergence (e.g., convergence of $\int_0^\infty F(t)e^{-zt}\, dt$ for each z with $\operatorname{Re} z > 0$) provided additional conditions are satisfied (e.g., G has an analytic extension to a neighborhood of each point on the imaginary axis). Tauber's original theorem can be found in *The Elements of Real Analysis* by R.G. Bartle.

7.3.2 Corollary

Let f be a nonnegative, piecewise continuous and nondecreasing function on $[1, \infty)$ such that $f(x) = O(x)$. Then its *Mellin transform*

$$g(z) = z \int_1^\infty f(x)x^{-z-1}\, dx$$

exists for $\operatorname{Re} z > 1$ and defines an analytic function g. Assume that for some constant c, the function

$$g(z) - \frac{c}{z-1}$$

has an analytic extension to a neighborhood of the line $\operatorname{Re} z = 1$. Then as $x \to \infty$,

$$\frac{f(x)}{x} \to c.$$

As stated earlier, we are first going to see how the prime number theorem follows from (7.3.1) and (7.3.2). To this end, let ψ be as above, namely

$$\psi(x) = \sum_{p \le x} \left[\frac{\ln x}{\ln p}\right] \ln p.$$

Then ψ is a nonnegative, piecewise continuous, nondecreasing function on $[1, \infty)$. Furthermore, by (7.2.2), $\psi(x) = O(x)$, so by (7.3.2) we may take $f = \psi$ and consider the Mellin transform

$$g(z) = z \int_1^\infty \psi(x) x^{-z-1} \, dx.$$

But by (7.1.4), actually $g(z) = -\zeta'(z)/\zeta(z)$, and by the discussion leading up to the statement of (7.1.4), $\frac{\zeta'(z)}{\zeta(z)} + \frac{1}{z-1}$ has an analytic extension to a neighborhood of each point of $\mathrm{Re}\, z = 1$, hence so does $g(z) - \frac{1}{z-1}$. Consequently, by (7.3.2), we can conclude that $\psi(x)/x \to 1$, which, by (7.2.1), is equivalent to the PNT. Thus we are left with the proof of (7.3.1) and its corollary (7.3.2).

Proof of (7.3.1)

Let F be as in the statement of the theorem. Then it follows just as in the proof of (7.1.2), the extension theorem for zeta, that F's Laplace transform G is defined and analytic on $\mathrm{Re}\, z > 0$. Assume that G has been extended to an analytic function on a region containing $\mathrm{Re}\, z \geq 0$. Since F is bounded we may as well assume that $|F(t)| \leq 1, t \geq 0$. For $0 < \lambda < \infty$, define

$$G_\lambda(z) = \int_0^\lambda F(t) e^{-zt} \, dt.$$

By (3.3.3), each function G_λ is entire, and the conclusion of our theorem may be expressed as

$$\lim_{\lambda \to \infty} G_\lambda(0) = G(0).$$

That is, the improper integral $\int_0^\infty F(t) \, dt$ exists and converges to $G(0)$. We begin the analysis by using Cauchy's integral formula to get a preliminary estimate of $|G_\lambda(0) - G(0)|$. For each $R > 0$, let $\delta(R) > 0$ be so small that G is analytic inside and on the closed path

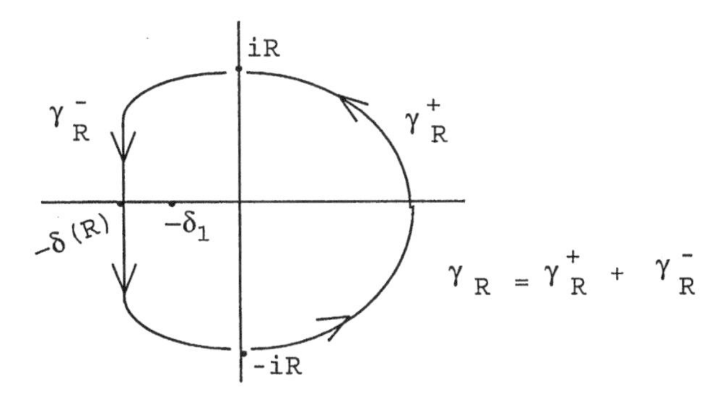

Figure 7.3.1

γ_R in Figure 7.3.1. (Note that since G is analytic on an open set containing $\mathrm{Re}\, z \geq 0$, such a $\delta(R) > 0$ must exist, although it may well be the case that $\delta(R) \to 0$ as $R \to +\infty$.) Let γ_R^+ denote that portion of γ_R that lies in $\operatorname{Re} z > 0$, and γ_R^- the portion that lies in $\operatorname{Re} z < 0$. By Cauchy's integral formula,

$$G(0) - G_\lambda(0) = \frac{1}{2\pi i} \int_{\gamma_R} (G(z) - G_\lambda(z)) \frac{1}{z}\, dz. \tag{1}$$

Let us consider the consequences of estimating $|G(0) - G_\lambda(0)|$ by applying the usual M-L estimates to the integral on the right hand side of (1) above. First, for $z \in \gamma_R^+$ and $x = \operatorname{Re} z$, we have

$$\begin{aligned}
\left|\frac{G(z) - G_\lambda(z)}{z}\right| &= \frac{1}{R}\left|\int_\lambda^\infty F(t)e^{-zt}\, dt\right| \\
&\le \frac{1}{R}\int_\lambda^\infty |F(t)|e^{-xt}\, dt \\
&\le \frac{1}{R}\int_\lambda^\infty e^{-xt}\, dt \\
&= \frac{1}{R}\frac{e^{-\lambda x}}{x} \qquad (2)\\
&\le \frac{1}{R}\frac{1}{x} = \frac{1}{R}\frac{1}{\operatorname{Re} z}.
\end{aligned}$$

But $1/\operatorname{Re} z$ is unbounded on γ_R^+, so we see that a more delicate approach is required to shows that $G(0) - G_\lambda(0) \to 0$ as $\lambda \to \infty$. Indeed, it is here that Newman's ingenuity comes to the fore, and provides us with a modification of the above integral representation for $G(0) - G_\lambda(0)$. This *will* furnish the appropriate estimate. Newman's idea is to replace the factor $1/z$ by $(1/z) + (z/R^2)$ in the path integral in (1). Since $(G(z) - G_\lambda(z))z/R^2$ is analytic, the value of the path integral along γ_R remains unchanged. We further modify (1) by replacing $G(z)$ and $G_\lambda(z)$ by their respective products with $e^{\lambda z}$. Since $e^{\lambda z}$ is entire and has the value 1 at $z = 0$, we can write

$$G(0) - G_\lambda(0) = \frac{1}{2\pi i} \int_{\gamma_R} (G(z) - G_\lambda(z))e^{\lambda z}\left(\frac{1}{z} + \frac{z}{R^2}\right) dz.$$

Note that for $|z| = R$ we have $(1/z) + (z/R^2) = (\overline{z}/|z|^2) + (z/R^2) = (2\operatorname{Re} z)/R^2$, so that if $z \in \gamma_R^+$, (recalling (2) above),

$$|(G(z) - G_\lambda(z))e^{\lambda z}\left(\frac{1}{z} + \frac{z}{R^2}\right)| \le \frac{1}{\operatorname{Re} z} e^{-\lambda \operatorname{Re} z} e^{\lambda \operatorname{Re} z} \frac{2\operatorname{Re} z}{R^2} = \frac{2}{R^2}.$$

Consequently,

$$\left|\frac{1}{2\pi i}\int_{\gamma_R^+} (G(z) - G_\lambda(z))e^{\lambda z}\left(\frac{1}{z} + \frac{z}{R^2}\right) dz\right| \le \frac{1}{R}$$

by the M-L theorem. Note that this estimate of the integral along the path γ_R^+ is independent of λ. Now let us consider the contribution to the integral along γ_R of the integral along γ_R^-. First we use the triangle inequality to obtain the estimate

$$\begin{aligned}&\left|\frac{1}{2\pi i}\int_{\gamma_R^-}(G(z)-G_\lambda(z))e^{\lambda z}(\frac{1}{z}+\frac{z}{R^2})\,dz\right|\\&\le\left|\frac{1}{2\pi i}\int_{\gamma_R^-}G(z)e^{\lambda z}(\frac{1}{z}+\frac{z}{R^2})\,dz\right|+\left|\frac{1}{2\pi i}\int_{\gamma_R^-}G_\lambda(z)e^{\lambda z}(\frac{1}{z}+\frac{z}{R^2})\,dz\right|\\&=|I_1(R)|+|I_2(R)|.\end{aligned}$$

First consider $I_2(R)$. Since $G_\lambda(z)$ is an entire function, we can replace the path of integration γ_R^- by the semicircular path from iR to $-iR$ in the left half plane. For z on this semicircular arc, the modulus of the integrand in $I_2(R)$ is

$$|(\int_0^\lambda F(t)e^{-zt}\,dt)e^{\lambda z}\frac{2\operatorname{Re}z}{R^2})|\le\frac{1}{|\operatorname{Re}z|}\frac{2|\operatorname{Re}z|}{R^2}=\frac{2}{R^2}.$$

(Note that $|F|\le 1$, we can replace the upper limit of integration by ∞, and $e^{\lambda x}\le 1$ for $x\le 0$.) This inequality also holds if $\operatorname{Re}z=0$ (let $z\to iy$). Thus by the M-L theorem we get $|I_2(R)|\le(1/2\pi)(2/R^2)(\pi R)=1/R$, again.

Finally, we consider $|I_1(R)|$. This will be the trickiest of all since we only know that on γ_R^-, G is an analytic *extension* of the explicitly defined G in the right half plane. To deal with this case, first choose a constant $M(R)>0$ such that $|G(z)|\le M(R)$ for $z\in\gamma_R^-$. Choose δ_1 such that $0<\delta_1<\delta(R)$ and break up the integral defining $I_1(R)$ into two parts, corresponding to $\operatorname{Re}z<-\delta_1$ and $\operatorname{Re}z\ge-\delta_1$. The first contribution is bounded in modulus by

$$\frac{1}{2\pi}M(R)e^{-\lambda\delta_1}(\frac{1}{\delta(R)}+\frac{1}{R})\pi R=\frac{1}{2}RM(R)(\frac{1}{\delta(R)}+\frac{1}{R})e^{-\lambda\delta_1},$$

which for fixed R and δ_1 tends to 0 as $\lambda\to\infty$. On the other hand, the second contribution is bounded in modulus by

$$\frac{1}{2\pi}M(R)(\frac{1}{\delta(R)}+\frac{1}{R})2R\arcsin\frac{\delta_1}{R},$$

the last factor arising from summing the lengths of two short circular arcs on the path of integration. Thus for fixed R and $\delta(R)$ we can make the above expression as small as we please by taking δ_1 sufficiently close to 0. So at last we are ready to establish the conclusion of this theorem. Let $\epsilon>0$ be given. Take $R=4/\epsilon$ and fix $\delta(R), 0<\delta(R)<R$, such that G is analytic inside and on γ_R. Then as we saw above, for all λ,

$$\left|\frac{1}{2\pi i}\int_{\gamma_R^+}(G(z)-G_\lambda(z))e^{\lambda z}(\frac{1}{z}+\frac{z}{R^2})\,dz\right|\le\frac{1}{R}=\frac{\epsilon}{4}$$

and also

$$\left|\frac{1}{2\pi i}\int_{\gamma_R^-}(G_\lambda(z)e^{\lambda z}(\frac{1}{z}+\frac{z}{R^2})\,dz\right|\le\frac{1}{R}=\frac{\epsilon}{4}.$$

Now choose δ_1 such that $0 < \delta_1 < \delta(R)$ and such that

$$\frac{1}{2\pi}M(R)(\frac{1}{\delta(R)} + \frac{1}{R})2R\arcsin\frac{\delta_1}{R} < \frac{\epsilon}{4}.$$

Since

$$\frac{1}{2}RM(R)(\frac{1}{\delta(R)} + \frac{1}{R})e^{-\lambda\delta_1} < \frac{\epsilon}{4}$$

for all λ sufficiently large, say $\lambda \geq \lambda_0$, it follows that

$$|G_\lambda(0) - G(0)| < \epsilon,\ \lambda \geq \lambda_0$$

which completes the proof. ♣

Proof of (7.3.2)

Let $f(x)$ and $g(z)$ be as in the statement of the corollary. Define F on $[0, +\infty)$ by

$$F(t) = e^{-t}f(e^t) - c.$$

Then F satisfies the first part of the hypothesis of the auxiliary Tauberian theorem, so let us consider its Laplace transform,

$$G(z) = \int_0^\infty (e^{-t}f(e^t) - c)e^{-zt}\,dt,$$

which via the change of variables $x = e^t$ becomes

$$\begin{aligned}
G(z) &= \int_1^\infty (\frac{1}{x}f(x) - c)x^{-z}\frac{dx}{x} \\
&= \int_1^\infty f(x)x^{-z-2}\,dx - c\int_1^\infty x^{-z-1}\,dx \\
&= \int_1^\infty f(x)x^{-z-2}\,dx - \frac{c}{z} \\
&= \frac{g(z+1)}{z+1} - \frac{c}{z} \\
&= \frac{1}{z+1}[g(z+1) - \frac{c}{z} - c].
\end{aligned}$$

It follows from the hypothesis that $g(z+1) - (c/z)$ has an analytic extension to a neighborhood of the line $\operatorname{Re} z = 0$, and consequently the same is true of the above function G. Thus the hypotheses of the auxiliary Tauberian theorem are satisfied, and we conclude that the improper integral $\int_0^\infty F(t)\,dt$ exists and converges to $G(0)$. In terms of f, this says that $\int_0^\infty (e^{-t}f(e^t) - c)\,dt$ exists, or equivalently (via the change of variables $x = e^t$ once more) that

$$\int_1^\infty (\frac{f(x)}{x} - c)\frac{dx}{x}$$

exists. Recalling that f is nondecreasing, we can infer that $f(x)/x \to c$ as $x \to \infty$. For let $\epsilon > 0$ be given, and suppose that for some $x_0 > 0, [f(x_0)/x_0] - c \geq 2\epsilon$. It follows that

$$f(x) \geq f(x_0) \geq x_0(c+2\epsilon) \geq x(c+\epsilon) \text{ for } x_0 \leq x \leq \frac{c+2\epsilon}{c+\epsilon}x_0.$$

Hence,

$$\int_{x_0}^{\frac{c+2\epsilon}{c+\epsilon}x_0} (\frac{f(x)}{x} - c)\frac{dx}{x} \geq \int_{x_0}^{\frac{c+2\epsilon}{c+\epsilon}x_0} \frac{\epsilon}{x}\, dx = \epsilon \ln(\frac{c+2\epsilon}{c+\epsilon}).$$

But $\int_{x_1}^{x_2} (\frac{f(x)}{x} - c)\frac{dx}{x} \to 0$ as $x_1, x_2 \to \infty$, because the integral from 1 to ∞ is convergent. Thus for all x_0 *sufficiently large*,

$$\int_{x_0}^{\frac{c+2\epsilon}{c+\epsilon}x_0} (\frac{f(x)}{x} - c)\frac{dx}{x} < \epsilon \ln(\frac{c+2\epsilon}{c+\epsilon}).$$

However, reasoning from the assumption that $[f(x_0)/x_0] - c \geq 2\epsilon$, we have just deduced the opposite inequality. We must conclude that for all x_0 sufficiently large, $[f(x_0)/x_0] - c < 2\epsilon$. Similarly, $[f(x_0)/x_0] - c > -2\epsilon$ for all x_0 sufficiently large. [Say $[f(x_0)/x_0] - c \leq -2\epsilon$. The key inequality now becomes

$$f(x) \leq f(x_0) \leq x_0(c-2\epsilon) \leq x(c-\epsilon) \text{ for } (\frac{c-2\epsilon}{c-\epsilon})x_0 \leq x \leq x_0$$

and the limits of integration in the next step are from $\frac{c-2\epsilon}{c-\epsilon}x_0$ to x_0.] Therefore $f(x)/x \to c$ as $x \to \infty$, completing the proof of both the corollary and the prime number theorem. ♣

The prime number theorem has a long and interesting history. We have mentioned just a few of the many historical issues related to the PNT in this chapter. There are several other *number theoretic* functions related to $\pi(x)$, in addition to the function $\psi(x)$ that was introduced earlier. A nice discussion of some of these issues can be found in Eric W. Weisstein, "Prime Number Theorem", from MathWorld—A Wolfram Web Resource, http://mathworld.wolfram.com/PrimeNumberTheorem.html. This source also includes a number of references on PNT related matters.

References

1. Karl Sabbach, "The Riemann Hypothesis-The Greatest Unsolved Problem in Mathematics," Farrar, Strass and Giroux, New York, 2002.

2. Julian Havil, "Gamma-Exploring Euler's Constant," Princeton University Press, Princeton and Oxford, 2003, Chapter 15.

Solutions

Chapter 1

1. $|z_1+z_2|^2+|z_1-z_2|^2=(z_1+z_2)(\overline{z}_1+\overline{z}_2)+(z_1-z_2)(\overline{z}_1-\overline{z}_2)=2|z_1|^2+2|z_2|^2$. A diagram similar to Fig. 1.1.1 illustrates the geometric interpretation that the sum of the squares of the lengths of the diagonals of a parallelogram equals twice the sum of the squares of the lengths of the sides.
2. Again, use a diagram similar to Fig. 1.1.1.
3. (a) Let $z_1=a+bi$, $z_2=c+di$; then $|z_1||z_2|\cos\theta$ is the dot product of the vectors (a,b) and (c,d), that is, $ac+bd=\operatorname{Re} z_1\overline{z}_2$. Also, $|z_1||z_2|\sin\theta$ is the length of the cross product of these vectors, that is, $|ad-bc|=|\operatorname{Im} z_1\overline{z}_2|$. [Strictly speaking, we should take the cross product of the 3-dimensional vectors $(a,b,0)$ and $(c,d,0)$.]

   (b) The area of the triangle is half the area of the parallelogram determined by z_1 and z_2. The area of the parallelogram is the length of the cross product of the vectors (a,b) and (c,d), which is $|\operatorname{Im} z_1\overline{z}_2|$.
4. Say $\frac{\partial g}{\partial x}$ exists near (x_0,y_0) and is continuous at (x_0,y_0), while $\frac{\partial g}{\partial y}$ merely exists at (x_0,y_0). Write, as in (1.4.1),

$$g(x,y)-g(x_0,y_0)=g(x,y)-g(x_0,y)+g(x_0,y)-g(x_0,y_0).$$

   Apply the mean value theorem to the first difference and the definition of $\frac{\partial g}{\partial y}(x_0,y_0)$ to the second difference to obtain

$$\frac{\partial g}{\partial x}(\overline{x},y)(x-x_0)+\frac{\partial g}{\partial y}(x_0,y_0)(y-y_0)+\epsilon(y)(y-y_0)$$

   where $\overline{x}$ is between x_0 and x and $\epsilon(y)\to 0$ as $y\to y_0$. In (1.4.1) we may take

$$A=\frac{\partial g}{\partial x}(x_0,y_0),\quad \epsilon_1(x,y)=\frac{\partial g}{\partial x}(\overline{x},y)-\frac{\partial g}{\partial x}(x_0,y_0),$$

$$B=\frac{\partial g}{\partial y}(x_0,y_0),\quad \epsilon_2(x,y)=\epsilon(y).$$

5. We have $u(x,y)=x$, $v(x,y)=-y$, hence $\frac{\partial u}{\partial x}=1, \frac{\partial v}{\partial y}=-1$. Thus the Cauchy-Riemann equations are never satisfied.

6. Since $u(x,y) = x^2 + y^2, v(x,y) = 0$, the Cauchy-Riemann equations are satisfied at $x = y = 0$, but nowhere else. The result follows from (1.4.2) and Problem 4. (Differentiability at $z = 0$ can also be verified directly, using the definition of the derivative.)

7. Since $u(0,y) = u(x,0) = 0$ for all x, y, $\frac{\partial u}{\partial x}(0,0) = \frac{\partial u}{\partial y}(0,0) = 0$. Take v to be identically 0. If u is real-differentiable at (0,0), then $f = u + iv$ is complex-differentiable at (0,0) by (1.4.2). Now differentiability of f at z_0 requires that $(f(z) - f(z_0))/(z - z_0)$ approach a unique limit as z approaches z_0 along an *arbitrary path*. In the present case, let $z \to 0$ along the line $y = x$. The difference quotient is

$$\frac{\sqrt{x^2}}{x + ix} = \begin{cases} \frac{1}{1+i} & \text{if } x > 0 \\ \frac{-1}{1+i} & \text{if } x < 0. \end{cases}$$

Therefore f is not complex-differentiable at the origin, hence u cannot be real-differentiable there.

8. Let $M_{ab} = \begin{bmatrix} a & b \\ -b & a \end{bmatrix}$, and let $h(a + bi) = M_{ab}$. Then h is 1-1 onto and $h(z_1 + z_2) = h(z_1) + h(z_2)$, $h(z_1 z_2) = h(z_1)h(z_2)$. The result follows.

9. By (b), either $1 \in P$ or $-1 \in P$. Since $i^2 = (-i)^2 = -1$, we have $-1 \in P$ by (a), hence $1 \in P$ by (a) again. But $-1 \in P$ and $1 \in P$ contradicts (b).

10. If $\alpha(z) < 0$, let $w^2 = z$; by (ii), $(\alpha(w))^2 = \alpha(z) < 0$, contradicting $\alpha(w) \in \mathbb{R}$. Thus $\alpha(z) \geq 0$ for all z. Since $\alpha(z^n) = [\alpha(z)]^n$ by (ii), it follows from (iii) that $\alpha(z) \leq 1$ for $|z| = 1$. By (i) and (ii), $|z|^2 = a(|z|^2) = \alpha(z)\alpha(\overline{z})$, so for z on the unit circle, $\alpha(z) < 1$ implies $\alpha(\overline{z}) > 1$, and therefore $\alpha(z) = 1$ for $|z| = 1$. Thus for arbitrary $z \neq 0$ we have $\alpha(z) = \alpha(z/|z|)\alpha(|z|) = \alpha(|z|) = |z|$.

11. As in Problem 10, $\alpha(z) \geq 0$ for all z. Also, $x^2 = \alpha(x^2) = \alpha(-x)\alpha(x)$, and consequently $\alpha(-x) = x, x \geq 0$. Thus $\alpha(x) = |x|$ for real x. If $z = x + iy$, then $\alpha(z) \leq \alpha(x) + \alpha(i)\alpha(y) = |x| + |y|$. (Note that $(\alpha(i))^2 = \alpha(i^2) = \alpha(-1) = 1$, so $\alpha(i) = 1$.) Therefore α is bounded on the unit circle, and the result follows from Problem 10.

12. Since $|z-\alpha|^2 = (z-\alpha)(\overline{z}-\overline{\alpha})$ and $|1-\overline{\alpha}z|^2 = (1-\overline{\alpha}z)(1-\alpha\overline{z})$, we have $|z-\alpha| = |1-\overline{\alpha}z|$ iff $z\overline{z} - \alpha\overline{z} - \overline{\alpha}z + \alpha\overline{\alpha} = 1 - \alpha\overline{z} - \overline{\alpha}z + \alpha\overline{\alpha}z\overline{z}$ iff $z\overline{z} - 1 = \alpha\overline{\alpha}(z\overline{z} - 1)$. Since $|\alpha| < 1$, this can happen iff $z\overline{z} = 1$, that is, $|z| = 1$.

13. If $z = r\cos\theta + ir\sin\theta$, then $1/z = (1/r)\cos\theta - i(1/r)\sin\theta$, so $z + 1/z = (r + 1/r)\cos\theta + i(r - 1/r)\sin\theta$, which is real iff $\sin\theta = 0$ or $r - 1/r = 0$. The result follows.

14. To show that u is harmonic, verify directly that $\partial^2 u/\partial x^2 + \partial^2 u/\partial y^2 = 0$. To find v, use the technique of (1.6.2). In part (i) we have

$$\frac{\partial v}{\partial x} = -\frac{\partial u}{\partial y} = -e^y \cos x, \qquad \frac{\partial v}{\partial y} = \frac{\partial u}{\partial x} = -e^y \sin x.$$

Thus (using calculus) $v(x,y) = -e^y \sin x$. In part (ii) we have

$$\frac{\partial v}{\partial x} = -\frac{\partial u}{\partial y} = -6xy, \qquad \frac{\partial v}{\partial y} = \frac{\partial u}{\partial x} = 2 - 3x^2 + 3y^2$$

so $v(x,y) = -3x^2y + 2y + y^3$. Note that if $z = x + iy$, then $u + iv$ can be written as $-z^3 + 2z$. After complex exponentials are studied further in Section 2.3, it will follow that in part (i), $u + iv = e^{-iz}$.

15. (i) Note that $|z - z_0| = r$ iff $|az + b - (az_0 + b)| = r|a|$.
(ii) $T(0) = 1 + i$, so $b = 1 + i$; $r|a| = |a| = 2$, so $T(z) = az + 1 + i$, $|a| = 2$.
(iii) Since $|-2 + 2i| > 2$, the desired result cannot be accomplished.
16. Since $u = e^x$, $v = 0$, the Cauchy-Riemann equations are never satisfied.
17. We have

$$\frac{g(z+h) - g(z)}{h} = \frac{\overline{f}(\overline{z} + \overline{h}) - \overline{f}(\overline{z})}{h} = \left[\overline{\frac{f(\overline{z} + \overline{h}) - f(\overline{z})}{\overline{h}}}\right].$$

Thus g is analytic at z iff f is analytic at $\overline{z}$, and in this case, $g'(z) = \overline{f'(\overline{z})}$. Since $z \in \Omega$ iff $\overline{z} \in \Omega$, the result follows.

18. The circle is described by $|z - z_0|^2 = r^2$, or, equivalently, $(z - z_0)(\overline{z} - \overline{z}_0) = r^2$; the result follows.
19. If $P(z) = 0$ for some $z \in D(0,1)$, then $(1 - z)P(z) = 0$, that is, $(1 - z)(a_0 + a_1 z + \cdots + a_n z^n) = 0$, which implies that

$$a_0 = (a_0 - a_1)z + (a_1 - a_2)z^2 + \cdots + (a_{n-1} - a_n)z^n + a_n z^{n+1}. \tag{1}$$

Since $a_i - a_{i+1} \geq 0$, the absolute value of the right side of (1) is at most $|z|(a_0 - a_1 + a_1 - a_2 + \cdots + a_{n-1} - a_n + a_n) = a_0|z|$. If $|z| < 1$, this is less than a_0, a contradiction.

20. If $P(z) = 0$ for some z with $|z| \leq 1$, then $|z| = 1$ by Problem 19. The only way for (1) in Problem 19 to be satisfied is if all terms $(a_0 - a_1)z, \ldots, (a_{n-1} - a_n)z^n, a_n z^{n+1}$ are nonnegative multiples of one another (cf. Problem 2), and this requires that z be real, i.e., $z = 1$. But $P(1) = a_0 + \cdots + a_n > 0$, so there are no roots in $\overline{D}(0,1)$.

Chapter 2

Section 2.1

1. We have $\gamma(t) = (1 - t)(-i) + t(1 + 2i) = t + i(3t - 1), 0 \leq t \leq 1$; thus

$$\int_\gamma y\,dz = \int_0^1 (\operatorname{Im}\gamma(t))\gamma'(t)\,dt = \int_0^1 (3t - 1)(1 + 3i)\,dt = \frac{1}{2} + i\frac{3}{2}.$$

2. We have $\gamma(t) = t + it^2, 1 \leq t \leq 2$; thus

$$\int_\gamma \overline{z}\,dz = \int_1^2 \overline{\gamma(t)}\,\gamma'(t)\,dt = \int_1^2 (t - it^2)(1 + i2t)\,dt = 9 + i\frac{7}{3}.$$

Intuitively,

$$\int_\gamma \overline{z}\,dz = \int_\gamma (x - iy)(dx + idy) = \int_\gamma x\,dx + y\,dy + i(x\,dy - y\,dx).$$

Since $y = x^2$ on γ^*, this becomes

$$\int_1^2 [x\,dx + x^2(2x\,dx) + ix(2x\,dx) - ix^2\,dx]$$

as above. Note also that, for example, $\int_\gamma x\,dy = \int_2^4 \sqrt{y}\,dy$.

3. The first segment may be parametrized as $(1-t)(-i)+t(2+5i) = 2t+i(6t-1), 0 \le t \le 1$, and the second segment as $(1-t)(2+5i) + t5i = 2 - 2t + 5i, 0 \le t \le 1$. Thus

$$\begin{aligned}\int_\gamma f(z)\,dz &= \int_0^1 [i\,\mathrm{Im}\,\gamma(t) + (\mathrm{Re}\,\gamma(t))^2]\gamma'(t)\,dt\\ &= \int_0^1 [i(6t-1) + 4t^2](2+6i)\,dt\\ &\quad + \int_0^1 [5i + (2-2t)^2](-2)\,dt\\ &= -\frac{28}{3} + 12i - \frac{8}{3} - 10i = -12 + 2i.\end{aligned}$$

4. Since γ is a path and h is continuously differentiable, it follows that γ_1 is a path. We have, with $s = h(t)$,

$$\begin{aligned}\int_{\gamma_1} f(z)\,dz &= \int_c^d f(\gamma_1(t))\gamma_1'(t)\,dt = \int_c^d f(\gamma(h(t)))\gamma'(h(t))h'(t)\,dt\\ &= f(\gamma(s))\gamma'(s)\,ds = \int_\gamma f(z)\,dz.\end{aligned}$$

(Strictly speaking, this argument is to be applied separately to the subintervals on which γ_1' is continuous.)

5. (a) By (2.1.6), $f(z_2) - f(z_1) = \int_{[z_1,z_2]} f'(w)\,dw$. If $w = (1-t)z_1 + tz_2$, we obtain $f(z_2) - f(z_1) = (z_2 - z_1)\int_0^1 f'((1-t)z_1 + tz_2)\,dt$. Since $\mathrm{Re}\,f' > 0$ by hypothesis, we have $\mathrm{Re}[(f(z_2) - f(z_1))/(z_2 - z_1)] > 0$. In particular, $f(z_1) \ne f(z_2)$.
(b) For $f(z) = z + 1/z$, we have $f'(z) = 1 - 1/z^2$, so in polar form, $\mathrm{Re}\,f'(re^{i\theta}) = 1 - (\cos 2\theta)/r^2$, which is greater than 0 iff $r^2 > \cos 2\theta$. By examining the graph of $r^2 = \cos 2\theta$ (a two-leaved rose), w see that for $a > 0$ and sufficiently large, and $\delta > 0$ and sufficiently small, we have $\mathrm{Re}\,f' > 0$ on $\Omega = \mathbb{C}\setminus A$, where A is the set of points inside or on the boundary of the infinite "triangle" determined by the rays $[a, (1-\delta)i, \infty)$ and $[a, (1-\delta)(-i), \infty)$. Now Ω is starlike and contains $\pm i$, with $f(i) = f(-i)$, which proves that (a) does not generalize to starlike regions.
(c) Since $f'(z_0) \ne 0$, either $\mathrm{Re}\,f'(z_0) \ne 0$ or $\mathrm{Im}\,f'(z_0) \ne 0$. If the real part is nonzero, then $\mathrm{Re}\,f'$ must be of constant sign (positive or negative) on a sufficiently small disk centered at z_0. The result then follows from (a). The remaining case is handled by observing that $\mathrm{Im}\,f' = \mathrm{Re}(-if') = \mathrm{Re}[(-if)']$.

Section 2.2

1. The statement about pointwise convergence follows because $\mathbb{C}$ is a complete metric space. If $f_n \to f$ uniformly on S, then $|f_n(z) - f_m(z)| \le |f_n(z) - f(z)| + |f(z) - f_m(z)|$, hence $\{f_n\}$ is uniformly Cauchy. Conversely, if $\{f_n\}$ is uniformly Cauchy, it is pointwise Cauchy and therefore converges pointwise to a limit function f. If $|f_n(z) - f_m(z)| \le \epsilon$ for all $n, m \ge N$ and all $z \in S$, let $m \to \infty$ to show that $|f_n(z) - f(z)| \le \epsilon$ for $n \ge N$ and all $z \in S$. Thus $f_n \to f$ uniformly on S.

2. This is immediate from (2.2.7).

3. We have $f'(x) = (2/x^3)e^{-1/x^2}$ for $x \ne 0$, and $f'(0) = \lim_{h\to 0}(1/h)e^{-1/h^2} = 0$. Since $f^{(n)}(x)$ is of the form $p_n(1/x)e^{-1/x^2}$ for $x \ne 0$, where p_n is a polynomial, an induction argument shows that $f^{(n)}(0) = 0$ for all n. If g is analytic on $D(0,r)$ and $g = f$ on $(-r,r)$, then by (2.2.16), $g(z) = \sum_{n=0}^{\infty} f^{(n)}(0)z^n/n!$, $z \in D(0,r)$. [Note that $f^{(n)}(0)$ is determined once f is specified on $(-r,r)$.] Thus g, hence f, is 0 on $(-r,r)$, a contradiction.

4. (a) The radius of convergence is at least $1/\alpha$. For if $\alpha = \infty$, this is trivial, and if $\alpha < \infty$, then for a given $\epsilon > 0$, eventually $|a_{n+1}/a_n| < \alpha + \epsilon$, say for $n \ge N$. Thus $|a_{N+k}z^{N+k}| \le |a_N||z|^N|(\alpha+\epsilon)z|^k, k = 0, 1, \dots$. By comparison with a geometric series, the radius of convergence is a least $1/(\alpha+\epsilon)$. Since ϵ is arbitrary, the result follows.
 Note that the radius of convergence may be greater than $1/\alpha$. for example, let $a_n = 2$ if n is even, and $a_n = 1$ if n is odd. The radius of convergence is 1, but $\limsup_{n\to\infty} |a_{n+1}/a_n| = 2$, so $1/\alpha = 1/2$.
 (b) The radius of convergence r is exactly $1/\alpha$. For $r \ge 1/\alpha$ by (a), and on the other hand we have $\lim_{n\to\infty} |a_{n+1}z^{n+1}/a_n z^n| = \alpha|z|$, which is greater than 1 if $|z| > 1/\alpha$. Thus $\lim_{n\to\infty} a_n z^n$ cannot be 0, and hence the series cannot converge, for $|z| > 1/\alpha$. [This is just the ratio test; see (2.2.2).]

5. Since $a_n = f^{(n)}(z_0)/n!$, we have $\limsup_{n\to\infty} |a_n|^{1/n} \ge \limsup_{n\to\infty}(b_n)^{1/n}$. The radius of convergence of the Taylor expansion bout z_0 is therefore 0, a contradiction.

6. (a) As in (2.2.16), write

$$f(z) = \frac{1}{2\pi i}\int_\Gamma \frac{f(w)}{w - z_0}\left[\frac{1}{1 - \frac{z-z_0}{w-z_0}}\right] dw.$$

 The term in brackets is

$$1 + \frac{z-z_0}{w-z_0} + \cdots + \left(\frac{z-z_0}{w-z_0}\right)^n + \frac{\left(\frac{z-z_0}{w-z_0}\right)^{n+1}}{1 - \frac{z-z_0}{w-z_0}}.$$

 By (2.2.11), $f(z) = \sum_{k=0}^{n}[f^{(k)}(z_0)(z-z_0)^k/k!] + R_n(z)$, where

$$R_n(z) = \frac{(z-z_0)^{n+1}}{2\pi i}\int_\Gamma \frac{f(w)}{(w-z)(w-z_0)^{n+1}}\, dw.$$

(b) If $|z - z_0| \le s < r_1$, then $|w - z| \ge r_1 - s$ for all $w \in \Gamma$, hence by (2.1.5),

$$|R_n(z)| \le \frac{|z - z_0|^{n+1}}{2\pi} \frac{M_f(\Gamma)}{(r_1 - s)r_1^{n+1}} 2\pi r_1 \le M_f(\Gamma)\frac{r_1}{r_1 - s}\left(\frac{s}{r_1}\right)^{n+1}.$$

7. We compute $\sum_{k=r}^{s} a_k \Delta b_k = a_r(b_{r+1} - b_r) + \cdots + a_s(b_{s+1} - b_s) = -a_r b_r + b_{r+1}(a_r - a_{r+1}) + \cdots + b_s(a_{s-1} - a_s) + a_s b_{s+1} - a_{s+1}b_{s+1} + a_{s+1}b_{s+1}$, and the result follows.
8. (a) If $|b_n| \le M$ for all n, then $\sum_{k=r}^{s} |b_{k+1}\Delta a_k| \le M \sum_{k=r}^{s} |\Delta a_k| = M(a_r - a_{r+1} + a_{r+1} - a_{r+2} + \cdots + a_s - a_{s+1}) = M(a_r - a_{s+1}) \to 0$ as $r, s \to \infty$. The result follows from Problem 7.
(b) By the argument of (a), $\sum_{k=r}^{s} a_k \Delta b_k(z) \to 0$ as $r, s \to \infty$, uniformly for $z \in S$.
9. (a) Let $a_n = 1/n, b_n(z) = \sum_{k=0}^{n-1} z^k = (1 - z^n)/(1 - z)$ if $z \ne 1$. For any fixed z with $|z| = 1, z \ne 1$, we have $|b_n(z)| \le 2/|1 - z| < \infty$ for all n, and the desired result follows from Problem 8(a).
(b) Let $a_n = 1/n$ and $b_n = \sum_{k=0}^{n-1} \sin kx = \operatorname{Im}(1 + e^{ix} + e^{i2x} + \cdots + e^{i(n-1)x}) = \operatorname{Im}[(1 - e^{inx})/(1 - e^{ix})]$ (if x is not an integral multiple of 2π; the series converges to 0 in that case). Now

$$\left|\frac{1 - e^{inx}}{1 - e^{ix}}\right|^2 = \frac{1 - \cos nx}{1 - \cos x} = \frac{\sin^2(nx/2)}{\sin^2(x/2)} \le \frac{1}{\sin^2(x/2)}$$

which is uniformly bounded on $\{x : 2k\pi + \delta \le x \le (2k + 2)\pi - \delta\}$. The result follows from Problem 8(b).
(c) Let $\sin nz = \sin n(x+iy) = (e^{inz} - e^{-inz})/2i = (e^{inx}e^{-ny} - e^{-inx}e^{ny})/2i$. If $y \ne 0$, then $(1/n)\sin nz \to \infty$ as $n \to \infty$, hence $\sum_n (1/n)\sin nz$ cannot converge.
10. If $z \notin \mathbb{C}^+ \cup \mathbb{R}$, then

$$\frac{f^*(z + h) - f^*(z)}{h} = \frac{\overline{f}(\overline{z} + \overline{h}) - \overline{f}(\overline{z})}{h} = \left[\overline{\frac{f(\overline{z} + \overline{h}) - f(\overline{z})}{\overline{h}}}\right] \to \overline{f'(\overline{z})} \text{ as } h \to 0.$$

Thus f^* is analytic on $\mathbb{C} \setminus \mathbb{R}$. On $\mathbb{R}$ we have $z = \overline{z}$ and $f(z) = \overline{f(z)} = \overline{f}(\overline{z})$, so f^* is continuous on $\mathbb{C}$.
11. The idea is similar to (2.1.12). If T is a triangle in $\mathbb{C}$, express $\int_T f^*(z)\,dz$ as a sum of integrals along polygons whose interiors are entirely contained in $\mathbb{C}^+$ or in the open lower half plane $\mathbb{C}^-$, and at worst have a boundary segment on $\mathbb{R}$. But, for example, $\int_{[a+i\delta, b+i\delta]} f^*(z)\,dz \to \int_{[a,b]} f^*(z)\,dz$ as $\delta \to 0$ (use the M-L theorem). It follows that $\int_T f^*(z)\,dz = 0$, and f^* is analytic on $\mathbb{C}$ by Morera's theorem.
12. (a) By (2.2.10), F is analytic on $\mathbb{C} \setminus C(z_0, r)$ and $F'(z) = \int_{C(z_0,r)} (w - z)^{-2}\,dw$. But for any fixed z, the function h given by $h(w) = 1/(w - z)^2$ has a primitive, namely $1/(w - z)$, on $\mathbb{C} \setminus \{z\}$. Thus by (2.1.6), $F'(z) = 0$. By (2.1.7b), F is constant on $D(z_0, r)$.
(b) We have $F(z_0) = 2\pi i$ [see the end of the proof of (2.2.9)] and thus by (a), $F(z) = 2\pi i$ for all $z \in D(z_0, r)$. As in the proof of (2.2.9),

$$\frac{1}{2\pi i}\int_{C(z_0,r)} \frac{f(w)}{w - z}\,dw = \frac{f(z)}{2\pi i}\int_{C(z_0,r)} \frac{1}{w - z}\,dw = f(z) \text{ by part (b).}$$

13. (a) This follows from (2.2.11), (2.1.4) and (2.1.2).
(b) By part (a) with $a = 0$, $|f^{(n)}(0)/n!| \le Mr^k/r^n \to 0$ as $r \to \infty$, if $n > k$. Thus the Taylor coefficient a_n is 0 for $n > k$, and the result follows.
(c) The statement and proof of (b) go through even if k is a nonnegative number, not necessarily an integer. Take $k = 3/2$ to obtain (c).

14. If $|z| < 1$, then $\sum_{n=0}^{\infty} |a_n z^n| < \infty$, hence the radius of convergence r is at least 1. If $r > 1$, then the series for $f'(z)$, namely $\sum_{n=1}^{\infty} n a_n z^{n-1}$, will converge absolutely when $|z| = 1$, so $\sum_{n=1}^{\infty} n|a_n| < \infty$, a contradiction. Thus $r = 1$.

15. Let T be a triangle such that $\hat{T} \subseteq \Omega$. By (2.1.8), $\int_T f_n(z)\,dz = 0$ for every n. Since $f_n \to f$ uniformly on T, we have $\int_T f(z)\,dz = 0$. By Morera's theorem, f is analytic on Ω.

Section 2.3

1. If $u + iv = \sin(x + iy)$, then $u = \sin x \cosh y$ and $v = \cos x \sinh y$. If $y = b$, then $(u^2/\cosh^2 b) + (v^2/\sinh^2 b) = 1$. Thus $\{x + iy : -\pi/2 < x < \pi/2, y > 0\}$ is mapped onto $\{u+iv : v > 0\}$, $\{x+iy : -\pi/2 < x < \pi/2, y < 0\}$ is mapped onto $\{u+iv : v < 0\}$, $\{x + iy : x = \pi/2, y \ge 0\}$ is mapped onto $\{u + iv : v = 0, u \ge 1\}$, and finally $\{x + iy : x = -\pi/2, y \le 0\}$ is mapped onto $\{u + iv : v = 0, u \le -1\}$, and the mapping is one-to-one in each case. Since $\sin(z + \pi) = -\sin z$, the statement of the problem follows.

2. If $\sin(x + iy) = 3$, then $\sin x \cosh y = 3$, $\cos x \sinh y = 0$. If $\sinh y = 0$ then $y = 0, \cosh y = 1, \sin x = 3$, which is impossible. Thus $\cos x = 0, x = (2n + 1)\pi/2$. If n is odd then $\sin x = -1, \cosh y = -3$, again impossible. Thus the solutions are $z = x + iy$ where $x = (4m + 1)\pi/2$, m an integer, y such that $\cosh y = 3$ (two possibilities, one the negative of the other).

3. Since $\sin z = z - z^3/3! + x^5/5! - \cdots$, the only nonzero contribution to the integral is the single term $-\int_{C(0,1)} dz/3!z = -2\pi i/6 = -\pi i/3$.

4. This follows from two observations:
(a) $1 + z + z^2/2! + \cdots + z^n/n! \to e^z$ as $n \to \infty$, uniformly for $|z| \le r$;
(b) $\min_{|z|\le r} |e^z| > 0$.

5. If $f(z) = \sum_{n=0}^{\infty} a_n z^n$, then since $f'' + f = 0$ we have $n(n-1)a_n + a_{n-2} = 0, n = 2, 3 \dots$. Since $f(0) = 0$ and $f'(0) = 1$, we have $a_0 = 0, a_1 = 1$, hence $a_2 = 0, a_3 = -1/3!, a_4 = 0, a_5 = 1/5!$, and so on. Thus $f(z) = z - z^3/3! + z^5/5! - z^7/7! + \cdots = \sin z$.

6. As in Problem 5, $na_n - a_{n-1} = 0, a_0 = 1$. Therefore $f(z) = 1 + z + z^2/2! + \cdots = e^z$.

Section 2.4

1. Take $f(z) = \sin(1/z), \Omega = \mathbb{C} \setminus \{0\}$; then f has zeros at $1/n\pi \to 0 \notin \Omega$.

2. If $f(z) = (z - z_0)^m g(z)$ on Ω with $g(z_0) \ne 0$, expand g in a Taylor series about z_0 to conclude that $a_j = 0$ for $j < m$ and $a_m \ne 0$. Conversely, if $a_0 = \cdots = a_{m-1} = 0, a_m \ne 0$, then $f(z) = \sum_{n=m}^{\infty} a_n(z - z_0)^n = (z - z_0)^m g(z)$ with $g(z_0) \ne 0$. (Strictly speaking, this holds only on some disk $D(z_0, r)$, but g may be extended to all of Ω by the formula $f(z)/(z - z_0)^m$.) The remaining statement of (2.4.5) follows from (2.2.16).

3. Let f be continuous on the region Ω.
(i) If f satisfies (b) of the maximum principle, f need not satisfy (a). For example, let $\Omega = D(0,2)$ and $f(z) = 1, |z| \le 1; f(z) = |z|, 1 < |z| < 2$.
(ii) If f satisfies (c), then f satisfies (b), hence (b) and (c) are equivalent (assuming Ω is bounded). This is because f must satisfy (c) if we take $M = \lambda$, and consequently f satisfies (b).
(iii) If Ω is bounded and f is continuous on $\overline{\Omega}$, then (d) implies (b), hence in this case (b), (c) and (d) are equivalent. For let z_0 be a point on the boundary of Ω such that $|f(z_0)| = M_0 = \max\{|f(z)| : x \in \partial\Omega\}$. Since $z_0 \in \partial\Omega$, there is a sequence of points $z_n \in \Omega$ with $z_n \to z_0$, hence $|f(z_n)| \to |f(z_0)| = M_0$. Thus $\lambda = \sup\{|f(z)| : z \in \Omega\} \ge M_0$. If $|f| < M_0$ on Ω, then $|f| < \lambda$ on Ω.

4. In a neighborhood of z_0, we have

$$\frac{f(z)}{g(z)} = \frac{a_m(z-z_0)^m + a_{m+1}(z-z_0)^{m+1} + \cdots}{b_n(z-z_0)^n + b_{n+1}(z-z_0)^{n+1} + \cdots}$$

where $a_m \ne 0, b_n \ne 0$ (that is, f has a zero of order m and g a zero of order n at z_0). Then

$$\lim_{z\to z_0} \frac{f(z)}{g(z)} = \lim_{z\to z_0} \frac{f'(z)}{g'(z)} = \begin{cases} a_m/b_m & \text{if } m = n \\ 0 & \text{if } m > n \\ \infty & \text{if } m < n \end{cases}.$$

(To handle the last case, apply the second case to g/f.)

5. Immediate from (2.4.12).

6. $\operatorname{Im} f = 0$ on ∂D, hence by part (d) of the maximum and minimum principles for harmonic functions (see (2.4.15) and its accompanying remark), $\operatorname{Im} f(z) = 0$ for all $z \in D$. Thus f is constant on D by the Cauchy-Riemann equations.

7. By the maximum principle, we need only consider ∂K. Now $\sin(x+iy) = \sin x \cosh y + i \cos x \sinh y$. If $x = 0$ or 2π, then $\sin(x+iy) = i \sinh y$. If $y = 0$, then $\sin(x+iy) = \sin x$. If $y = 2\pi$, then $\sin(x+iy) = \cosh 2\pi \ \sin x + i \sinh 2\pi \ \cos x$. Since $\cosh 2\pi > \sinh 2\pi > 1$, it follows that the maximum modulus is attained at $x = \pi/2$ or $3\pi/2, y = 2\pi$, and $\max|f| = \cosh 2\pi$.

8. Choose $z_0 \in K$ such that $|f(z_0)| = \max\{|f(z)| : z \in K\}$. If $z_0 \in \partial K$, we are finished, so assume that $z_0 \in K^0$. By (2.4.12a), f is constant on the component Ω_0 of Ω that contains z_0, which proves the "furthermore" part. To see that $|f(z_0)| = \max\{|f(z)| : z \in \partial K\}$, note that by continuity, f must also be constant on $\overline{\Omega}_0 \subseteq \overline{K}_0 \subseteq K$. Since Ω_0 is bounded, its boundary is not empty. Choose any $z_1 \in \partial\Omega_0$. Then $f(z_0) = f(z_1)$ since $z_1 \in \overline{\Omega}_0$, so $|f(z_1)| = \max\{|f(z)| : z \in K\}$. But $z_1 \in K$ and z_1 is not an interior point of K. (If $z_1 \in K^0$ then $D(z_1, r) \subseteq K^0$ for some $r > 0$, and it would not be possible for z_1 to be a boundary point of a component of K^0.) Consequently, $z_1 \in \partial K$, and

$$\max\{|f(z)| : z \in K\} = |f(z_1)| \le \max\{|f(z)| : z \in \partial K\} \le \max\{|f(z)| : z \in K\}.$$

The result follows.

9. By Problem 8, $\max_{z\in\overline{\Omega}} |f(z)| = \max_{z\in\partial\overline{\Omega}} |f(z)|$. But $\partial\overline{\Omega} = \overline{\Omega}\setminus\Omega = \partial\Omega$, and the result follows.

10. Take $u = \operatorname{Im} f$ where f is a nonconstant entire function that is real-valued on $\mathbb{R}$. For example, $f(z) = e^z, u(x+iy) = e^x \sin y$; or $f(z) = z, u(x+iy) = y$.

11. If Ω is disconnected and A is a component of Ω, let $f(z) = 1$ if $z \in A$, and $f(z) = 0$ if $z \notin A$. Let $g(z) = 0$ if $z \in A$, and $g(z) = 1$ if $z \notin A$. Then $fg \equiv 0$ but $f \not\equiv 0, g \not\equiv 0$. Assume Ω connected, and let f, g be analytic on Ω with $fg \equiv 0$. If $f(z_0) \neq 0$, then f is nonzero on some disk $D(z_0, r)$, hence $g \equiv 0$ on $D(z_0, r)$. By (2.4.8), $g \equiv 0$ on Ω.

12. The given function can be extended to a function f^* analytic on $S \cup \{z : \operatorname{Im} z < 0\}$ by the technique of the Schwarz reflection principle (2.2.15). Since $f^*(z) = z^4 - 2z^2$ for $z \in (0,1)$, the identity theorem (2.4.8) implies that this relation holds for all $z \in S \cup \{z : \operatorname{Im} z < 0\}$. Thus if $z \in S$ we have $f(z) = z^4 - 2z^2$, and in particular, $f(i) = 3$.

13. Apply Liouville's theorem to $1/f$.

14. No, by the identity theorem. If S is an uncountable set, then infinitely many points of S must lie in some disk $D(0, r)$, hence S has a limit point.

15. Fix the real number α. Then $\sin(\alpha+\beta) - \sin\alpha\cos\beta - \cos\alpha\sin\beta$ is an analytic function of β, and is zero for real β, hence is identically zero by the identity theorem. A repetition of this argument with fixed β and variable α completes the demonstration.

16. If $f = u + iv$, then $|e^{if}| = e^{-v} \leq 1$ (because $v \geq 0$ by hypothesis). By Liouville's theorem, e^{if} is constant, hence $|e^{if}| = e^{-v}$ is constant, so v is constant. But then by the Cauchy-Riemann equations, u is constant, so f is constant.

17. We have $(f/g)' = (gf' - fg')/g^2$, and by the identity theorem, $gf' - fg'$ is identically zero on $D(0,1)$. The result follows.

18. By Liouville's theorem, $f(z) - e^z \sin z = c$ where $|c| < 4$. Since $f(0) = 0$ we have $c = 0$, so $f(z) = e^z \sin z$.

19. By the maximum and minimum principles for harmonic functions, $\operatorname{Re}(f - g)$ is identically zero. Therefore $f - g$ is constant.

20. If f is never 0 and $\{f(z_n)\}$ is unbounded whenever $|z_n| \to 1$, then $1/f(z) \to 0$ as $|z| \to 1$. By the maximum principle, $1/f \equiv 0$, a contradiction.

21. Let $f = u + iv$ with f analytic on $\mathbb{C}$. Then $|e^f| = e^u \geq e^0 = 1$, hence $|e^{-f}| \leq 1$. By Liouville's theorem, e^{-f} is constant. But then $|e^{-f}|$, hence $|e^f|$, is constant. Since $|e^f| = e^u$, the result follows.

22. If f is never 0 in $D(0,1)$ then by the maximum principle, $\max_{|z|\leq 1} |1/f(z)| < 1$, hence $|1/f(0)| < 1$. This contradicts $f(0) = i$.

23. If $z = x + iy$ then $u = \operatorname{Re} z^3 = x^3 - 3xy^2$. By the maximum principle, it suffices to consider u on each of the four line segments forming the boundary of the square. By elementary calculus we find that the maximum value is 1 and occurs at $x = 1, y = 0$.

24. If K is a compact subset of $D(0,1)$, then $K \subseteq \overline{D}(0,r)$ for some $r \in (0,1)$. If $M = \max\{|f(z)| : |z| \leq r\}$, then $[|f(rz)|/M] \leq 1$ on $\overline{D}(0,1)$. By (2.4.16), $[|f(rz)|/M] \leq |z|$ on $\overline{D}(0,1)$. Make the substitution $z = w/r$ to obtain $|f(w)| \leq M|w|/r$ on $\overline{D}(0,r)$,

hence on K. The result now follows from the uniform convergence of the series $\sum |w|^n, |w| \le r$.

25. If $z_n \to z \in C(0,1)$, then for some k, $e^{ik\beta}z \in A_1$, so $f(e^{ik\beta}z_n) \to 0$, and therefore $F(z_n) \to 0$. By (2.4.12c), $F \equiv 0$. Now for any $z \in D(0,1), F(z) = 0$, so $f(e^{ih\beta}z) = 0$ for some $h = 0, 1, \dots, n$. Thus f has uncountably many zeros, hence a limit point of zeros, in $D(0,1)$. By the identity theorem, $f \equiv 0$.

26. (a) By Problem 9, $\{f_n\}$ is uniformly Cauchy on $\overline{\Omega}$, hence by (2.2.4), $\{f_n\}$ converges uniformly on $\overline{\Omega}$. By (2.2.17), f is analytic on Ω (and continuous on $\overline{\Omega}$ by the uniform convergence). The proof of (2.2.17), in particular the formula (2.2.11), may be adapted to show that each derivative $f_n^{(k)}$ extends to a continuous function on $\overline{\Omega}$, and that $f_n^{(k)} \to f^{(k)}$ uniformly on $\overline{\Omega}$ for all k.
(b) If $p_1, p_2, \dots$ are polynomials and $p_n \to f$ uniformly on $C(0,1)$, then by (a), p_n converges uniformly on $\overline{D}(0,1)$ to a limit function g, where g is analytic on $D(0,1)$ and continuous on $\overline{D}(0,1)$ (and of course $g = f$ on $C(0,1)$. Conversely, if f is the restriction to $C(0,1)$ of such a function g, then f can be uniformly approximated by polynomials. To see this, let $\{r_n\}$ be an increasing sequence of positive reals converging to 1, and consider $g_n(z) = g(r_n z), |z| < 1/r_n$. Since g_n is analytic on $D(0, 1/r_n)$, there is a (Taylor) polynomial p_n such that $|p_n(z) - g_n(z)| < 1/n$ for $|z| \le 1$. But g_n converges uniformly to g on $\overline{D}(0,1)$ by uniform continuity of g on $\overline{D}(0,1)$. The result follows.

Chapter 3

Sections 3.1 and 3.2

1. (a) This follows because $\log_\alpha$ is discontinuous on the ray $\mathbb{R}_\alpha$ [see (3.1.2b)].
(b) Let U be as indicated in Figure S3.2.1, and define

$$g(z) = \begin{cases} \ln|z| + i\theta(z), 0 \le \theta < 2\pi, & \text{for } z \in \Omega_1 \\ \ln|z| + i\theta(z), \pi \le \theta < 3\pi, & \text{for } z \in \Omega_2. \end{cases}$$

Locally, $g(z)$ coincides with one of the elementary branches of $\log z$, hence g is an analytic version of $\log z$ on Ω.

2. First, we show that f does not have an analytic logarithm on Ω. For $f'(z)/f(z) = [1/(z-a)] + [1/(z-b)]$, so that if γ describes a circle enclosing both a and b, (3.2.3) yields $\int_\gamma [f'(z)/f(z)]\,dz = 2\pi i(n(\gamma,a) + n(\gamma,b)) = 4\pi i \ne 0$. By (3.1.9), f does not have an analytic logarithm on Ω. However, f has an analytic square root. For if θ_0 is the angle of $[a,b]$ (see Figure S3.2.2), then define

$$(z-a)^{1/2} = |z-a|^{1/2} \exp(i\frac{1}{2} \arg(z-a))$$

$$(z-b)^{1/2} = |z-b|^{1/2} \exp(i\frac{1}{2} \arg(z-b))$$

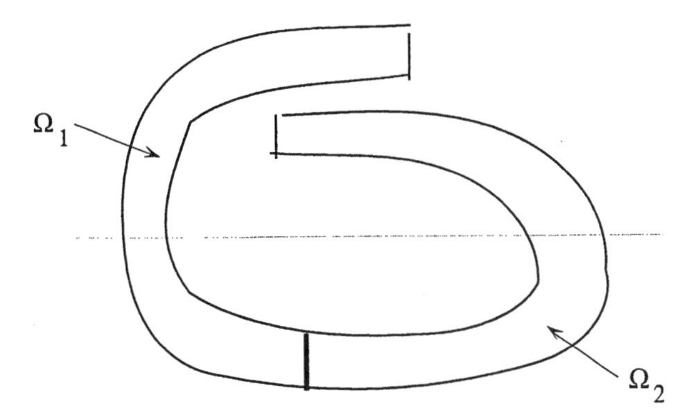

Figure S3.2.1

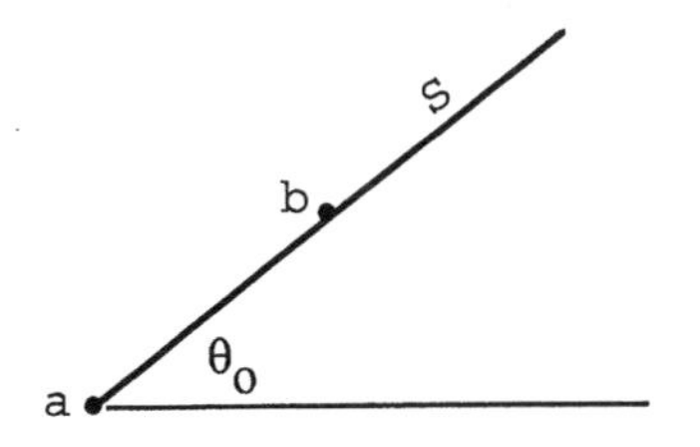

Figure S3.2.2

where the angles are chosen in the interval $[\theta_0, \theta_0+2\pi)$. Then $g(z) = (z-a)^{1/2}(z-b)^{1/2}$ is the desired analytic square root. The key intuitive point is that if z traverses a circle enclosing both a and b, then the arguments of $(z-a)^{1/2}$ and $(z-b)^{1/2}$ each change by π, so that $g(z)$ returns to its initial value. This shows that the set $S = \{z : z-a = re^{i\theta_0}, r > |b-a|\}$ is not a barrier to analyticity.

Remark

$f(z) = z^2$ on $D(0,1)\setminus\{0\}$ gives an easier example of an analytic function that is never 0 and has an analytic square root, but not an analytic logarithm.

3. By (3.1.11) we have (a) implies (b), and (b) implies (c) is obvious. To prove that (c) implies (a), let g_k be analytic on Ω with $g_k^k = f$. Then $f'/f = kg_k'/g_k$, so that if γ is a closed path in Ω, we have

$$\frac{1}{2\pi i}\int_\gamma \frac{g_k'(z)}{g_k(z)}\,dz = \frac{1}{2\pi ik}\int_\gamma \frac{f'(z)}{f(z)}\,dz \to 0$$

as $k \to \infty$ through an appropriate subsequence. By (3.2.3), $n(g_k \circ \gamma, 0) \to 0$ as $k \to \infty$. Since the index must be an integer, $n(g_k \circ \gamma, 0) = 0$ for large k. Therefore $\int_\gamma \frac{f'(z)}{f(z)}\,dz = 0$, and the result follows from (3.1.9).

4. As in (3.2.4d), $0 \notin \gamma_1^* \cup \gamma_2^*$. If $\gamma = \gamma_2/\gamma_1$, then $|1-\gamma| < 1 + |\gamma|$, which implies that $\gamma(t)$ can never be real and negative. Thus $\operatorname{Arg}\circ\gamma$ is a continuous argument of γ, hence $n(\gamma, 0) = 0$. As in (3.2.4d), $n(\gamma_1, 0) = n(\gamma_2, 0)$.

   The hypothesis is satisfied by all possible values of $\gamma_1(t)$ and $\gamma_2(t)$ except those lying on a line through the origin, with $\gamma_1(t)$ and $\gamma_2(t)$ on opposite sides of the origin. Thus if initially the angle between $\gamma_1(t)$ and $\gamma_2(t)$ (visualizing a complex number z as a vector in the plane pointing from 0 to z) is less than π and remains less than π for all t, then γ_1 and γ_2 have the same net number of revolutions about 0.

   The interpretation of the hypothesis is that the length of the leash is always less than the sum of the distances of man and dog from the tree.

5. Suppose θ is a continuous argument of f. Let $\gamma(t) = e^{it}, 0 \le t \le 2\pi$. Since $z = |z|e^{i\theta(z)} = e^{i\theta(z)}$ when $|z| = 1$, we have

$$e^{it} = e^{i\theta(e^{it})}, 0 \le t \le 2\pi.$$

   Thus t and $\theta(e^{it})$ are each continuous arguments of γ, so by (3.1.6c), $\theta(e^{it}) = t + 2\pi k$ for some integer k. Let $t \to 0$ to obtain $\theta(1) = 2\pi k$, and let $t \to 2\pi$ to obtain $\theta(1) = 2\pi + 2\pi k$, a contradiction.

   Note that $\theta(z), z \in S$, if it is to exist, must be a continuous function of z, that is, a continuous function of *position in the plane*, as opposed to $\theta(e^{it}), 0 \le t \le 1$, which is a continuous function of the "time parameter" t. If we specify $\theta(1) = 0$ and move around the circle, continuity requires that $\theta(1) = 2\pi$, which produces a contradiction. ("A function is a function is a function.")

6. Since f is (uniformly) continuous on S, $m = \min\{|f(z)| : z \in S\} > 0$, and for some $\delta > 0$ we have $|f(z) - f(z')| < m$ whenever $z, z' \in S, |z - z'| < \delta$. Break up S into closed squares whose diagonal is less than δ. Let A be a particular square, and pick any $z_0 \in A$. Then $|z - z_0| < \delta$ for every $z \in A$, hence $|f(z) - f(z_0)| < m$. Thus $f(z) \in D(f(z_0), m)$, and it follows just as in (3.1.7) that f has a continuous logarithm on A.

   Now let A and B be adjacent squares having a common side. If $e^{g_1} = f$ on A and $e^{g_2} = f$ on B, then for some integer k, $g_1 - g_2 = 2\pi ik$ on the common side. If we replace g_2 by $g_2 + 2\pi ik$, we produce a continuous logarithm of f on $A \cup B$. Continuing in this fashion, we may construct a continuous logarithm on each horizontal strip of S, and then piece the horizontal strips together to cover all of S. Formal details are not difficult to supply.

   Remark: The same technique works if S is an infinite rectangular strip.

7. If D is a disk contained in Ω, then f has an analytic logarithm h on D. Thus $g - h$ is constant on D, so g is analytic on D, hence on all of Ω.

8. We will show that f and g are entire functions such that $f^2 + g^2 = 1$ iff for some entire function h, we have $f = \cos h$ and $g = \sin h$. The "if" part is immediate, so consider the "only if" assertion. Since $f + ig$ is entire and never 0, $f + ig$ has an analytic logarithm h_0 on $\mathbb{C}$. If $h = -ih_0$, then $f + ig = e^{ih}$ and $f - ig = (f + ig)^{-1} = e^{-ih}$. Consequently, $f = (e^{ih} + e^{-ih})/2 = \cos h$ and $g = (e^{ih} - e^{-ih})/2i = \sin h$.

9. (a) Since $(f/g)^n = 1$, f/g is a continuous map of S into $\{e^{i2\pi k/n} : k = 0, 1, \dots, n-1\}$. Since S is connected, the image must be connected. Therefore the image consists of a single point, so $f/g = e^{i2\pi k/n}$ for some fixed k.
   (b) Take $S = [-1, 1]$ and let $f(x) = g(x) = \sqrt{x}, 0 \le x \le 1$; $f(x) = i\sqrt{|x|}, -1 \le x \le 0$; $g(x) = -i\sqrt{|x|}, -1 \le x \le 0$. Then $f^2(x) = g^2(x) = x$ for all $x \in S$.

Section 3.3

1. By (i) of (3.3.1), $\int_\gamma [\frac{f(w)-f(z)}{w-z}]\, dw = 0$, and the result follows from (3.2.3).

2. In order to reproduce the proof in the text, two key observations must be made.
   (a) Theorem 3.2.3 holds when γ is a cycle [this was noted in (3.3.5)].
   (b) For any cycle $\gamma = k_1\gamma_1 + \dots + k_m\gamma_m$, we have $n(\gamma, z) = 0$ for all sufficiently large $|z|$. This holds because if $|z|$ is large enough, then for each j, z will be in the unbounded component of $\mathbb{C} \setminus \gamma_j^*$. Thus $n(\gamma, z) = 0$ by (3.2.5).

   With these modifications, the proof in the text goes through.

3. By (3.2.5), $n(\gamma, z)$ is locally constant, and the result then follows from (2.2.10) and (ii) of (3.3.1).

4. By partial fraction expansion,

$$\frac{1}{z^2 - 1} = \frac{1}{(z-1)(z+1)} = \frac{1/2}{z-1} - \frac{1/2}{z+1}.$$

By the Cauchy integral formula [(ii) of (3.3.1)],

$$\int_\gamma \frac{1}{z^2-1}\,dz = 2\pi i(\frac{1}{2}-\frac{1}{2}) = 0.$$

5. Apply Problem 3 with $k=3$, $f(w) = e^w + \cos w$, $z=0$, to obtain

$$\int_{\gamma_j} \frac{e^z+\cos z}{z^4}\,dz = \frac{2\pi i}{3!}n(\gamma_j,0)f^{(3)}(0) = \frac{\pi i}{3}n(\gamma_j,0).$$

Since $n(\gamma_1,0) = -1, n(\gamma_2,0) = -2$, the integral on γ_1 is $-\pi i/3$ and the integral on γ_2 is $-2\pi i/3$.

6. By (3.3.7), we may replace γ by $\gamma_0(t) = \cos t + i\sin t$, hence $\int_\gamma dz/z = \int_{\gamma_0} dz/z = 2\pi i$. But

$$\int_\gamma \frac{dz}{z} = \int_0^{2\pi} \frac{-a\sin t + ib\cos t}{a\cos t + ib\sin t}\,dt.$$

Take imaginary parts to obtain

$$2\pi = ab\int_0^{2\pi} \frac{\cos^2 t + \sin^2 t}{a^2\cos^2 t + b^2\sin^2 t}\,dt,$$

and the result follows.

Section 3.4

1. (a) Let $\Omega = \mathbb{C}\setminus\{0\}$. Then $\hat{\mathbb{C}}\setminus\Omega = \{0,\infty\}$, which is not connected. If $f(z) = 1/z$ on Ω and γ describes any circle with center at 0, then $\int_\gamma f(z)\,dz = 2\pi i \neq 0$.
(b) Let Ω be the union of two disjoint disks D_1 and D_2. Then Ω is disconnected, but $\hat{\mathbb{C}}\setminus\Omega$ is connected.

2. No. For example, the situation illustrated in Figure 3.4.4 can occur even if Ω is connected. In this case, there is no way to replace the cycle γ by a single closed path.

3. (a) Since $1-z$ is analytic and never 0 on the simply connected open set $\mathbb{C}\setminus\Gamma_1$, it has an analytic square root f. If we specify that $\sqrt{1} = 1$, then f is determined uniquely, by Problem 9(a) of Section 3.2. A similar analysis applies to g.
(b) Since $f^2 = g^2$ and $f(0) = g(0) = 1$, $f = g$ on any connected open set containing 0, by Problem 9(a) of Section 3.2. In particular, $f = g$ below Γ. Suppose $f = g$ above Γ. Since f is analytic on $\Gamma_2\setminus\{1\}$ and g is analytic on $\Gamma_1\setminus\{1\}$, f can be extended to a function analytic on $\mathbb{C}\setminus\{1\}$. Thus $1-z$ has an analytic square root on $\mathbb{C}\setminus\{1\}$, so that z has an analytic square root on $\mathbb{C}\setminus\{0\}$, a contradiction. (If $h^2(z) = z$ with h continuous on $\mathbb{C}\setminus\{0\}$, then $h(e^{it}) = e^{it/2}k(t)$, where $k(t) = \pm 1$. A connectedness argument shows that either $k(t) \equiv 1$ or $k(t) \equiv -1$, and in either case we obtain a contradiction by letting $t \to 0$ and $t \to 2\pi$.) It follows that $f(z_0) = -g(z_0)$ for at least one point z_0 above Γ, and as above, we must have $f = -g$ at all points above Γ.
(c) The function h may be obtained by expanding g in a Taylor series on $D(0,1)$. Thus $h(z) = f(z), z \in D(0,1)$, z below Γ_1, and $h(z) = -f(z), z \in D(0,1)$, z above Γ_1.

4. It follows from part (a) of (3.4.3) that if $\Omega \subseteq \mathbb{C}$, then Ω is open in $\mathbb{C}$ iff Ω is open in $\hat{\mathbb{C}}$. If K is a compact subset of $\mathbb{C}$, then $\hat{\mathbb{C}} \setminus K$ is open in $\hat{\mathbb{C}}$. On the other hand, if $\infty \in V$, where V is open in $\hat{\mathbb{C}}$, then $\hat{\mathbb{C}} \setminus V$ is a closed and bounded (hence compact) subset of $\mathbb{C}$. Thus the open sets in $\hat{\mathbb{C}}$ are of two types: (i) open subsets of $\mathbb{C}$, and (ii) complements of compact subsets of $\mathbb{C}$. Consequently, $\hat{\mathbb{C}}$ (with the topology induced by the chordal metric) is homeomorphic (via the identity map) to the one point compactification of $\mathbb{C}$.

Chapter 4

Section 4.1

1. (a) If f has a removable singularity at z_0, then as in the proof of (4.1.5), f can be defined or redefined at z_0 so as to be analytic on $D(z_0, r)$ for some $r > 0$. Thus f is bounded on $D'(z_0, \delta)$ for some $\delta > 0$, in fact f is bounded on $D(z_0, \delta)$. Conversely, if f is bounded on $D'(z_0, \delta)$, let $g(z) = (z - z_0)f(z)$. Then $g(z) \to 0$ as $z \to z_0$, so by the first equivalence of (4.1.5a), g has a removable singularity at z_0. Since the Laurent expansion of g has only nonnegative powers of $z - z_0$, it follows that f has either a removable singularity or a pole of order 1 at $z = z_0$. The second case is impossible by the first equivalence of (4.1.5b), and the result follows.
(b) If f has a pole of order m at z_0, then $(z - z_0)^m f(z) \to K \neq 0$ as $z \to z_0$, so $|f(z)| \to \infty$. Conversely, if $|f(z)| \to \infty$ as $z \to z_0$, then by (4.1.5a) and (4.1.6) (which we use instead of (4.1.5c) to avoid circularity), f cannot have a removable or essential singularity at z_0, so f must have a pole.

2. (a) Since $\lim_{z\to n\pi}(z - n\pi)z/\sin z = n\pi/\cos n\pi = (-1)^n n\pi$, there are, by (4.1.5), simple poles at $z = n\pi$, n a nonzero integer. Since $z/\sin z \to 1$ as $z \to 0$, there is a removable singularity at $z = 0$. Now $f(1/z) = 1/z \sin 1/z$ has poles at $z = 1/n\pi, n = \pm 1, \pm 2, \dots$, so 0 is a nonisolated singularity of $f(1/z)$, hence ∞ is a nonisolated singularity of $f(z)$.
(b) There is an isolated essential singularity at 0 since $e^{1/x} \to \infty$ as $x \to 0^+, e^{1/x} \to 0$ as $x \to 0^-$. There is a removable singularity at ∞ since e^z is analytic at 0.
(c) There is an isolated essential singularity at 0 since $z \cos 1/z = z(1 - 1/2!z^2 + 1/4!z^4 - \cdots), z \neq 0$. There is a simple pole at ∞ because $(1/z)\cos z$ has a simple pole at 0.
(d) There is a pole of order 2 at 0 since $z^2 f(z) \to 1$ as $z \to 0$. There are poles of order 1 at $z = i2n\pi, n = \pm 1, \pm 2, \dots$ since $(z - i2n\pi)/z(e^z - 1) \to (1/i2n\pi)(1/e^{i2n\pi}) = 1/i2n\pi$ as $z \to i2n\pi$.
(e) There are simple poles at $z = n\pi, n = 0, \pm 1, \pm 2, \dots$ since $(z - n\pi)\cos z/\sin z \to \cos n\pi/\cos n\pi = 1$ as $z \to n\pi$. There is a non-isolated singularity at ∞ because ∞ is a limit point of poles.

3. We have $f(z) = \frac{1}{z} - \frac{3}{z+1} + \frac{2}{z-2}$, and

$$\frac{1}{z} = \frac{1}{(z+1) - 1} = \frac{-1}{1 - (z+1)} = -\sum_{n=0}^{\infty}(z+1)^n, \quad |z+1| < 1;$$

$$\frac{2}{z-2} = \frac{2}{(z+1)-3} = \frac{-2/3}{1-\frac{1}{3}(z+1)} = -\frac{2}{3}\sum_{n=0}^{\infty}(1/3)^n(z+1)^n, \quad |z+1|<3.$$

Thus

$$f(z) = -\frac{3}{z+1} - \sum_{n=0}^{\infty}\left[1+\frac{2}{3^{n+1}}\right](z+1)^n, \quad 0<|z+1|<1.$$

We may obtain a Laurent expansion for $1<|z+1|<3$ by modifying the expansion of $1/z$, as follows:

$$\frac{1}{z} = \frac{1}{(z+1)-1} = \frac{\frac{1}{z+1}}{1-\frac{1}{z+1}} = \frac{1}{z+1}\sum_{n=0}^{\infty}\frac{1}{(z+1)^n}, \quad |z+1|>1.$$

Therefore

$$f(z) = -\frac{2}{z+1} + \sum_{k=1}^{\infty}\frac{1}{(z+1)^k} - \frac{2}{3}\sum_{n=0}^{\infty}(1/3)^n(z+1)^n, \quad 1<|z+1|<3.$$

For $|z+1|>3$, the expansion $1/z = \sum_{n=0}^{\infty} 1/(z+1)^{n+1}$ is acceptable, but the expansion of $2/(z-2)$ must be modified:

$$\frac{2}{z-2} = \frac{2}{(z+1)-3} = \frac{\frac{2}{z+1}}{1-\frac{3}{z+1}} = \frac{2}{z+1}\sum_{n=0}^{\infty}3^n(z+1)^{-n}.$$

Thus

$$f(z) = -\frac{3}{z+1} + \sum_{n=1}^{\infty}\frac{1}{(z+1)^n} + 2\sum_{n=1}^{\infty}\frac{3^{n-1}}{(z+1)^n} = \sum_{n=2}^{\infty}\left[\frac{1+2(3^{n-1})}{(z+1)^n}\right], \quad |z+1|>3.$$

4. We have

$$\frac{1}{z+2} = \frac{1/2}{1+z/2} = \sum_{n=0}^{\infty}(-1)^n z^n/2^{n+1}, \quad |z|<2$$

and

$$\frac{1}{z+2} = \frac{1/z}{1+2/z} = \sum_{n=0}^{\infty}(-1)^n 2^n/z^{n+1}, \quad |z|>2.$$

Now $1/(1-z) = \sum_{n=0}^{\infty} z^n$, $|z|<1$, and therefore by differentiation, $1/(1-z)^2 = \sum_{n=1}^{\infty} nz^{n-1}$, $|z|<1$. Also

$$\frac{1}{1-z} = \frac{-1/z}{1-1/z} = -\sum_{n=0}^{\infty}\frac{1}{z^{n+1}}, \quad |z|>1,$$

hence

$$\frac{1}{(1-z)^2} = \sum_{n=0}^{\infty} \frac{n+1}{z^{n+2}}, \quad |z| > 1.$$

Thus

$$\begin{aligned} f(z) &= \frac{1}{z} + \sum_{n=0}^{\infty} [n+1+(-1)^n 2^{-(n+1)}] z^n, \quad 0 < |z| < 1 \\ &= \frac{1}{z} + \sum_{n=0}^{\infty} \frac{n+1}{z^{n+2}} + \sum_{n=0}^{\infty} (-1)^n 2^{-(n+1)} z^n, \quad 1 < |z| < 2 \\ &= \frac{2}{z} + \sum_{n=2}^{\infty} [n-1+(-1)^{n-1} 2^{n-1}] \frac{1}{z^n}, \quad |z| > 2. \end{aligned}$$

Remark

The coefficients of the Taylor expansion of $f(1/z)$ about $z = 0$ are the same as the coefficients of the Laurent expansion of $f(z)$ valid for $|z| > 2$, that is, in a neighborhood of ∞. For this reason, the expansion of $f(z)$ for $|z| > 2$ may be called the "Taylor expansion of f about ∞."

5. Since $g(z) = z/(e^z - e^{-z}) \to 1/2$ as $z \to 0$, g has a removable singularity at $z = 0$. We may compute the derivatives of g at $z = 0$ to form the Taylor expansion $g(z) = (1/2) - (1/12)z^2 + (7/720)z^4 - \cdots$, $0 < |z| < \pi$. Thus

$$f(z) = \frac{1}{z^2(e^z - e^{-z})} = \frac{1}{2z^3} - \frac{1}{12z} + \frac{7}{720}z + \cdots, 0 < |z| < \pi.$$

Alternatively,

$$g(z) = \frac{1}{2}\left(1 + \frac{z^2}{3!} + \frac{z^4}{5!} + \cdots\right)^{-1} = \sum_{n=0}^{\infty} a_n z^n,$$

and the Taylor coefficients may be found by ordinary long division, or by matching coefficients in the equation

$$\left(1 + \frac{z^2}{3!} + \frac{z^4}{5!} + \cdots\right)(a_0 + a_1 z + a_2 z^2 + \cdots) = \frac{1}{2}.$$

6. The function

$$\frac{1}{\sin z} - \frac{1}{z} + \frac{1}{z-\pi} + \frac{1}{z+\pi}$$

is analytic for $|z| < 2\pi$, hence has a Taylor expansion $\sum_{n=0}^{\infty} a_n z^n$. Also,

$$\frac{1}{z} - \frac{1}{z-\pi} - \frac{1}{z+\pi}$$

has a Laurent expansion $\sum_{n=-\infty}^{\infty} b_n z^n$, $\pi < |z| < 2\pi$; the expansion may be found by the procedure illustrated in Problems 3 and 4. Addition of these two series gives the Laurent expansion of $1/\sin z, \pi < |z| < \pi$.

7. Expand $R(z)$ in a Laurent series about $z = z_1$:

$$R(z) = \frac{A_{1,0}}{(z-z_1)^{n_1}} + \frac{A_{1,1}}{(z-z_1)^{n_1-1}} + \cdots + \frac{A_{1,n_1-1}}{z-z_1} + R_1^*(z),$$

where R_1^* is analytic at z_1 and the representation is valid in some deleted neighborhood of z_1. Define

$$R_1(z) = R(z) - \sum_{i=0}^{n_1-1} \frac{A_{1,i}}{(z-z_1)^{n_1-i}}.$$

Then R_1 is a rational function whose poles are at $z_2, \ldots, z_k$ with orders $n_2, \ldots, n_k$, and R_1 has a removable singularity at z_1 since $R_1 = R_1^*$ near z_1. Similarly, expand $R_1(z)$ in a Laurent series about z_2 to obtain

$$R_2(z) = R_1(z) - \sum_{i=0}^{n_2-1} \frac{A_{2,i}}{(z-z_2)^{n_2-i}},$$

where R_2 is a rational function with poles at $z_3, \ldots, z_k$ with orders $n_3, \ldots, n_k$. Continue in this fashion until we reach R_k:

$$R_k(z) = R_{k-1}(z) - \sum_{i=0}^{n_k-1} \frac{A_{k,i}}{(z-z_k)^{n_k-i}} = R(z) - \sum_{i=1}^{k} B_i(z).$$

Now R_{k-1} has a pole only at z_k, so R_k is a rational function with no poles, that is, a polynomial. But $R(z) \to 0$ as $z \to \infty$ by hypothesis ($\deg P < \deg Q$), and $B_i(z) \to 0$ as $z \to \infty$ by construction. Thus $R_k(z) \equiv 0$. Finally,

$$\lim_{z \to z_j} \frac{d^r}{dz^r}[(z-z_j)^{n_j} B_m(z)] = 0, \quad m \neq j,$$

and when $m = j$, the limit is

$$\lim_{z \to z_j} \frac{d^r}{dz^r} \sum_{i=0}^{n_j-1} A_{j,i}(z-z_j)^i = r! A_{j,r}.$$

Hence

$$r! A_{j,r} = \lim_{z \to z_j} \frac{d^r}{dz^r}[(z-z_j)^{n_j} R(z)]$$

as desired. Now

$$\frac{1}{z(z+i)^3} = \frac{A}{z} + \frac{B}{(z+i)^3} + \frac{C}{(z+i)^2} + \frac{D}{z+i}, \text{ where}$$

$$A = [zR(z)]_{z\to 0} = \frac{1}{i^3} = i$$

$$B = (z+i)^3 R(z)]_{z\to -i} = \frac{1}{-i} = i$$

$$C = \left[\frac{d}{dz}[(z+i)^3 R(z)]\right]_{z\to -i} = \left[\frac{-1}{z^2}\right]_{z\to -i} = 1$$

$$D = \frac{1}{2!}\left[\frac{d^2}{dz^2}[(z+i)^3 R(z)]\right]_{z\to -i} = \left[\frac{1}{z^3}\right]_{z\to -i} = -i.$$

Thus

$$\frac{1}{z(z+i)^3} = \frac{i}{z} + \frac{i}{(z+i)^3} + \frac{1}{(z+i)^2} - \frac{i}{z+i}.$$

8. The series converges absolutely on $U = \{x+iy : -1 < y < 1\}$, uniformly on $\{x+iy : -1+\epsilon \le y \le 1-\epsilon\}$ for every $\epsilon > 0$, hence uniformly on compact subsets of U. The series diverges for $z \notin U$. For

$$\sum_{n=0}^{\infty} e^{-n} e^{inz} = \sum_{n=0}^{\infty} e^{(iz-1)n} = \frac{1}{1-e^{iz-1}}$$

if $|e^{iz-1}| < 1$, that is, $e^{-(y+1)} < 1$, or $y > -1$; this series diverges if $|e^{iz-1}| \ge 1$, that is, $y \le -1$. The convergence is uniform for $|e^{iz-1}| \le r < 1$, that is, $y \ge -1-\epsilon$. Similarly,

$$\sum_{n=0}^{\infty} e^{-n} e^{-inz} = \frac{1}{1-e^{-iz-1}}$$

if $|e^{-iz-1}| < 1$, that is, $y < 1$, with uniform convergence for $y \le 1-\epsilon$. The result follows; explicitly, we have

$$\sum_{n=0}^{\infty} e^{-n} \sin nz = \frac{1}{2i}\left[\frac{1}{1-e^{iz-1}} - \frac{1}{1-e^{-iz-1}}\right], \quad z \in U.$$

9. (a) Since $\hat{\mathbb{C}}$ is compact, f is bounded. The result follows from Liouville's theorem.
(b) If $f(z) = \sum_{m=0}^{\infty} b_m z^m, z \in \mathbb{C}$, then $g(z) = f(1/z) = \sum_{m=0}^{\infty} b_m z^{-m}, z \in \mathbb{C}, z \ne 0$, a Laurent expansion of g about $z = 0$. By (4.1.3),

$$b_m = \frac{1}{2\pi i}\int_{|z|=1/r} g(w) w^{m-1}\, dw,$$

hence

$$|b_m| \le \max\{|g(z)| : |z| = 1/r\}(1/r)^m = \max\{|f(z)| : |z| = r\}(1/r)^m,$$

which approaches 0 as $r \to \infty$ if $m > k$. Thus $b_m = 0$ for $m > k$, and the result follows.

(c) The argument is the same as in (b), except now we know by hypothesis that b_m is nonzero for only finitely many m, so it is not necessary to use (4.1.3).

(d) Let the poles of f in $\mathbb{C}$ be at $z_1, \dots, z_k$, with orders $n_1, \dots, n_k$. (If f had infinitely many poles, there would be a nonisolated singularity somewhere in $\hat{\mathbb{C}}$. Let $g(z) = f(z)\prod_{j=1}^{k}(z-z_j)^{n_j}$. Then g is analytic on $\mathbb{C}$ and has a nonessential singularity at ∞. By (c), g is a polynomial, hence f is a rational function.

10. (a) Pole of order 2 at $z = 0$, isolated essential singularity at ∞.

(b) Isolated essential singularity at $z = 0$, pole of order 1 at $z = -1$, removable singularity at ∞.

(c) $z \csc z \to 1$ as $z \to 0$, hence $\csc z - k/z$ has poles of order 1 at $z = n\pi, n = \pm 1, \pm 2, \dots$, and a pole of order 1 at $z = 0$ as long as $k \neq 1$. If $k = 1$, there is a removable singularity at $z = 0$. The point at ∞ is a nonisolated singularity.

(d) If z is real and near $2/[(2n+1)\pi], n = 0, \pm 1, \pm 2, \dots$, then $\exp[\sin(1/z)/\cos(1/z)]$ will be near ∞ or 0 depending on the sign of $z - [2/(2n+1)\pi]$. By (4.1.5), $\exp[\tan(1/z)]$ has an isolated essential singularity at $z = [2/(2n+1)\pi]$. There is a nonisolated singularity at 0 and a removable singularity at ∞.

(e) $\sin(x+iy) = n\pi$ when $\sin x \cosh y + i \cos x \sinh y = n\pi + i0$. Thus if $n = 1, 2, \dots$, then $y = \cosh^{-1} n\pi, x = (4k+1)\pi/2$, k an integer. ($\cosh^{-1} n\pi$ refers to the *two* numbers u and $-u$ such that $\cosh u = n\pi$.) If $n = -1, -2, \dots$, then $y = \cosh^{-1}(-n\pi), x = (4k+3)\pi/2$, k an integer. If $n = 0$, then $x = k\pi, y = 0$, k an integer. If $z_0 = x_0 + iy_0$ is any of these points, then by Problem 4 of Section 2.4,

$$\lim_{z \to z_0} \frac{z - z_0}{\sin(\sin z)} = \frac{1}{\cos(\sin z_0)\cos z_0} = \frac{1}{\cos n\pi \cos z_0}.$$

Now $\cos(x_0 + iy_0) = \cos x_0 \cosh y_0 - i \sin x_0 \sinh y_0$, and this is nonzero, by the above argument. Thus all the points are poles of order 1. The point at ∞ is a nonisolated singularity.

11. Let $f(z) = (z-a)/(z-b)$ and $U = \mathbb{C} \setminus [a, b]$. For any closed path γ in U,

$$\frac{1}{2\pi i}\int_\gamma \frac{f'(z)}{f(z)}\,dz = \frac{1}{2\pi i}\int_\gamma \left(\frac{1}{z-a} - \frac{1}{z-b}\right) dz = n(\gamma, a) - n(\gamma, b) = 0$$

because a and b lie in the same component of $\mathbb{C} \setminus \gamma^*$. (Note that $\gamma^* \subseteq U$, hence $[a, b] \cap \gamma^* = \emptyset$.) By (3.1.9), f has an analytic logarithm g on U. Now $g' = f'/f$ [see (3.1.9)], hence

$$\begin{aligned} g'(z) &= \frac{1}{z-a} - \frac{1}{z-b} \\ &= \sum_{n=0}^{\infty} \frac{(a^n - b^n)}{z^{n+1}}, \quad |z| > \max(|a|., |b|) \\ &= \sum_{n=0}^{\infty} \left(\frac{1}{b^{n+1}} - \frac{1}{a^{n+1}}\right) z^n, \quad |z| < \min(|a|, |b|). \end{aligned}$$

Thus

$$g(z) = \log\left(\frac{z-a}{z-b}\right) = k + \sum_{n=1}^{\infty} \frac{b^n - a^n}{nz^n}, \qquad |z| > \max(|a|,|b|), z \in U$$

and

$$g(z) = k' + \sum_{n=1}^{\infty} \frac{1}{n}\left(\frac{1}{b^n} - \frac{1}{a^n}\right), \qquad |z| < \min(|a|,|b|), z \in U$$

where k is any logarithm of 1 and k' is any logarithm of a/b.

12. If $f(\mathbb{C})$ is not dense in $\mathbb{C}$, then there is a disk $D(z_0, r)$ such that $D(z_0,r) \cap f(\mathbb{C}) = \emptyset$. Thus for all $z \in \mathbb{C}$, $|f(z) - z_0| \geq r$, and the result now follows from Liouville's theorem applied to $1/[f(z) - z_0]$.

13. If $P(z) = \sum_{j=0}^{n} a_j z^j$, then $P(f(z)) = a_n[f(z)]^n + \cdots + a_1 f(z) + a_0$. By hypothesis, $(z-\alpha)^m f(z)$ approaches a finite nonzero limit as $z \to \alpha$, hence so does $(z-\alpha)^{mn}[f(z)]^n$. But if $j < n$, then $(z-\alpha)^{mn} f(z)^j = (z-\alpha)^{mn} f(z)^n / [f(z)]^{n-j} \to 0$ as $z \to \alpha$; the result follows.

Section 4.2

1. By (4.2.7), $n(f \circ \gamma, 0) = -1$. Geometrically, as z traverses γ once in the positive sense, the argument of $z - 1$ changes by 2π, the argument of $z + 2i$ also changes by 2π, and the argument of $z - 3 + 4i$ has a net change of 0. Thus the total change in the argument of $f(z)$ is $2\pi - 2(2\pi) = -2\pi$, hence $n(f \circ \gamma, 0) = -1$.

2. Let γ describe the contour of Figure S4.2.1, with r "very large". Now $f(z) = z^3 - z^2 + 3z + 5$, so $f(iy) = 5 + y^2 + i(3y - y^3)$. Thus $f \circ \gamma$ is as indicated in Figure S4.2.2. Note that in moving from B to C, the argument of z changes by π. Since $f(z) = z^3(1 - z^{-1} + 3z^{-2} + 5z^{-3}) = z^3 g(z)$ where $g(z) \to 1$ as $z \to \infty$, the argument of $f(z)$ changes by approximately 3π. Note also that $f(\overline{z}) = \overline{f(z)}$, so that $f \circ \gamma$ is symmetrical about the real axis. It follows that $n(f \circ \gamma, 0) = 2$, so that f has two roots in the right half plane. (In fact $f(z) = (z+1)[(z-1)^2 + 4]$, with roots at $-1, 1 + 2i, 1 - 2i$.)

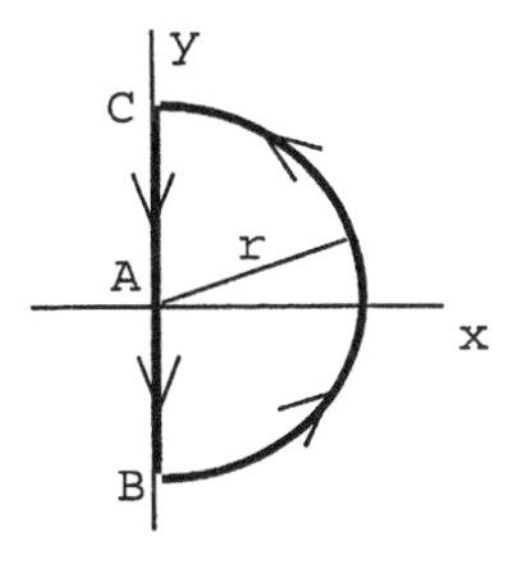

Figure S4.2.1

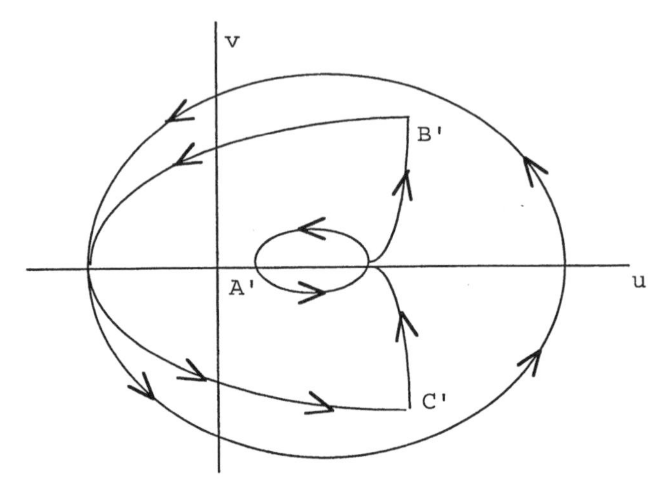

Figure S4.2.2

3. Let $f(z) = a_n z^n + \cdots + a_1 z + a_0$, $a_n \neq 0$, $g(z) = a_n z^n$. Then if γ describes a sufficiently large circle centered at the origin, $|f - g| < |g|$ on γ^*, so by Rouché's theorem, f has exactly n zeros inside γ, counting multiplicity.

4. (a) Integrate $f(z) = ze^{iaz}/(z^4 + 4)$ around the contour γ indicated in Figure S4.2.3. Then $\int_\gamma f(z)\,dz = 2\pi i \sum$ residues of f at poles in the upper half plane. The poles of f are at $\sqrt{2}e^{i\pi/4}$, $\sqrt{2}e^{i3\pi/4}$, $\sqrt{2}e^{i5\pi/4}$, $\sqrt{2}e^{i7\pi/4}$. The residue at $z = z_0$ is

$$\lim_{z \to z_0} \frac{(z - z_0)ze^{iaz}}{z^4 + 4} = \frac{z_0 e^{iaz_0}}{4z_0^3}.$$

Thus

$$\int_\gamma f(z)\,dz = \frac{2\pi i}{4}\left[\frac{\exp(ia\sqrt{2}e^{i\pi/4})}{2e^{i\pi/2}} + \frac{\exp(ia\sqrt{2}e^{i3\pi/4})}{2e^{i3\pi/2}}\right]$$

which reduces to

$$\frac{\pi}{4}[\exp(ia\sqrt{2}(\sqrt{2}/2 + i\sqrt{2}/2)) - \exp(ia\sqrt{2}(-\sqrt{2}/2 + i\sqrt{2}/2))]$$

$$= \frac{\pi}{4}e^{-a}(e^{ia} - e^{-ia}) = \frac{\pi}{2}ie^{-a}\sin a.$$

An application of (2.1.5) shows that the integral of f around the semicircle approaches 0 as $r \to \infty$. Thus in the expression

$$\int_{-r}^{r} f(x)\,dx + \int_{\substack{z=re^{it},\\ 0\leq t\leq\pi}} f(z)\,dz = \frac{\pi}{2}ie^{-a}\sin a,$$

we may let $r \to \infty$ to obtain

$$\int_{-\infty}^{\infty} \frac{xe^{iax}}{x^4+4}\,dx = \frac{\pi}{2} ie^{-a}\sin a.$$

Take imaginary parts to obtain

$$\int_{-\infty}^{\infty} \frac{x\sin ax}{x^4+4}\,dx = \frac{\pi}{2} e^{-a}\sin a.$$

(b) By the analysis of (a), the integral is $2\pi i$ times the sum of the residues in the

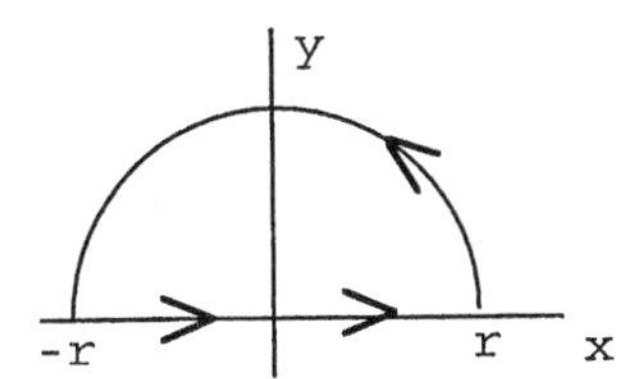

Figure S4.2.3

upper half plane of

$$\frac{z}{(z^2+1)(z^2+2z+2)} = \frac{z}{(z-i)(z+i)(z-(-1+i))(z-(-1-i))}.$$

The residue at $z = i$ is

$$\frac{i}{2i(i^2+2i+2)} = \frac{1}{2(1+2i)} = \frac{1-2i}{10}.$$

The residue at $-1+i$ is

$$\frac{-1+i}{[(-1+i)^2+1]}\frac{1}{2i} = \frac{-1+i}{4+2i} = \frac{-1+3i}{10}.$$

Thus the integral is $2\pi i(i/10) = -\pi/5$.

(c) The integral is $2\pi i \sum$ residues of $1/(z^2-4z+5)^2$ in the upper half plane. Now $z^2-4z+5 = (z-2)^2+1$, so there are poles of order 2 at $2+i$ and $2-i$. By (4.2.2d), he residue at $2+i$ is

$$\frac{d}{dz}\left[\frac{1}{(z-(2-i))^2}\right]_{z=2+i} = \left[\frac{-2}{(z-(2-i))^3}\right]_{z=2+i} = \frac{-2}{8i^3}.$$

The integral is $2\pi i/4i = \pi/2$.

(d) The integral is

$$\int_0^{2\pi} \frac{e^{i\theta}+e^{-i\theta}}{2(5+2(e^{i\theta}+e^{-i\theta}))}\,d\theta = \int_{|z|=1} \frac{z+z^{-1}}{2(5+2(z+z^{-1}))}\frac{dz}{iz}$$

which is $2\pi i$ times the sum of the residues of the integrand inside the unit circle. Multiply numerator and denominator of the integrand by z to get

$$\frac{z^2+1}{2iz(2z^2+5z+2)} = \frac{z^2+1}{2iz(2z+1)(z+2)}.$$

The residue at $z=0$ is $1/4i$, and the residue at $z=-1/2$ is

$$\frac{5/4}{2i(-1/2)2(3/2)} = \frac{-5}{12i}.$$

Thus the integral is $2\pi i(-2/12i) = -\pi/3$.

(e) Since the integrand is an even function, we may integrate from $-\infty$ to ∞ and divide by 2 to get $\pi i \sum$ residues of $1/(z^4+a^4)$ in the upper half plane. By a computation similar to (a), the residue at $ae^{i\pi/4}$ is $1/(4a^3 e^{i3\pi/4})$, and the residue at $ae^{i3\pi/4}$ is $1/(4a^3 e^{i9\pi/4})$. Thus the integral is

$$\frac{\pi i}{4a^3}(e^{-i\pi/4} + e^{-i3\pi/4}) = \frac{\pi}{4a^3}(\sin\frac{\pi}{4} + \sin\frac{3\pi}{4}) = \sqrt{2}\frac{\pi}{4a^3}.$$

(f) We may integrate $e^{ix}/(x^2+1)$ from $-\infty$ to ∞, divide by 2, and take the real part to get $\text{Re}(\pi i \sum$ residues of $e^{iz}/(z^2+1)$ in the upper half plane). The residue at $z=i$ is $e^{-1}/2i$, hence the integral is $\text{Re}(\pi i e^{-1}/2i) = \pi/2e$.

(g) The integral is

$$\int_0^{2\pi} \left(\frac{e^{i\theta}-e^{-i\theta}}{2i}\right)^{2n} d\theta = \int_{|z|=1} \left(\frac{z-z^{-1}}{2i}\right)^{2n} \frac{dz}{iz}$$

which is $2\pi i$ times the residue of $(z^2-1)^{2n}/(i2^{2n}z^{2n+1})(-1)^n$ at $z=0$. But the Taylor expansion of $(1-z^2)^{2n}$ is

$$1 - 2nz^2 + \binom{2n}{2}z^4 - \binom{2n}{3}z^6 + \cdots + (-1)^n\binom{2n}{n}z^{2n} + \cdots + z^{4n}.$$

Thus the coefficient of $1/z$ in the Laurent expansion of $(z^2-1)^{2n}/z^{2n+1}$ is $(-1)^n\binom{2n}{n}$. Therefore

$$\int_0^{2\pi} (\sin\theta)^{2n}\, d\theta = \frac{2\pi\binom{2n}{n}}{2^{2n}} = \frac{2\pi(2n)!}{(2^n n!)^2}.$$

Remark: In these examples [except for (d) and (g)] we needed a result of the form

$$\int_{\substack{z=re^{it} \\ 0\le t\le\pi}} f(z)\, dz \to 0 \text{ as } r \to \infty.$$

By (2.1.5), this will hold if $zf(z) \to 0$ as $z \to \infty$ in the upper half plane.

5. When $z = i(2n+1)\pi$, n an integer, $1 + e^z = 0$. These are simple poles of $f(z) = (\operatorname{Log} z)/(1+e^z)$ with residues $\operatorname{Log}(i(2n+1)\pi)/e^{i(2n+1)\pi} = -\operatorname{Log} i(2n+1)\pi$. Since $n(\gamma, -i\pi) = -1$, the integral is

$$2\pi i[\operatorname{Res}(f, i3\pi) - \operatorname{Res}(f, -i\pi)] = 2\pi i[-\operatorname{Log}(i3\pi) + \operatorname{Log}(-i\pi)]$$

$$= 2\pi i[-\ln 3 - i\pi] = 2\pi^2 - i2\pi \ln 3.$$

6. (a) The Taylor expansion of $\sin^2 z$ has no term of degree 3, so the residue is 0.
(b) The Taylor expansion of $z^3 \sin z^2$ is

$$z^5[1 - \frac{z^4}{3!} + \frac{z^8}{5!} - \frac{z^{12}}{7!} + \cdots]$$

and by long division, the reciprocal of the expression in brackets has a z^4 term with coefficient $1/3! = 1/6$. The residue is therefore $1/6$.
(c) We have

$$z \cos \frac{1}{z} = z \left[1 - \frac{1}{2!z^2} + \frac{1}{4!z^4} - \cdots\right]$$

and the residue is therefore -1/2.

7. We have

$$\sin(\frac{e^z}{z}) = \frac{e^z}{z} - \frac{e^{3z}}{3!z^3} + \frac{e^{5z}}{5!z^5} - \cdots = \sum_{n=1}^{\infty} f_n(z), z \neq 0,$$

where

$$f_n(z) = [(-1)^{(n-1)/2}]\frac{e^{nz}}{n!z^n}, \quad n \text{ odd}$$

and $f_n(z) = 0$ for n even. Now for n odd,

$$f_n(z) = \frac{(-1)^{(n-1)/2}}{n!z^n}(1 + nz + \frac{n^2z^2}{2!} + \cdots) = \sum_{k=-\infty}^{\infty} a_{kn}z^k, \ z \neq 0,$$

where $a_{kn} = 0$, $k < -n$, and the series is the Laurent expansion of f_n about $z = 0$. Now

$$\sum_{n=1}^{\infty} \sum_{k=-\infty}^{\infty} |a_{kn}z^k| = \sum_{\substack{n=1 \\ n \text{ odd}}}^{\infty} \frac{e^{n|z|}}{n!|z|^n} = \sinh\left(\frac{e^{|z|}}{|z|}\right) < \infty.$$

Thus we may reverse the order of summation to obtain

$$\sin(e^z/z) = \sum_{k=-\infty}^{\infty} \left(\sum_{n=1}^{\infty} a_{kn}\right) z^k, \quad z \neq 0.$$

This is the Laurent expansion of $\sin(e^z/z)$ about $z = 0$. The residue at $z = 0$ is therefore $\sum_{n=1}^{\infty} a_{-1,n}$. But $a_{-1,n} = 0$ for n even, and for n odd we have $a_{-1,n} = (-1)^{(n-1)/2}n^{n-1}/(n-1)!n!$. Thus the residue is

$$\sum_{n=1,3,5,\cdots} \frac{(-1)^{(n-1)/2}n^{n-1}}{(n-1)!n!}.$$

8. (a) Since $\sin\theta$ lies above the line segment joining (0,0) to $(\pi/2, 1)$, we have $\sin\theta \geq 2\theta/\pi, 0 \leq \theta \leq \pi/2$. Thus

$$\int_0^{\pi/2} e^{-r\sin\theta}\, d\theta \leq \int_0^{\pi/2} e^{-2r\theta/\pi}\, d\theta = \frac{\pi}{2r}(1 - e^{-r}).$$

(b) Let γ be the path of Figure 4.2.3, traversed in the positive sense; γ consists of a radial path γ_1 away from z_0, followed by γ_ϵ, and completed by a radial path γ_2 toward z_0. If $g(z) = f(z) - [k/(z - z_0)], k = \text{Res}(f, z_0)$, then g is analytic at z_0, so $\int_\gamma g(z)\, dz = 0$ by Cauchy's theorem. Now

$$\begin{aligned}\int_{\gamma_\epsilon} f(z)\, dz &= \int_{\gamma_\epsilon} g(z)\, dz + k\int_{\gamma_\epsilon} \frac{dz}{z - z_0} \\ &= -\int_{\gamma_1} g(z)\, dz - \int_{\gamma_2} g(z)\, dz + k\int_{\gamma_\epsilon} \frac{dz}{z - z_0}.\end{aligned}$$

Since the integrals along γ_1 and γ_2 approach 0 as $\epsilon \to 0$ by (uniform) continuity of g, we must show that $\int_{\gamma_\epsilon} \frac{dz}{z - z_0}\, dz \to \alpha i$. In fact, if θ_0 is the angle between γ_1 and the horizontal, then

$$\int_{\gamma_\epsilon} \frac{dz}{z - z_0} = \int_{\theta_0}^{\theta_0 + \alpha} \frac{i\epsilon e^{i\theta}}{\epsilon e^{i\theta}}\, d\theta = \alpha i.$$

9. (a) By Problem 8a, the integral around the large semicircle approaches 0 as the radius approaches ∞. By Problem 8b, the integral around the small semicircle approaches $-i\pi \,\text{Res}(e^{iz}/z, 0) = -i\pi$ as the radius approaches 0. It follows that $\int_{-\infty}^{\infty}(e^{ix}/x)\, dx - i\pi = 0$ (where the integral is interpreted as a Cauchy principal value), or $\int_{-\infty}^{\infty}[(\sin x)/x]\, dx = \pi$.
(b) By Cauchy's theorem,

$$0 = \int_0^r e^{ix^2}\, dx + \int_0^{\pi/4} \exp(ir^2 e^{i2t}) ire^{it}\, dt + \int_r^0 \exp(is^2 e^{i\pi/2}) e^{i\pi/4}\, ds.$$

The second integral is, in absolute value, less than or equal to

$$r\int_0^{\pi/4} e^{-r^2 \sin 2t}\, dt \to 0 \text{ as } r \to \infty$$

($\sin\theta \geq 2\theta/\pi, 0 \leq \theta \leq \pi/2$; see Problem 8 for details). Thus

$$\int_0^\infty e^{ix^2}\, dx = \int_0^\infty e^{-s^2} e^{i\pi/4}\, ds = \frac{1}{2}\sqrt{\pi} e^{i\pi/4},$$

and therefore $\int_0^\infty \cos x^2\,dx = \int_0^\infty \sin x^2\,dx = \frac{1}{2}\sqrt{\frac{\pi}{2}}$.

(c) The integral of $[\mathrm{Log}(z+i)]/(z^2+1)$ on γ is $2\pi i$ times the residue at $z = i$ of $[\mathrm{Log}(z+i)]/(z^2+1)$, which is $2\pi i(\mathrm{Log}\,2i)/2i = \pi\ln 2 + i\pi^2/2$. (Note that $\mathrm{Log}(z+i)$ is analytic except for $z = -i - x, x \geq 0$.) Thus

$$\int_{-r}^{0} \frac{\mathrm{Log}(x+i)}{x^2+1}\,dx + \int_0^r \frac{\mathrm{Log}(x+i)}{x^2+1}\,dx \to \pi\ln 2 + \frac{i\pi^2}{2}.$$

(The integral around the large semicircle approaches 0 as $r \to \infty$, by the M-L theorem.) Now let $x' = -x$ in the first integral to obtain

$$\int_0^r \frac{[\mathrm{Log}(i-x) + \mathrm{Log}(i+x)]}{x^2+1}\,dx \to \pi\ln 2 + i\frac{\pi^2}{2}.$$

But $\mathrm{Log}(i-x) + \mathrm{Log}(i+x) = \ln[|i-x||i+x|] + i(\theta_1+\theta_2) = \ln(x^2+1) + i\pi$ (see Figure S4.2.4). Hence

$$\int_0^\infty \frac{\ln(x^2+1)}{x^2+1}\,dx + i\pi\int_0^\infty \frac{dx}{x^2+1} = \pi\ln 2 + \frac{i\pi^2}{2}$$

or

$$\int_0^\infty \frac{\ln(x^2+1)}{x^2+1}\,dx = \pi\ln 2.$$

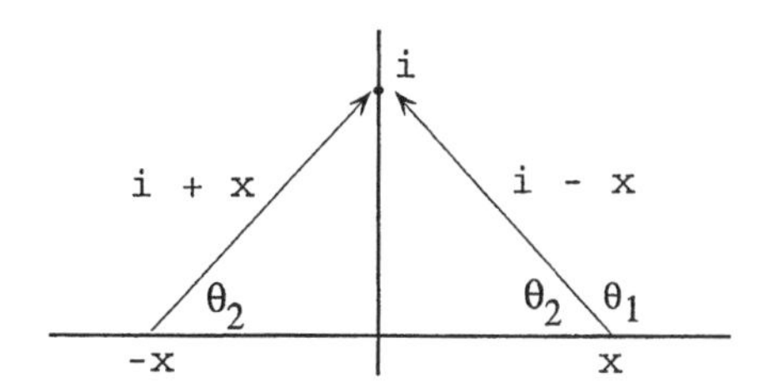

Figure S4.2.4

(d) In (c) let $x = \tan\theta$ to obtain

$$\pi\ln 2 = \int_0^{\pi/2} \frac{\ln(\tan^2\theta+1)}{\tan^2\theta+1}\sec^2\theta\,d\theta = -2\int_0^{\pi/2} \ln\cos\theta\,d\theta$$

so

$$\int_0^{\pi/2} \ln\cos\theta\,d\theta = -\frac{\pi}{2}\ln 2.$$

Set $\theta = \frac{\pi}{2} - x$ to get

$$-\frac{\pi}{2}\ln 2 = -\int_{\pi/2}^{0} \ln\cos(\frac{\pi}{2} - x)\,dx = \int_0^{\pi/2} \ln\sin x\,dx.$$

10. Let $f(z) = z^4, g(z) = z^4 + 6z + 3$. Then $|f(z) - g(z)| = |6z + 3|$; if $|z| = 2$, this is less than or equal to $12 + 3 < |z|^4 = |f(z)|$. Since f has all its zeros inside $\{z : |z| = 2\}$, so does g.

    Now let $f(z) = 6z, g(z) = z^4 + 6z + 3$. Then $|f(z) - g(z)| = |z^4 + 3| \leq 4 < 6|z|$ for $|z| = 1$. Thus g has one root inside $\{z : |z| = 1\}$, hence there are 3 roots in $\{z : 1 < |z| < 2\}$. (Since $|f - g| < |f|$ when $|z| = 1$, g cannot be 0 when $|z| = 1$.)

11. Apply Rouché's theorem to $f(z) - z^n$ and $-z^n$. We have $|f(z) - z^n + z^n| = |f(z)| < |-z^n|$ when $|z| = 1$. Since $-z^n$ has n zeros inside the unit circle, so does $f(z) - z^n$.

12. Apply the hexagon lemma (3.4.5) to the compact set $K_0 = \{z : |f(z)+g(z)| = |f(z)|+|g(z)|\}$. If $\gamma_1, \dots, \gamma_m$ are the polygonal paths given by the lemma, let $\gamma = \sum_{j=1}^m \gamma_j$. Then $\gamma^* \subseteq \Omega \setminus K_0$, so $|f + g| < |f| + |g|$ on γ^*. Since $Z(f) \cup Z(g) \subseteq K_0$, we have $n(\gamma, z) = 1$ for each $z \in Z(f) \cup Z(g)$. Again by (3.4.5), γ is Ω-homologous to 0. The result now follows from (4.2.9).

13. First let $u \geq 0$. Then the integral is $2\pi i$ times the residue of $e^{iuz}/[\pi(1 + z^2)]$ at $z = i$, which is $2\pi i e^{-u}/2\pi i = e^{-u}$. Now let $u < 0$. Then $|e^{iu(x+iy)}| = e^{-uy}$ is bounded on $\{x + iy : y \leq 0\}$ but *not* on $\{x + iy : y \geq 0\}$. Thus we must complete the contour in the *lower* half plane, as indicated in Figure S4.2.5. Therefore the integral is $-2\pi i$ times the residue of $e^{iuz}/[\pi(1 + z^2)]$ at $z = -i$, which is $-2\pi i e^u/-2\pi i = e^u$.

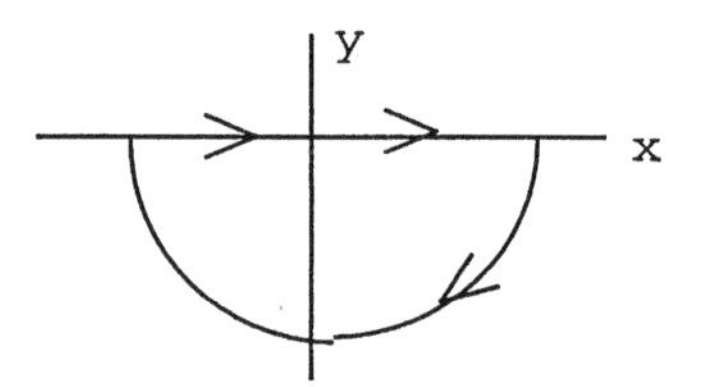

Figure S4.2.5

14. We have

$$e^{\sin 1/z} = \sum_{k=0}^{\infty} \frac{[\sin(1/z)]^k}{k!}.$$

Since $|\sin 1/z|$ is bounded on $\{z : |z| = 1\}$, the Weierstrass M-test shows that the series converges uniformly on this set and may therefore be integrated term by term. The residue theorem yields

$$\int_{|z|=1} e^{\sin 1/z}\, dz = 2\pi i \sum_{n=0}^{\infty} \frac{1}{n!} \operatorname{Res}(\sin^n 1/z, 0).$$

But

$$\sin \frac{1}{z} = \frac{1}{z} - \frac{1}{3!z^3} + \frac{1}{5!z^5} - \cdots, z \neq 0,$$

and thus all residues of $(\sin 1/z)^n$ at $z = 0$ are 0 except for $n = 1$, in which case the residue is 1. The integral is therefore $2\pi i$.

15. (a) Near z_0 we have

$$\frac{f(z)}{g(z)} = \frac{(z-z_0)^k[a_k + a_{k+1}(z-z_0) + \cdots]}{(z-z_0)^{k+1}[b_{k+1} + b_{k+2}(z-z_0) + \cdots]}.$$

The residue is therefore

$$\frac{a_k}{b_{k+1}} = \frac{f^{(k)}(z_0)/k!}{g^{(k+1)}(z_0)/(k+1)!} = (k+1)\frac{f^{(k)}(z_0)}{g^{(k+1)}(z_0)}.$$

(b) Near z_0 we have $f(z)/g(z) = [a_0 + a_1(z-z_0) + \cdots]/[(z-z_0)^2 h(z)]$ where $h(z) = b_0 + b_1(z-z_0) + \cdots$. The residue is therefore

$$\frac{d}{dz}[f(z)/h(z)]_{z=z_0} = \frac{h(z_0)f'(z_0) - f(z_0)h'(z_0)}{h^2(z_0)} = [f'(z_0)/b_0] - [f(z_0)b_1/b_0^2].$$

But $g(z) = b_0(z-z_0)^2 + b_1(z-z_0)^3 + \cdots$, so $b_0 = g''(z_0)/2!$ and $b_1 = g'''(z_0)/3!$, and the result follows.

16. let $f(z) = 3z, g(z) = 3z - e^{-z}$. then

$$|f(z) - g(z)| = |e^{-z}| = |e^{-(x+iy)}| = e^{-x} \le e < 3 = |f(z)| \text{ for } |z| = 1.$$

The result follows from Rouché's theorem.

17. If $w \in D(0,\epsilon)$, we must show that $w = f(z)$ for some $z \in D(0,r)$, that is, $f - w$ has a zero in $D(0,r)$. Now when $|z| = r$ we have $|(f(z) - w) - f(z)| = |w| < \epsilon \le |f(z)|$ by hypothesis. By Rouché's theorem, $f - w$ and f have the same number of zeros in $D(0,r)$. But f has at least one zero in $D(0,r)$ since $f(0) = 0$, and the result follows.

18. The analytic function $1/e^z$ contributes zero to the integral, as does $\cos 1/z$, whose residue at 0 is 0. Since $+i$ is inside the circle $C(1+i,2)$ but $-i$ is outside, the integral is $2\pi i$ times the residue of $e^{\pi z}/[(z-i)(z+i)]$ at $z = i$. Thus the integral is $2\pi i(e^{i\pi}/2i) = -\pi$.

19. Let γ_r be the contour formed by traveling from $-r$ to r along the real axis, and then returning to $-r$ on the semicircle $S(0,r)$ (in the upper half plane) with center at 0 and radius r. The integral of $P(z)/Q(z)$ on the semicircle approaches 0 as $r \to \infty$, by the M-L theorem. For r sufficiently large, γ_r encloses all the poles of P/Q in the upper half plane, so

$$\int_{\gamma_r} \frac{P(z)}{Q(z)}\,dz = \int_{-r}^{r} \frac{P(x)}{Q(x)}\,dx + \int_{S(0,r)} \frac{P(z)}{Q(z)}\,dz$$

and we may let $r \to \infty$ to get the desired result. For the specific example, note that the poles of $z^2/(1+z^4)$ in the upper half plane are at $z = e^{i\pi/4}$ and $e^{i3\pi/4}$. The residues are

$$\lim_{z\to e^{i\pi/4}} \frac{(z-e^{i\pi/4})z^2}{z^4+1} = e^{i\pi/2} \lim_{z\to e^{i\pi/4}} \frac{(z-e^{i\pi/4})}{z^4+1} = \frac{e^{i\pi/2}}{4e^{i3\pi/4}} = \frac{1}{4}e^{-i\pi/4}$$

and

$$\lim_{z\to e^{i3\pi/4}} \frac{(z - e^{i3\pi/4})z^2}{z^4+1} = \frac{e^{i3\pi/2}}{4e^{i9\pi/4}} = \frac{1}{4}e^{-i3\pi/4}.$$

Thus the integral is

$$\frac{2\pi i}{4}(e^{-i\pi/4} + e^{-i3\pi/4}) = \frac{\pi i}{2}(-i\sin\frac{\pi}{4} - i\sin\frac{3\pi}{4}) = \frac{\pi i}{2}(-i\sqrt{2}) = \frac{1}{2}\pi\sqrt{2}.$$

20. Apply Rouché's theorem with $f(z) = az^n$ and $g(z) = az^n - e^z$. Then for $|z| = 1$, $|f(z) - g(z)| = |e^z| \le e^{|z|} = e < |a| = |f(z)|$, and the result follows.
21. Let $f(z) = 2z, g(z) = 2z + 1 - e^z$. then for $|z| = 1$,
$$|f(z) - g(z)| = |e^z - 1| = |z + \frac{z^2}{2!} + \frac{z^3}{3!} + \cdots|$$
so
$$|f(z) - g(z)| \le 1 + \frac{1}{2!} + \frac{1}{3!} + \cdots = e - 1 < 2 = |f(z)|$$
and Rouché's theorem applies.
22. Let $g(z) = -5z^4$. If $|z| = 1$, then $|f(z) - g(z)| = |z^7 + z^2 - 2| \le 1 + 1 + 2 < |g(z)|$ and Rouché's theorem applies.
23. If $g(z) = z^5$, then for $|z| = 2$ we have $|f(z) - g(z)| = |15z + 1| \le 31 < 2^5 = |g(z)|$. If $h(z) = 15z$, then for $|z| = 1/2, |f(z) - h(z)| = |z^5 + 1| \le (1/2)^5 + 1 < 15/2 = |h(z)|$. The result follows from Rouché's theorem.
24. Apply Rouché's theorem with $f(z) = z^5, g(z) = z^5 + z + 1$. We have, for $|z| = 5/4, |f(z) - g(z)| = |z + 1| \le (5/4) + 1 = 9/4$. But $|f(z)| = (5/4)^5 = 3.05 > 9/4$, and the result follows.
25. If $f_n(z_n) = 0$ for all n and $z_n \to 0$, then $|f(z_0)| \le |f(z_0) - f(z_n)| + |f(z_n) - f_n(z_n)| + |f_n(z_n)| \to 0$ as $n \to \infty$ by the uniform convergence of f_n on compact subsets and the continuity of f at z_0. Thus $f(z_0) = 0$. Conversely, assume $f(z_0) = 0$. Since f is not identically zero, there is a disk $D(z_0, r)$ containing no zero of f except z_0. Let $\delta = \min\{|f(z)| : |z - z_0| = r\} > 0$. For sufficiently large m, $|f(z) - f_m(z)| < \delta$ for all $z \in \overline{D}(z_0, r)$, hence on $C(z_0, r)$ we have $|f(z) - f_m(z)| < |f(z)|$. By Rouché's theorem, f_m has a zero in $D(z_0, r)$, say at z_m. We may repeat this process using the disks $D(z_0, 2^{-n}r), n = 1, 2, 3, \ldots$ to find the desired subsequence.
26. (a) This is a direct calculation.
(b) By hypothesis, p must have $n - k$ zeros in $|z| > 1$, and the result follows from (a).
(c) This follows from (a) if we note that for $|z| = 1$, we have $z\overline{z} = 1$, hence $1/\overline{z} = z$.
(d) Assume $|a_0| > |a_n|$. If $g(z) = \overline{a}_0 p(z)$, then $|f(z) - g(z)| = |a_n q(z)| < |\overline{a}_0 p(z)|$ by part (c), so $|f(z) - g(z)| < |g(z)|$. By Rouché's theorem, f has k zeros in $|z| < 1$. Now assume $|a_0| < |a_n|$. If $h(z) = -a_n q(z)$, then for $|z| = 1, |f(z) - h(z)| = |\overline{a}_0 p(z)| < |-a_n q(z)| = |h(z)|$. By Rouché's theorem and part (b), f has $n - k$ zeros in $|z| < 1$.
(e) If $|a_0| > |a_n|$ and p has no zeros in $|z| > 1$, then p has n zeros in $|z| < 1$, hence so does f, by (d). If $|a_0| < |a_n|$ and p has no zeros in $|z| < 1$, then by (d), f has n zeros in $|z| < 1$. In either case there is a contradiction, because f is a polynomial of degree at most $n - 1$.

Section 4.3

1. Near z_0 we have $f(z) = \sum_{n=-1}^{\infty} a_n(z-z_0)^n$ and $g(z) = \sum_{m=0}^{\infty} b_m(z-z_0)^m$. The Laurent expansion of $g(z)f(z)$ is found by multiplying the two series, and $\text{Res}(gf, z_0) = b_0 a^{-1} = g(z_0)\,\text{Res}(f, z_0)$, as desired. For the counterexample, take $z_0 = 0$, $f(z) = (1/z^2) + (1/z)$, $g(z) = 1 + z$. Then $\text{Res}(gf, 0) = 2$; on the other hand, $g(0) = 1, \text{Res}(f, 0) = 1$.

2. If $f(z_0) = w_0$, then since f is one-to-one, $k = \min_{z \in C(z_0,r)} |f(z) - w_0| > 0$. Thus if $|w - w_0| < k$, we may expand

$$\frac{1}{f(z) - w} = \frac{1}{f(z) - w_0 - (w - w_0)} = \frac{1}{f(z) - w_0}\left[\frac{1}{1 - \frac{w - w_0}{f(z) - w_0}}\right]$$

in a geometric series. Term by term integration shows that f^{-1} is analytic at w_0.

3. Let $z_0 \in P$. If r is sufficiently small, then $V = \{1/f(z) : z \in D(z_0, r)\}$ is open in $\hat{\mathbb{C}}$ by (4.3.1). Also, $W = \{1/z : z \in V\}$ is open in $\hat{\mathbb{C}}$ because the image under $1/z$ of a disk containing 0 is a neighborhood of ∞. But $W = f(D(z_0, r))$, and the result follows.

4. By the residue theorem, the integral is $\sum_{j=1}^{n} \text{Res}(gf'/f, a_j)$. Since $\text{Res}(gf'/f, a_j) = m(f, a_j)g(a_j)$ by (4.2.2e) and Problem 1 of this section, the result follows.

5. If $z_0 \in \Omega$ and $D(z_0, r) \subseteq \Omega$, then the image of $D(z_0, r)$ under f will contain a disk $D(f(z_0), s)$. Since $D(f(z_0), s)$ will contain points w_1, w_2, w_3 such that $|w_1| > |f(z_0)|, \text{Re}\, w_2 > \text{Re}\, f(z_0)$, and $\text{Im}\, w_3 > \text{Im}\, f(z_0)$, it follows that $|f|, \text{Re}\, f$, and $\text{Im}\, f$ cannot take on a local maximum at z_0.

Sections 4.4 and 4.5

1. For the inverse, solve $w = (az + b)/(cz + d)$ for z. For the composition, consider $w = (au + b)/(cu + d)$, $u = (\alpha z + \beta)/(\gamma z + \delta)$ and substitute. Alternatively, use the fact that a linear fractional transformation is a composition of maps of types (i)-(iv) of (4.4.1).

2. (a) If $w = (1 + z)/(1 - z)$ then $z = (w - 1)/(w + 1)$, so $T^{-1}(w) = (w - 1)/(w + 1)$.
(b) It is easier to deal with T^{-1}. Figure $S4.5.1$ shows that T^{-1} maps $\text{Re}\, w > 0$ onto $|w| < 1$, $\{\text{Re}\, w = 0\} \cup \{\infty\}$ onto $|w| = 1$, and $\text{Re}\, w < 0$ onto $|w| > 1$; the result follows.

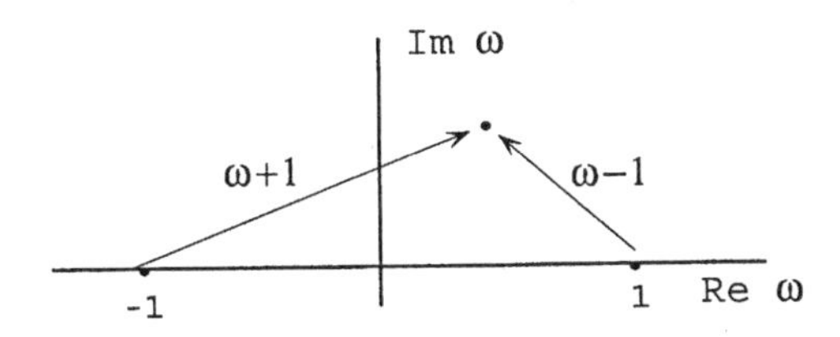

Figure S4.5.1

3. (a) Possibly motivated by the analysis of Problem 2, try $T(z) = k(z-i)/(z+i)$. Since $T(1) = 1$ we have $k = (1+i)/(1-i)$, and this does yield $T(-1) = -1$, as desired.
(b) The desired transformation is accomplished by an inversion followed by a 180 degree rotation, in other words, $T(z) = -1/z$.

4. (a) T must be of the form $T(z) = k(z - z_1)/(z - z_3)$. Since $T(z_2) = 1$ we have $1 = k(z_2 - z_1)/(z_2 - z_3)$, which determines k uniquely.
(b) If $z_1 = \infty$ then $T(z) = (z_2 - z_3)/(z - z_3)$. If $z_2 = \infty$ then $T(z) = (z - z_1)/(z - z_3)$, and if $z_3 = \infty$ then $T(z) = (z - z_1)/(z_2 - z_1)$.
(c) If T_1 is the unique linear fractional transformation mapping z_1, z_2, z_3 to $0, 1, \infty$, and T_2 is the unique linear fractional transformation mapping w_1, w_2, w_3 to $0, 1, \infty$, then $T = T_2^{-1} \circ T_1$. (If T^* is another linear fractional transformation mapping z_1, z_2, z_3 to w_1, w_2, w_3, then $T_2 \circ T^*$ maps z_1, z_2, z_3 to $0, 1, \infty$. Thus $T_2 \circ T^* = T_1$, hence $T^* = T_2^{-1} \circ T_1 = T$, proving T unique.)

5. (a) This follows from the fact that f is one-to-one.
(b) This is a consequence of the open mapping theorem for meromorphic functions (Section 4.3, Problem 3).
(c) Let $w \in f(D(0,1))$, which is open in $\hat{\mathbb{C}}$ by part (b). If ∞ is an essential singularity, then by the Casorati-Weierstrass theorem we find $z_n \to \infty$ with $f(z_n) \to w$. Thus for large n, $z_n \notin D(0,1)$ but $f(z_n) \in f(D(0,1))$, contradicting the assumption that f is one-to-one.
(d) If f is analytic on $\hat{\mathbb{C}}$, then f is constant by Liouville's theorem. Thus by part (a), there is only one remaining case to consider, in which f has poles at ∞ and at $z_0 \in \mathbb{C}$. As in (b), $f(D(z_0,1))$ and $f(\hat{\mathbb{C}} \setminus \overline{D}(z_0,1))$ are disjoint open sets in $\hat{C}$. Since $\infty \in f(D(z_0,1))$ (because $f(z_0) = \infty$), $f(\hat{C} \setminus \overline{D}(z_0,1))$is a bounded set, that is, f is bounded on the complement of $\overline{D}(z_0,1)$. This contradicts the assumption that ∞ is a pole.
(e) If $z_0 = \infty$, then by Problem 9(c) of Section 4.1, f is a polynomial, and $\deg f = 1$ because f is one-to-one. If $z_0 \in \mathbb{C}$, then since f has a pole at z_0, g is analytic at z_0. By the open mapping theorem (4.3.1), $g'(z_0) \neq 0$. (If $g'(z_0) = 0$ then g, hence f, is not one-to-one.
(f) If $z_0 = \infty$, this follows from (e), so assume $z_0 \in \mathbb{C}$. By (e), $g'(z_0) \neq 0$, hence $(z - z_0)f(z) = (z - z_0)/(g(z) - g(z_0)) \to 1/g'(z_0)$ as $z \to z_0$. By part (b) of (4.1.5), f has a simple pole at z_0.
(g) Let $h(z) = f(z) - [\text{Res}(f, z_0)/(z - z_0)]$. By (4.2.2d), $\lim_{z \to z_0}(z - z_0)h(z) = 0$. Thus $h(z)$ has only removable singularities in $\hat{\mathbb{C}}$ and is therefore constant.

Section 4.6

1. By (4.6.3i),

$$\left|\frac{f(z) - f(a)}{z - a}\right| \leq \left|\frac{1 - \overline{f(a)}f(z)}{1 - \overline{a}z}\right|;$$

let $z \to a$ to obtain (4.6.3ii).

2. Since $\operatorname{Re} z > 0$, we have $|w - f(0)| < |w - (-\overline{f(0)})|$ for $\operatorname{Re} w > 0$ (draw a picture). Thus T maps $\{w : \operatorname{Re} w > 0\}$ into $D(0,1)$, so $T \circ f$ is an analytic map of $D(0,1)$ into itself. Since $T(f(0)) = 0$, Schwarz's lemma implies that $|T(f(z)| \leq |z|, z \in D(0,1)$, that is, $|f(z) - f(0)| \leq |z||f(z) + \overline{f(0)}|$. Thus both $|f(z)| - |f(0)|$ and $|f(0)| - |f(z)|$ are less than or equal to $|z|[|f(z)| + |f(0)|]$. This yields the first statement of the problem. Now

$$\frac{d}{dz}T(f(z)) = \frac{f(z) + \overline{f(0)} - (f(z) - f(0))}{[f(z) + \overline{f(0)}]^2} f'(z),$$

and this is at most 1 in absolute value when $z = 0$, by Schwarz's lemma. Thus

$$\frac{|2 \operatorname{Re} f(0)|}{|2 \operatorname{Re} f(0)|^2} |f'(0)| \leq 1$$

and the result follows.

3. If $f(z_0) = z_0$ and $f(a) = a$, with $z_0 \neq a$, then equality holds at z_0 in (4.6.3i). In this case $b = f(a) = a$, so $f = \varphi_a^{-1} \circ \lambda\varphi_a$ with $|\lambda| = 1$. Now $z_0 = f(z_0) = \varphi_a^{-1}(\lambda\varphi_a(z_0))$, hence $\varphi_a(z_0) = \lambda\varphi_a(z_0))$. Since $z_0 \neq a$, we have $\varphi_a(z_0) \neq 0$, so $\lambda = 1$ and $f = \varphi_a^{-1} \circ \varphi_a$, the identity function.

4. (a) The function f must have the form given in (4.6.6) in $D(0,1)$, hence on $\mathbb{C}$ by the identity theorem. Since f is entire, the only possibility is $n = 1, a_1 = 0$, so $f(z) = \lambda z^k$ for some unimodular λ and nonnegative integer k.
(b) Let the poles of f in $D(0,1)$ be at $b_1, \ldots, b_m$, with orders $l_1, \ldots, l_m$ respectively. Then by (4.6.6), f is of the form

$$f(z) = \frac{\lambda \prod_{j=1}^{n} \left(\frac{z - a_j)}{1 - \overline{a}_j z}\right)^{k_j}}{\prod_{j=1}^{m} \left(\frac{z - b_j)}{1 - \overline{b}_j z}\right)^{l_j}}$$

with $|\lambda| = 1; a_j, b_j \in D(0,1); k_j, l_j = 0, 1, \ldots$.
(Note that $f(z)$ times the denominator of the above fraction has only removable singularities in $D(0,1)$.)

5. The function g satisfies the hypothesis of (4.6.3), so by (4.6.3i),

$$\left|\frac{g(z) - g(a)}{1 - \overline{g(a)}g(z)}\right| \leq \left|\frac{z - a}{1 - \overline{a}z}\right|, \quad a, z \in D,$$

that is,

$$\left|\frac{M(f(Rz) - f(Ra))}{M^2 - \overline{f(Ra)}f(Rz)}\right| \leq \left|\frac{z - a}{1 - \overline{a}z}\right|.$$

Let $w = Rz, w_0 = Ra$, to obtain

$$\left|\frac{M(f(w) - f(w_0))}{M^2 - \overline{f(w_0)}f(w)}\right| \leq \left|\frac{R(w - w_0)}{R^2 - \overline{w}_0 w}\right|, \quad w, w_0 \in D(0, R)$$

which is the desired generalization of (i). By (4.6.3ii), $|g'(a)| \le (1-|g(a)|^2)/(1-|a|^2)$, that is,

$$\frac{R}{M}|f'(Ra)| \le \frac{1-[|f(Ra)|^2/M^2]}{1-|a|^2}.$$

Thus

$$|f'(w_0)| \le \frac{(M/R)-[|f(w_0)|^2/MR]}{1-|w_0/R|^2},$$

or

$$|f'(w_0)| \le \frac{R(M^2-|f(w_0)|^2)}{M(R^2-|w_0|^2)}$$

which generalizes (ii).

6. Let

$$g(z) = \frac{f(z)}{\prod_{j=1}^{n}\left(\frac{z-z_j}{1-\overline{z}_j z}\right)^{k_j}}.$$

Then g is analytic on $D(0,1)$, continuous on $\overline{D}(0,1)$, and $|g(z)| = |f(z)| \le 1$ when $|z| = 1$. The assertion now follows from the maximum principle. If equality holds at some point z_0 in $D(0,1)$ (other than the z_j), then $|g(z_0)| = 1$, so g is constant by the maximum principle. Thus

$$f(z) = c\prod_{j=1}^{n}\left(\frac{z-z_j}{1-\overline{z}_j z}\right)^{k_j}$$

where c is a constant with $|c| \le 1$.

Section 4.7

1. If $|z| < 1$, then $(2\pi i)^{-1}\int_{|w|=1}\frac{w+z}{w(w-z)}\,dw = -1+2 = 1$ by the residue theorem. Thus $(2\pi)^{-1} = \int_{-\pi}^{\pi}\frac{e^{it}+z}{e^{it}-z}\,dt = 1$, as desired.
2. Since $-1 \le \cos(\theta-t) \le 1$, we have

$$\frac{1-r}{1+r} = \frac{1-r^2}{(1+r)^2} \le P_r(\theta-t) \le \frac{1-r^2}{(1-r)^2} = \frac{1+r}{1-r}.$$

The result now follows from (4.7.8) and the observation that by (4.7.9),

$$u(0) = \frac{1}{2\pi}\int_0^{2\pi} u(e^{it})\,dt.$$

3. If $\overline{D}(z_0,R) \subseteq \Omega$, then by (4.7.8) with $r=0$, $u_n(z_0) = (2\pi)^{-1}\int_0^{2\pi} u_n(z_0+Re^{it})\,dt$. Let $n\to\infty$ to obtain $u(z_0) = (2\pi)^{-1}\int_0^{2\pi} u(z_0+Re^{it})\,dt$. By (4.7.10), u is harmonic on Ω.

4. It is sufficient to consider the case where u is continuous on $\overline{D}(0,1)$ and analytic on $D(0,1)$. Then by (4.7.8), $u(z) = (2\pi)^{-1}\int_0^{2\pi} P_z(t)u(e^{it})\,dt$, $|z| < 1$. Let $f(z) = (2\pi)^{-1}\int_0^{2\pi} Q_z(t)u(e^{it})\,dt$. Then f is analytic on $D(0,1)$ by (3.3.3), and $\operatorname{Re} f = u$ by (4.7.2), as desired.
5. (i) We have

$$[z_0, z, \infty) = ([z_0, z, \infty) \cap \Omega) \cup ([z_0, z, \infty) \cap \partial\Omega) \cup ([z_0, z, \infty) \cap (\mathbb{C} \setminus \overline{\Omega})).$$

The first and third sets on the right are nonempty, relatively open subsets of $[z_0, z, \infty)$. Since $[z_0, z, \infty)$ is connected, $[z_0, z, \infty) \cap \partial\Omega \neq \emptyset$. Let β be any point in $[z_0, z, \infty) \cap \partial\Omega$. It follows from (a) and (b) that $[z_0, \beta) \subseteq \Omega$. (See Figure 4.7.1 to visualize this.)

Now either $z \in (z_0, \beta)$ or $\beta \in (z_0, z)$. If $\beta \in (z_0, z)$, we can repeat the above argument with z_0 replaced by β to get $\beta_1 \in \partial\Omega$ such that $\beta_1 \in (\beta, z, \infty)$. But then (a) and (b) imply that $\beta \in \Omega$, a contradiction. Thus $z \in (z_0, \beta)$, hence $[z_0, z] \subseteq [z_0, \beta) \subseteq \Omega$.
(ii) We have $\int_{\gamma_\delta} f(w)\,dw = 0$ by (3.3.1). Since $|\gamma(t) - \gamma_\delta(t)| = (1-\delta)|\gamma(t) - z_0| \to 0$ as $\delta \to 1$, uniformly in t, it follows from the uniform continuity of f on compact sets that we may let $\delta \to 1$ to obtain $\int_\gamma f(w)\,dw = 0$. The result $n(\gamma, z)f(z) = (2\pi i)^{-1}\int_\gamma [f(w)/(w-z)]\,dw$ is obtained similarly. (Note that $n(\gamma_\delta, z) = n(\gamma, z)$ for all δ sufficiently close to 1, by (3.2.3) and (3.2.5).]
6. The two equations given in the outline follow immediately from Problem 5. Subtract the second equation from the first to obtain

$$f(z) = \frac{1}{2\pi i}\int_\gamma f(w)\left[\frac{1}{w-z} - \frac{1}{w-\overline{z}}\right]dw.$$

If $z = x + iy$, $w = t + i\beta$, then

$$\frac{1}{w-z} - \frac{1}{w-\overline{z}} = \frac{z-\overline{z}}{(w-z)(w-\overline{z})} = \frac{2iy}{[t-x+i(\beta-y)][t-x+i(\beta+y)]}.$$

If w is real, so that $\beta = 0$, this becomes $2iy/[(t-x)^2 + y^2]$. Thus

$$f(z) = \frac{1}{\pi}\int_{-R}^{R}\frac{yf(t)}{(t-x)^2+y^2}\,dt + \frac{1}{\pi}\int_{\Gamma_R}\frac{yf(w)}{(w-z)(w-\overline{z})}\,dw$$

where Γ_R is the semicircular part of the contour. Let $M_f(R)$ be the maximum value of $|f|$ on Γ_R. By the M-L theorem, for large R the integral around Γ_R is bounded in absolute value by a constant times $M_f(R)/R$, so that if $M_f(R)/R \to 0$ as $R \to \infty$, we obtain

$$f(z) = \lim_{R\to\infty}\frac{1}{\pi}\int_{-R}^{R}\frac{yf(t)}{(t-x)^2+y^2}\,dt.$$

If $|f(z)|/|z|^{1-\delta} \to 0$ as $z \to \infty$ for some $\delta > 0$, then we may write

$$f(z) = \frac{1}{\pi}\int_{-\infty}^{\infty}\frac{yf(t)}{(t-x)^2+y^2}\,dt$$

where the integral exists in the improper Riemann sense, not simply as a Cauchy principal value. Take real parts to get the desired result.

Section 4.8

1. We may write $f(z) = cz^k + a_{k+1}z^{k+1} + a_{k+2}z^{k+2} + \cdots$ where $c \neq 0$. Apply Jensen's formula (4.8.2b) to $f(z)/z^k$ to obtain

$$\ln|c| = \sum_{j=1}^{n(r)} \ln\left|\frac{a_j}{r}\right| + \frac{1}{2\pi}\int_0^{2\pi} \ln\left|\frac{f(re^{it})}{r^k}\right| dt.$$

Thus

$$k\ln r + \ln|c| = \sum_{j=1}^{n(r)} \ln\left|\frac{a_j}{r}\right| + \frac{1}{2\pi}\int_0^{2\pi} \ln|f(re^{it})|\, dt.$$

But $c = f^{(k)}(0)/k!$, and the result follows.

2. The statement is

$$\ln|f(0)| = \sum_{j=1}^{n} k_j \ln\left|\frac{a_j}{R}\right| - \sum_{j=1}^{m} l_j \ln\left|\frac{b_j}{R}\right| + \frac{1}{2\pi}\int_0^{2\pi} \ln|f(Re^{it})|\, dt.$$

To prove the statement, note that we may write $f = g/h$, where g has zeros at $a_1, \dots, a_n$, h has zeros at $b_1, \dots, b_m$, and g and h each satisfy the hypothesis of (4.8.1). Since $\ln|f| = \ln|g| - \ln|h|$, the result follows.

3. First note that if $0 < r < R$, then $n(t)$ is a step function on $[0, r]$ which is left continuous, having jumps only at the radii of those circles that pass through zeros of f. To avoid cumbersome notation, we illustrate the ideas with a concrete example Suppose $0 < |a_1| = |a_2| = |a_3| < |a_4| < |a_5| = |a_6| < |a_7| < r \le |a_8|$. Then the graph of $n(t), 0 \le t \le r$, is shown in Figure S4.8.1. Since $n(t)$ is constant between jumps and $\int(1/t)\, dt = \ln t$, we have

$$\begin{aligned}\int_0^r \frac{n(t)}{t}\, dt &= n(|a_3|)\ln|a_3| + n(|a_4|)(\ln|a_4| - \ln|a_3|)\\ &\quad + n(|a_6|)(\ln|a_6| - \ln|a_4|) + n(|a_7|)(\ln|a_7| - \ln|a_6|)\\ &\quad + n(r)(\ln r - \ln|a_7|).\end{aligned}$$

If we observe that $|a_7| < r \le |a_8|$, so that $n(r) - n(|a_8|)$, we may write

$$\begin{aligned}\int_0^r \frac{n(t)}{t}\, dt &= -\ln|a_3|[n(|a_4|) - n(|a_3|)]\\ &\quad - \ln|a_4|[n(|a_6|) - n(|a_4)] - \ln|a_6|[n(|a_7|) - n(|a_6|)]\\ &\quad - \ln|a_7|[n(|a_8|) - n(|a_7|)] + n(r)\ln r.\end{aligned}$$

Now

$$-\ln|a_6|[n(|a_7|) - n(|a_6|)] = -2\ln|a_6| = -\ln|a_5| - \ln|a_6|$$

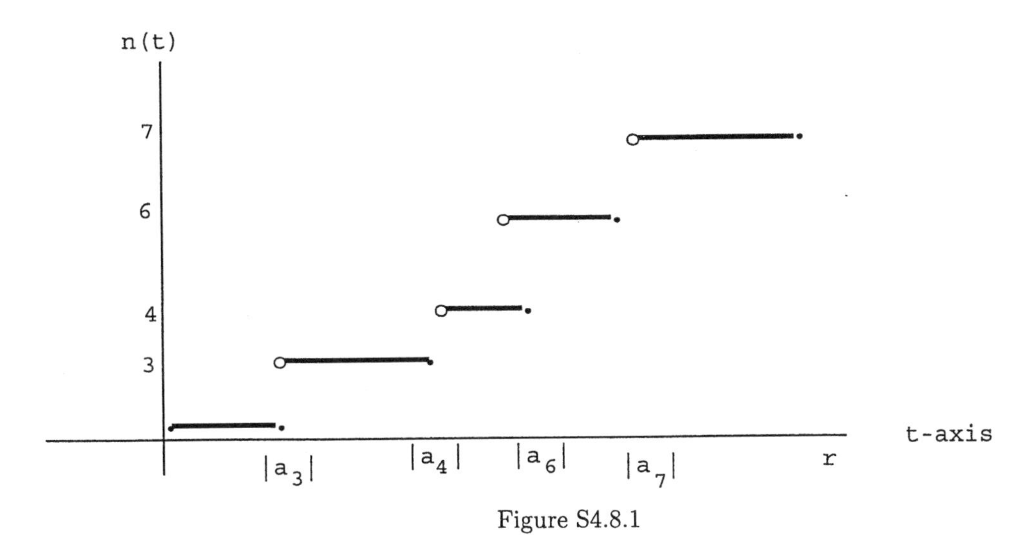

Figure S4.8.1

and similarly for the other terms. Thus

$$\int_0^r \frac{n(t)}{t}\,dt = -\sum_{j=1}^{n(r)} \ln|a_j| + n(r)\ln r = \sum_{j=1}^{n(r)} \ln\frac{r}{|a_j|}$$

as desired.

4. By Problem 3,

$$\int_0^r \frac{n(t)}{t}\,dt = \sum_{j=1}^{n(r)} \ln\frac{r}{|a_j|}.$$

Also,

$$\ln[|f^{(k)}(0)|r^k/k!] = k\ln r + \ln[|f^{(k)}(0)|/k!].$$

The result now follows from (4.8.5) if we observe that

$$\frac{1}{2\pi}\int_0^{2\pi} \ln|f(re^{it}|\,dt \le \frac{1}{2\pi}\int_0^{2\pi} \ln M(r)\,dt = \ln M(r).$$

5. By (4.8.5) and Problem 3,

$$\frac{1}{2\pi}\int_0^{2\pi} \ln|f(re^{it}|\,dt = k\ln r + \ln[|f^{(k)}(0)/k!] + \int_0^r \frac{n(t)}{t}\,dt,$$

which is a continuous, increasing function of r. Each time $n(t)$ has a jump, say a jump of size c at $t = r_0$ (see Figure S4.8.1), $\int_0^r [n(t)/t]\,dt$ contributes a term of the form $c(\ln r - \ln r_0), r > r_0$.

Section 4.9

1. If $z = re^{i2\pi p/q}$, then $z^{n!} = r^{n!}e^{i2\pi n!p/q} = r^{n!}$ if $n \geq q$, and $r^{n!} \to 1$ as $r \to 1$. It follows that any analytic function that agrees with f on $D(0,1)$ cannot approach a finite limit as z approaches a point on $C(0,1)$ of the form $e^{i2\pi p/q}$. Since these points are dense in $C(0,1)$, there can be no extension of f to a function analytic on $D(0,1) \cup D(w,\epsilon), |w| = 1$.

2. (a) Let $S = \{x + iy : y = 0, x \leq 0\}$. If $z_1 \notin S$, let $D_1, \dots, D_n$ be disks such that $D_i \cap D_{i+1} \neq \emptyset, i = 1, \dots, n-1, 1 \in D_1, z_1 \in D_n$, and $D_i \cap S = \emptyset, i = 1, \dots, n$. Let $f_i(z) = \operatorname{Log} z, z \in D_i$. Then (f_n, D_n) is a continuation of $(f_1, D_1) = (f, D)$ relative to Ω, so $f_n, D_n) \in \Phi$.
If, say, z_1 is in the second quadrant, then $\operatorname{Log} z_1 = \ln|z_1| + i\theta(z_1)$ where θ can be be chosen in the interval $[0, 2\pi)$. If $z_2 \in S, z_2 \neq 0$, let $E_1, \dots, E_m$ be disks such that $E_1 = D_n$ (so $z_1 \in E_1$), $E_i \cap E_{i+1} \neq \emptyset, i = 1, \dots, m-1, z_2 \in E_m$, and $E_i \cap T = \emptyset, i = 1, \dots, m$, where $T = \{x+iy : y = 0, x \geq 0\}$. Let $g_i(z) = \log z = \ln|z| + i\theta(z), 0 \leq \theta < 2\pi, z \in E_i$. Then (g_m, E_m) is a continuation of $(g_1, E_1) = f_n, D_n)$ relative to Ω, so $(g_m, E_m) \in \Phi$.
(b) By the argument of (a), if there were such an h, then $h(z) = \operatorname{Log} z, z \notin S$, and hence h must be discontinuous on the negative real axis, a contradiction.

3. The reasoning beginning with "since power series converge absolutely" is faulty. If $\sum_{k=0}^{\infty} b_k(z - z_1)^k$ converges absolutely at some point $z \notin \overline{D}(z_0, r)$, this does not imply that the original series converges at z. For

$$\sum_{k=0}^{\infty} |b_k|\,|z - z_1|^k = \sum_{k=0}^{\infty} \left| \sum_{n=k}^{\infty} \binom{n}{k} a_n (z_1 - z_0)^{n-k} \right| |z - z_1|^k < \infty$$

does not imply that

$$\sum_{k=0}^{\infty} \sum_{n=k}^{\infty} |a_n|\,|z_1 - z_0|^{n-k}\,|z - z_1|^k < \infty,$$

and the latter is what is needed to reverse the order of summation.

4. If $g_1, \dots, g_k$ are analytic on Ω, so is $h(z) = F(z, g_1(z), \dots, g_k(z)), z \in \Omega$. (The derivative of h may be calculated explicitly by the chain rule.) It follows that if $h_j(z) = F(z, f_{1j}(z), \dots, f_{kj}(z))$, then h_j is analytic on $D_j, j = 1, \dots, n$. Thus $(h_1, D_1), \dots, (h_n, D_n)$ forms a continuation. But $D_1 = D$ and $h_1 = 0$ on D, by hypothesis. By successive application of the identity theorem (2.4.8), we have $h_n = 0$ on D_n, as desired.

5. If (f_{i+1}, D_{i+1}) is a direct continuation of (f_i, D_i), then $f_i = f_{i+1}$ on $D_i \cap D_{i+1}$, hence $f_i' = f_{i+1}'$ on $D_i \cap D_{i+1}$. Therefore (f_{i+1}', D_{i+1}) is a direct continuation of (f_i', D_i), and the result follows.

Chapter 5

Section 5.1

1. Since $|f(z)| \leq \frac{1+|z|}{1-|z|}|f(0)|$, $\mathcal{F}$ is bounded, hence $\overline{\mathcal{F}}$ is closed and bounded, and therefore compact. thus $\mathcal{F}$ is relatively compact. To show that $\mathcal{F}$ is not compact, let $f_n(z) =$

$\frac{1}{n}\frac{1+z}{1-z}$, $f(z) \equiv 0$. By Section 4.5, Problem 2, $f_n \in \mathcal{F}$; since $f_n \to f$ uniformly on compact subsets of $D(0,1)$ but $f \notin \mathcal{F}$, $\mathcal{F}$ is not closed, and therefore not compact.

2. We may take $a = 0$ (if not, consider $f - a$). Since $1/|f(z)| \le 1/r$ for all $f \in \mathcal{F}$ and $z \in \Omega$, by (5.1.10) we have a subsequence $\{f_{n_k}\}$ such that $1/f_{n_k} \to g \in A(\Omega)$, uniformly on compact subsets. If g is not identically 0, then g is never 0 by (5.1.4), and it follows that $f_{n_k} \to 1/g$ uniformly on compact subsets. If $g \equiv 0$, then $f_{n_k} \to \infty$ uniformly on compact subsets.

3. (a) If $\mathcal{F}$ is relatively compact then $\overline{\mathcal{F}}$ is compact, so if $f_n \in \mathcal{F}, n = 1, 2, \ldots$, there is a subsequence $\{f_{n_k}\}$ converging to a limit in $\overline{\mathcal{F}}$ (not necessarily in $\mathcal{F}$). Conversely, if each sequence in $\mathcal{F}$ has a convergent subsequence, the same is true for $\overline{\mathcal{F}}$. (If $f_n \in \overline{\mathcal{F}}$, choose $g_n \in \mathcal{F}$ with $d(f_n, g_n) < 1/n$; if the subsequence $\{g_{n_k}\}$ converges, so does $\{f_{n_k}\}$). Thus $\overline{\mathcal{F}}$ is compact.
(b) $\mathcal{F}$ is bounded iff $\overline{\mathcal{F}}$ is bounded (by definition of boundedness), iff $\overline{\mathcal{F}}$ is closed and bounded (since $\overline{\mathcal{F}}$ is always closed), iff $\overline{\mathcal{F}}$ is compact (by the first statement of (5.1.11)), iff $\mathcal{F}$ is relatively compact.

4. Let $\mathcal{F}$ be relatively compact. If $f \in \mathcal{F}$ and $f(z) = \sum_{n=0}^{\infty} a_n z^n$, then by (2.4.1), $|a_n| \le r^{-n} \max\{|f(z)| : |z| = r\}, 0 < r < 1$. But by compactness, $\max\{|f(z)| : |z| = r\}$ is bounded by a constant $M(r)$ independent of the particular $f \in \mathcal{F}$. Thus

$$M_n = \sup\{|a_n(f)| : f \in \mathcal{F}\} \le M(r)/r^n.$$

Consequently, $\sum M_n z^n$ converges if $|z| < r$, so by (2.2.7), $(\limsup_{n\to\infty} M_n^{1/n})^{-1} \ge r$. Let $r \to 1$ to obtain $\limsup_{n\to\infty} M_n^{1/n} \le 1$. Conversely, if the desired M_n exist, then if $f \in \mathcal{F}$ and $|z| \le r < 1$, we have $|f(z)| \le |a_n||z|^n \le \sum_{n=0}^{\infty} M_n r^n < \infty$. Thus $\mathcal{F}$ is bounded, hence relatively compact.

5. (a) Apply Cauchy's formula for a circle to the function f^2 to get, for $0 \le r < R$,

$$f^2(a) = \frac{1}{2\pi}\int_0^{2\pi} f^2(a + re^{it})\,dt$$

(the mean value of f^2). Thus

$$|f(a)|^2 = |f^2(a)| \le \frac{1}{2\pi}\int_0^{2\pi} |f(a + re^{it})|^2\,dt.$$

Now multiply on both sides by r and integrate with respect to r from 0 to R to obtain

$$\frac{R^2}{2}|f(a)|^2 \le \frac{1}{2\pi}\int_0^R r\int_0^{2\pi} |f(a + re^{it})|^2\,dt\,dr$$

and the result follows.
(b) By part (a), $\mathcal{F}$ is bounded, and the result follows from (5.1.10).

6. Let $f \to H(f)$ be the suggested map. Since $|f| \le 1$ on Ω and $f = 0$ on the boundary of K, the integral over K is greater than 0 and H is well defined. If $f_n \in \mathcal{F}$ and $f_n \to f$, that is, $d(f_n, f) \to 0$, then $f_n \to f$ uniformly on K, hence $H(f_n) \to H(f)$, so that H

is continuous. If $\mathcal{F}$ were compact, then $H(\mathcal{F})$ would be a compact, hence bounded, subset of the reals.

If $0 < r < R$, let f be a continuous function from Ω to $[0,1]$ such that $f = 1$ on $\overline{D} = \overline{D}(a,r)$ and $f = 0$ off K (Urysohn's lemma). Then

$$\int_K\int |f(x+iy)|\,dx\,dy \geq \int_{\overline{D}}\int 1\,dx\,dy \to \int_K\int 1\,dx\,dy$$

as $r \to R$. Thus $H(\mathcal{F})$ is unbounded, a contradiction.

7. If z is a point on the open radial line S from 0 to $e^{i\theta}$, then $e^{i\theta} + (1/n)(z - e^{i\theta}) = (1 - 1/n)e^{i\theta} + (1/n)z$ also lies on S, and approaches $e^{i\theta}$ as $n \to \infty$. By hypothesis, f_n converges pointwise on S. Since S certainly has a limit point in $S(\theta, \alpha)$, Vitali's theorem implies that f_n converges uniformly on compact subsets. Given $\epsilon > 0$ there exists $\delta > 0$ such that if $z \in S(\theta, \alpha)$ and $|z - e^{i\theta}| < \delta$, then $|z - w| < \epsilon$ for some $w \in S$. It follows that by choosing z sufficiently close to $e^{i\theta}$, we can make $f(z)$ as close as we wish to L, as desired.

8. If k is a complex number, then k will also be used to denote the function that is identically k. Since $L(1) = L(1^2) = L(1)L(1)$, $L(1)$ must be 0 or 1. But if $L(1) = 0$, then for any $f \in A(\Omega), L(f) = L(f1) = L(f)L(1) = 0$, hence $L \equiv 0$, a contradiction. Thus $L(1) = 1$, so $L(k) = L(k1) = kL(1) = k$. Now let $z_0 = L(I)$. If $z_0 \notin \Omega$, then $h(z) = 1/(z - z_0)$ gives $h \in A(\Omega)$. Thus $h(I - z_0) = 1$, hence $L(h)(z_0 - z_0) = 1$, a contradiction. Therefore $z_0 \in \Omega$. If $f \in A(\Omega)$ and g is as defined in the outline, then $g \in A(\Omega)$ and $g(I - z_0) = f - f(z_0)$. It follows that $L(f) - f(z_0) = L(g)(L(I) - z_0) = L(g)(z_0 - z_0) = 0$.

9. Define A_n as suggested. Then each A_n is a closed subset of Ω, and since for each $z \in \Omega$, $f_k(z)$ converges to a finite limit as $k \to \infty$, we have $\cup_{n=1}^{\infty} A_n = \Omega$. By the Baire category theorem, some A_n contains a disk D. The f_k are uniformly bounded on D, hence by Vitali's theorem, $f_n \to f$ uniformly on compact subsets of D. (Note that D is connected, although Ω need not be.) Thus f is analytic on D. Finally, let U be the union of all disks $D \subseteq \Omega$ such that $f_n \to f$ uniformly on compact subsets of D. Then U is an open subset of Ω and $f_n \to f$ uniformly on any compact $K \subseteq U$ (because K is covered by finitely many disks). If W is an open subset of Ω, the first part of the proof shows that W contains one of the disks D whose union is U. Thus U is dense in Ω.

Section 5.2

1. For $j = 1, 2$, let g_j be the unique analytic map of Ω_j onto D such that $g_j(z_j) = 0$ and $g_j'(z_j) > 0$ (5.2.3d). Then $f = g_2^{-1} \circ g_1$ satisfies $f(z_1) = z_2$ and $f'(z_1) > 0$. If h is another such map, then $g_2 \circ h = g_1$ by (5.2.3d), so $h = f$.

2. From the definition, h is a continuous map of $\mathbb{C}$ into $D(0,1)$. To prove that h is one-to-one and onto, note that $h(re^{i\theta}) = re^{i\theta}/(1+r)$. If $h(z_n) \to h(z)$, then $h(z) \in D(0, r/(1+r))$ for r sufficiently close to 1. But since h maps $\overline{D}(0,r)$ one-to-one onto $\overline{D}(0, r/(1+r))$, it follows by compactness that h is a homeomorphism of these sets. Thus $z_n \to z$, so h^{-1} is continuous.

3. By convexity, $H(t,s) \in \Omega$ for all $t \in [a,b]$ and all $s \in [0,1]$. Since $H(t,0) = \gamma(t)$ and $H(t,1) = \gamma(a)$, the result follows.
4. Proceed as in Problem 3, with the initial point $\gamma(a)$ replaced by the star center, to obtain an Ω-homotopy of the given curve γ to a point (namely the star center).
5. Let $\Omega_n = \hat{\mathbb{C}} \setminus K_n$. By definition of K_n, we have

$$\Omega_n = \{\infty\} \cup \{z : |z| > n\} \cup \bigcup_{w \in \mathbb{C}\setminus\Omega} D(w, 1/n).$$

Now consider any component T of Ω_n. Since T is a maximal connected subset of Ω_n, it follows that $T \supseteq \{\infty\} \cup \{z : |z| > n\}$ or $T \supseteq D(w, 1/n)$ for some $w \in \mathbb{C}\setminus\Omega$. In either case, T meets $\hat{\mathbb{C}} \setminus \Omega$. Since $\Omega_n \supseteq \hat{\mathbb{C}} \setminus \Omega$, T must contain any component of $\hat{\mathbb{C}} \setminus \Omega$ that it meets, and such a component exists by the preceding sentence.

6. (a) Form the sets K_n as in (5.1.1), and find by (5.2.8) a rational function R_n with poles in S such that $|f - R_n| < 1/n$ on K_n. For any compact subset K of Ω, $K \subseteq K_n$ for sufficiently large n, so that $R_n \to f$ uniformly on compact subsets of Ω.
(b) By Problem 5, each component of $\hat{\mathbb{C}} \setminus K_n$ contains a component of $\hat{\mathbb{C}} \setminus \Omega$, so if Ω is simply connected, i.e., $\hat{\mathbb{C}} \setminus \Omega$ is connected, then $\hat{\mathbb{C}} \setminus K_n$ is connected for all n. Therefore in part (a), the R_n can be taken to be polynomials. Conversely, assume that for every $f \in A(\Omega)$ there is a sequence of polynomials P_n converging to f uniformly on compact subsets of Ω. If γ is a closed path in Ω, then $\int_\gamma P_n(z)\,dz = 0$ for all n, hence $\int_\gamma f(z)\,dz = 0$ because γ^* is compact. Thus Ω is simply connected.
7. (a) By Runge's theorem (see part (b) of Problem 6) there are polynomials p_n such that $|p_n(z) - f_n(z)| < 1/n$ for all $z \in K_n \cup L_n \cup M_n$. Then $p_n \to 0$ pointwise. But if K is any compact set containing all the B_n, then p_n cannot approach 0 uniformly on K because $\sup\{|p_n(z)| : z \in B_n\} \geq 1 - \frac{1}{n} \to 1$.
(b) Choose polynomials p_n such that $|p_n(z) - g_n(z)| < 1/n$ for all $z \in K_n \cup M_n$. Then $p_n \to g$ pointwise, where $g(z) = 1$ for $\operatorname{Re} z > 0$ and $g(z) = 0$ for $\operatorname{Re} z \leq 0$.

Section 5.3

1. Let f be a homeomorphism of $\overline{\Omega}$ onto $\overline{D}$ such that f is a one-to-one analytic map of Ω onto D; f exists by (5.3.9) and (5.2.2). If $g = f^{-1}$ and $u^* = u_0 \circ (g|_{\partial D})$, then u^* is real-valued and continuous on ∂D, so by (4.7.6), u^* extends to a function that is continuous on $\overline{D}$ and harmonic on D. Let $u = u^* \circ f$; then $u = u_0$ on $\partial\Omega$ and u is continuous on $\overline{\Omega}$. If $h = u^* + iv^*$ is analytic on D, then $h \circ f$ is analytic on Ω and $\operatorname{Re} h \circ f = u^* \circ f = u$, hence u is harmonic on Ω.
2. (a) Let u be the unique argument of z in $[-\pi, \pi)$; see (3.1.2).
(b) Apply (5.2.2) and (5.3.9).
(c) Note that $u(f(z)) = \operatorname{Im} \log_\pi(f(z))$, and $\log_\pi f(z)$ is analytic on D by (3.1.2).
(d) Suppose $u(f(z)) + iV(z)$ is analytic on D. Write $V(z) = v(f(z))$ where v is harmonic on Ω. Then $iu(f(z)) - v(f(z))$ is analytic on D, so by (3.1.6), $\ln|f(z)| = -v(f(z)) + 2\pi ik$ for some integer k. Consequently, $e^{-v(f(z))} = |f(z)|$. If V is bounded, so is v, which yields a contradiction. (Examine $f(z)$ near z_0, where $f(z_0) = 0$.)
3. Apply (5.3.9), along with Problems 3.2.6 and 3.2.7.

Chapter 6

Section 6.1

1. If $f(z) = 0$, then since $f_n(z) \to 1$ as $n \to \infty$, it follows that for sufficiently large N, the infinite product $\prod_{n=N}^{\infty} f_n(z)$ converges. Thus $f(z) = [\prod_{k=1}^{N-1} f_k(z)]g(z)$ where g is analytic at z and $g(z) \neq 0$. Hence $m(f,z) = \sum_{k=1}^{N-1} m(f_k, z) = \sum_{n=1}^{\infty} m(f_n, z)$.
2. The first statement is immediate from the power series expansion of $-\ln(1-x)$, namely
$$x + \frac{x^2}{2} + \frac{x^3}{3} + \cdots = x + x^2(\frac{1}{2} + \frac{x}{3} + \frac{x^2}{4} + \cdots).$$
Now if $\sum_n a_n$ converges, then $-\ln(1 - a_n) = \sum_n [a_n + g(a_n)a_n^2]$ where $g(a_n) \to 1/2$ as $n \to \infty$. By (6.1.1), $\prod_n (1 - a_n)$ converges to a nonzero limit iff $\sum_n a_n^2 < \infty$. The remaining statement of the problem follows similarly.
3. (a) Absolutely convergent by (6.1.2).
(b) Does not converge to a nonzero limit by Problem 2, since $\sum_n (n+1)^{-2} < \infty$, $\sum_n (n+1)^{-1} = \infty$. In fact,
$$\prod_{k=1}^{n} (1 - \frac{1}{k+1}) = \frac{1}{2} \cdot \frac{2}{3} \cdots \frac{n}{n+1} = \frac{1}{n+1} \to 0.$$
(c) Does not converge to a nonzero limit by Problem 2. Here, $a_n = (-1)^{n+1}/\sqrt{n}$, hence $\sum_n a_n$ converges but $\sum_n a_n^2 = \infty$.
(d) Absolutely convergent by (6.1.2).
4. (a) See Problem 3(c).
(b) Take $a_{2n-1} = 1/\sqrt{n}$ and $a_{2n} = (-1/\sqrt{n}) + (1/n)$.
Remark: This is also an example of an infinite product that is convergent but not absolutely convergent.
5. (a) Since $\sum_{n=1}^{\infty} |a^n z|$ converges uniformly on compact subsets, the result follows from (6.1.7).
(b) Restrict z to a compact set K. For sufficiently large n (positive or negative),
$$\begin{aligned} \operatorname{Log}\left[(1 - \frac{z}{n})e^{z/n}\right] &= \operatorname{Log}\left[(1 - \frac{z}{n})\right] + \operatorname{Log} e^{z/n} \\ &= -\left[\frac{(z/n)^2}{2} + \frac{(z/n)^3}{3} + \cdots\right] \\ &= \frac{z^2}{n^2} g(z/n) \end{aligned}$$
where $g(w) \to -1/2$ as $w \to 0$. Since K is bounded, there is a constant M such that
$$\left|\operatorname{Log}\left[(1 - \frac{z}{n})e^{z/n}\right]\right| \leq \frac{M}{n^2}$$
for all $z \in K$. Thus $\sum_n \operatorname{Log}[(1 - z/n)e^{z/n}]$ converges uniformly on K. As in the proof of (6.1.6), the infinite product converges uniformly on K, so that the resulting function

is entire.

(c) Since $\sum_{n=2}^{\infty} \frac{1}{n(\ln n)^2}$ converges, $\sum_{n=2}^{\infty} \frac{|z|}{n(\ln n)^2}$ converges uniformly on compact subsets and (6.1.6) applies.

6. If we try to prove that the convergence of $\sum z_n$ implies the convergence of $\sum z_n g(z_n)$, we run into difficulty. We would like to argue that $|\sum_{k=n}^{m} z_k g(z_k)| \leq \sum_{k=n}^{m} |z_k g(z_k)| \to 0$ as $n, m \to \infty$, but this requires the *absolute* convergence of $\sum z_n$. A similar difficulty occurs in the converse direction. [Note that $\prod_n (1+z_n)$ converges absolutely iff $\sum_n z_n$ converges absolutely, by (6.1.2).]

Section 6.2

1. (a) We have $m = 0$, and the canonical product is $\prod_{n=1}^{\infty}(1 - z/2^n)$.

   (b) The canonical product is $\prod_{n=1}^{\infty} E_m(z/z_n)$ where m is the least integer strictly greater than $(1/b) - 1$.

   (c) We have $m = 0$, and the canonical product is $\prod_{n=1}^{\infty}[1 - z/n(\ln n)^2]$.

2. We may proceed exactly as in (6.2.5), using (6.2.6) in place of (6.2.3).

Section 6.3

1. By (6.3.7), the result holds for $n = 2$. For if d is a gcd of $\{f_1, f_2\}$, then f_1/d and f_2/d are relatively prime. If $(f_1 g_1/d) + (f_2 g_2/d) = 1$, then $f_1 g_1 + f_2 g_2 = d$. To go from $n-1$ to n, let d be a gcd for $\{f_1, \ldots, f_n\}$ and d_1 a gcd for $\{f_1, \ldots, f_{n-1}\}$. Then d is a gcd for $\{d_1, f_n\}$ (by definition of gcd). By the induction hypothesis, we have $g_1, \ldots, g_{n-1} \in A(\Omega)$ such that $f_1 g_1 + \cdots + f_{n-1} g_{n-1} = d_1$, and by (6.3.7) there exist $h, g_n \in A(\Omega)$ such that $d_1 h + f_n g_n = d$. But then $f_1 g_1 h + \cdots + f_{n-1} g_{n-1} h + f_n g_n = d$.

2. Let $\{a_n\}$ be a sequence of points in Ω with no limit point in Ω. By (6.2.6) or (6.2.3), there exists $f_n \in A(\Omega)$ such that $Z(f_n) = \{a_n, a_{n+1}, \ldots\}$ and $m(f_n, a_j) \neq 0, j \geq n$. Let I be the ideal generated by $f_1, f_2, \ldots$, that is, I is the set of all finite linear combinations of the form $g_{i_1} f_{i_1} + \cdots + g_{i_k} f_{i_k}, k = 1, 2, \ldots, g_{i_j} \in A(\Omega)$. If I were principal, it would be generated by a single f. But then $Z(f) \subseteq Z(h)$ for each $h \in I$, in particular, $Z(f) \subseteq Z(f_n)$ for all n. It follows that f has no zeros, so $1 = f(1/f) \in I$. By definition of I, $1 = g_1 f_1 + \cdots + g_n f_n$ for some positive integer n and $g_1, \ldots, g_n \in A(\Omega)$. Since $f_1(a_1) = f_2(a_n) = \cdots = f_n(a_n) = 0$, we reach a contradiction.

List of Symbols

Index